New Methods for the Study of Biomolecular Complexes

NATO ASI Series

Advanced Science Institutes Series

A Series presenting the results of activities sponsored by the NATO Science Committee, which aims at the dissemination of advanced scientific and technological knowledge, with a view to strengthening links between scientific communities.

The Series is published by an international board of publishers in conjunction with the NATO Scientific Affairs Division

A Life Sciences **B Physics**	Plenum Publishing Corporation London and New York
C Mathematical and Physical Sciences **D Behavioural and Social Sciences** **E Applied Sciences**	Kluwer Academic Publishers Dordrecht, Boston and London
F Computer and Systems Sciences **G Ecological Sciences** **H Cell Biology** **I Global Environmental Change**	Springer-Verlag Berlin, Heidelberg, New York, London, Paris and Tokyo

PARTNERSHIP SUB-SERIES

1. Disarmament Technologies	Kluwer Academic Publishers
2. Environment	Springer-Verlag / Kluwer Academic Publishers
3. High Technology	Kluwer Academic Publishers
4. Science and Technology Policy	Kluwer Academic Publishers
5. Computer Networking	Kluwer Academic Publishers

The Partnership Sub-Series incorporates activities undertaken in collaboration with NATO's Cooperation Partners, the countries of the CIS and Central and Eastern Europe, in Priority Areas of concern to those countries.

NATO-PCO-DATA BASE

The electronic index to the NATO ASI Series provides full bibliographical references (with keywords and/or abstracts) to more than 50000 contributions from international scientists published in all sections of the NATO ASI Series.
Access to the NATO-PCO-DATA BASE is possible in two ways:

– via online FILE 128 (NATO-PCO-DATA BASE) hosted by ESRIN, Via Galileo Galilei, I-00044 Frascati, Italy.

– via CD-ROM "NATO-PCO-DATA BASE" with user-friendly retrieval software in English, French and German (© WTV GmbH and DATAWARE Technologies Inc. 1989).

The CD-ROM can be ordered through any member of the Board of Publishers or through NATO-PCO, Overijse, Belgium.

Series C: Mathematical and Physical Sciences – Vol. 510

New Methods for the Study of Biomolecular Complexes

edited by

W. Ens

K. G. Standing

The University of Manitoba,
Winnipeg, Manitoba, Canada

and

I. V. Chernushevich

PE SCIEX,
Concord, Ontario, Canada

Kluwer Academic Publishers

Dordrecht / Boston / London

Published in cooperation with NATO Scientific Affairs Division

Proceedings of the NATO Advanced Research Workshop on
New Methods for the Study of Molecular Aggregates
The Lodge at Kananaskis Village, Alberta, Canada
16–20 June 1996

A C.I.P. Catalogue record for this book is available from the Library of Congress.

ISBN 0-7923-5003-0

Published by Kluwer Academic Publishers,
P.O. Box 17, 3300 AA Dordrecht, The Netherlands.

Sold and distributed in the U.S.A. and Canada
by Kluwer Academic Publishers,
101 Philip Drive, Norwell, MA 02061, U.S.A.

In all other countries, sold and distributed
by Kluwer Academic Publishers,
P.O. Box 322, 3300 AH Dordrecht, The Netherlands.

Printed on acid-free paper

Printed in the Netherlands

This book contains the proceedings of a NATO Advanced Research Workshop held within the programme of activities of the NATO Special Programme on Supramolecular Chemistry as part of the activities of the NATO Science Committee.

Other books previously published as a result of the activities of the Special Programme are:

WIPFF, G. (Ed.), *Computational Approaches in Supramolecular Chemistry.* (ASIC 426) 1994. ISBN 0-7923-2767-5

FLEISCHAKER, G.R., COLONNA, S. and LUISI, P.L. (Eds.), *Self-Production of Supramolecular Structures.* From Synthetic Structures to Models of Minimal Living Systems. (ASIC 446) 1994. ISBN 0-7923-3163-X

FABBRIZZI, L., POGGI, A. (Eds.), *Transition Metals in Supramolecular Chemistry.* (ASIC 448) 1994. ISBN 0-7923-3196-6

BECHER, J. and SCHAUMBURG, K. (Eds.), *Molecular Engineering for Advanced Materials.* (ASIC 456) 1995. ISBN 0-7923-3347-0

LA MAR, G.N. (Ed.), *Nuclear Magnetic Resonance of Paramagnetic Macromolecules.* (ASIC 457) 1995. ISBN 0-7923-3348-9

SIEGEL, JAY S. (Ed.), *Supramolecular Stereochemistry.* (ASIC 473) 1995. ISBN 0-7923-3702-6

WILCOX, C.S. and HAMILTON A.D. (Eds.), *Molecular Design and Bioorganic Catalysis.* (ASIC 478) 1996. ISBN 0-7923-4024-8

MEUNIER, B. (Ed.), *DNA and RNA Cleavers and Chemotherapy of Cancer and Viral Diseases.* (ASIC 479) 1996. ISBN 0-7923-4025-6

KAHN, O. (Ed.), *Magnetism: A Supramolecular Function.* (ASIC 484) 1996. ISBN 0-7923-4153-8

ECHEGOYEN, L., KAIFER ANGEL E. (Eds.), *Physical Supramolecular Chemistry.* (ASIC 485) 1996. ISBN 0-7923-4181-3

DESVERGNE J.P., CZARNIK A.W. (Eds.), *Chemosensors of Ion and Molecule Recognition.* (ASIC 492) 1997. ISBN 0-7923-4555-X

MICHL J., *Modular Chemistry.* (ASIC 499) ISBN 0-7923-4730-7

TABLE OF CONTENTS

Matrix-Assisted Laser Desorption/Ionization Mass Spectrometry

X-Ray Crystallography and Mass Spectrometry

Other Techniques

Preface

A NATO Advanced Research Workshop entitled New Methods for the Study of Molecular Aggregates was held at The Lodge at Kananaskis Village, Alberta, Canada from 16 - 20 June 1996. In fact the meeting was entirely concerned with the problem of analyzing biomolecular complexes, so the title of these proceedings has been altered to give a more precise description of the content. The workshop was hosted by the time-of-flight group of the Department of Physics at the University of Manitoba, and was attended by 64 participants from around the world. Twenty-one invited talks were given and 27 papers were presented as posters. Of the 48 contributions, 22 papers (12 orals, 10 posters) are included in these proceedings.

The subject of the conference was the investigation of noncovalent biomolecular complexes, with particular focus on the application of mass spectrometry to their characterization. Two new ionization techniques introduced in the late 1980s, electrospray ionization (ESI) and matrix-assisted laser desorption/ionization (MALDI), resulted in a breakthrough in mass spectrometry, enabling its use in molecular weight and primary structure determination of biopolymers larger than 100 kDa. Recently it has been discovered that ESI mass spectrometry may also be used to characterize complexes containing noncovalent interactions, thus opening new perspectives for supramolecular chemistry. ESI mass spectrometry has the advantage that the sample is introduced from a homogenous solution which can be maintained at near physiological conditions of pH, concentration, and temperature. Analysis of noncovalent complexes by MALDI mass spectrometry is more difficult because the sample is normally dissolved in a solution with an acidic matrix and the ions are desorbed from the solid phase. Both of these steps can potentially cause complexes to denature. Still, the simplicity of the MALDI spectra and its greater tolerance to impurities has motivated considerable effort to solve the problems and some progress has been made.

The first section of these proceedings contains papers which indicate the role mass spectrometry can play in combination with chemical methods in the investigation of biomolecular complexes. The next two sections focus more on the specifics of ESI and MALDI mass spectrometry as new methods to characterize these interesting entities. The final two sections describe other techniques including some emerging methods as well as the established methods like x-ray crystallography and NMR.

The Lodge at Kananaskis provided a relaxed, informal environment suitable for lively and productive discussions during the sessions as well as between them. We are grateful to NATO for providing generous financial support for the invited speakers and for a number of young researchers.

Werner Ens
Ken Standing
Winnipeg, Nov. 1997

A GENERAL APPROACH TO DECIPHERING HOW PROTEINS WORK USING SYNTHETIC CHEMISTRY AND MASS SPECTROMETRY

MICHAEL C. FITZGERALD and STEPHEN B. H. KENT
The Scripps Research Institute
10666 N. Torrey Pines Rd.
La Jolla, CA 92037

1. Introduction

Understanding the molecular basis of protein function is currently a major focus of biological research. As chemists and biologists we would like to understand, for example, how protein domains recognize their target ligands or how enzymes are capable of catalyzing chemical reactions at nearly diffusion limited rates which in some cases are 10^6-10^{12} times faster than that of the uncatalyzed reaction. One way to get answers to questions like these is to use chemical synthesis to introduce specific mutations into the protein molecule and determine their effect on function [1].

Chemical synthesis is an especially powerful approach for studying protein structure and function. The preparation of proteins by total chemical synthesis permits the construction of protein analogues with a wide variety of both natural and unnatural modifications. Therefore by using total chemical synthesis it becomes possible to ask - and answer - very specific questions about how a protein works.

Here we describe how chemical synthesis can be used, in conjunction with mass spectrometry, to study the molecular basis of protein function. Currently, solid phase peptide synthesis (SPPS) methods permit the routine preparation of polypeptide chains up to ~40 to 60 amino acids [2]. The total chemical synthesis of larger protein constructs (up to 200 amino acid residues in length) has also been possible using recently developed chemical ligation strategies for the chemoselective ligation of unprotected peptide segments [3]. Thus, small proteins and enzymes in the 40-200 amino acid size range are accessible by total chemical synthesis. The ability to synthesize proteins of this size is of quite general utility because the fundamental building blocks of the protein world are 'functional domains' that are typically 110 $\pm$ 40 amino acids. As an example of how total chemical synthesis can be used to study the structure and function of an enzyme, we highlight our work on 4-oxalocrotonate tautomerase (4OT).

W. Ens et al. (eds.), New Methods for the Study of Biomolecular Complexes, 1–16.

2. Probing the Function of an Enzyme by Total Chemical Synthesis.

4OT is an enzyme used by certain soil bacteria to help degrade aromatic hydrocarbons to provide intermediates for the Krebs cycle [4,5]. The 4OT enzyme, which contains no cysteine residues, is a hexamer of identical 62 amino acid polypeptide chains and catalyzes the 1,3-allylic isomerization of 2-oxo-4-*(E)*-hexenedioate (**1**) to 2-oxo-3-*(E)*-hexenedioate (**3**) through the intermediate 2-hydroxymuconate (**2**) (Figure 1A). Recent X-ray crystallographic data on a 4OT isozyme has shown that the homo-hexameric, enzyme complex is a trimer of dimers; each dimer contains a 4 stranded beta sheet region and 2 anti-parallel helices [6]. Results from inhibition studies with an active site-directed irreversible inhibitor of 4OT have suggested that the enzyme complex contains six active sites and have implicated the N-terminal proline as the catalytic base [7]. It has also been hypothesized from the NMR solution structure and the three dimensional crystal structure of 4OT that two Arg residues (Arg^{11} of one subunit and Arg^{39} of an adjacent subunit) are important for binding the dicarboxylic acid substrate (**1**) in the active site of the enzyme (Figure 1B) [6,8].

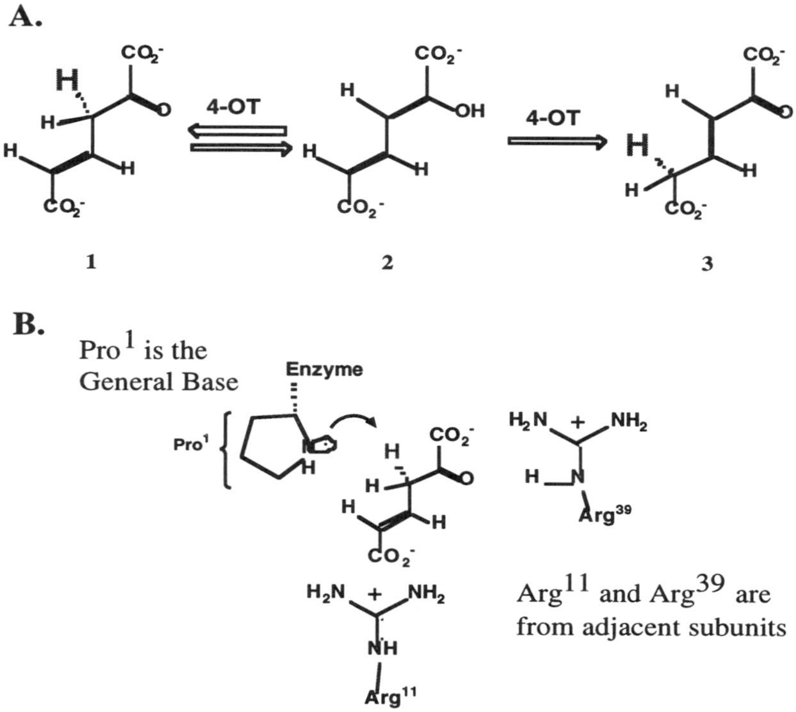

Figure 1. 4OT-catalyzed isomerization reaction (A) and proposed mechanism of action (B).

The X-ray crystallographic results and biochemical studies on 4OT that are referenced above have each helped identify catalytically important residues in the enzyme; however, the molecular basis for the function of these residues is not well understood. Currently, we are using total chemical synthesis to prepare 4OT analogues with various chemical modifications designed to address specific questions about the catalytic mechanism of this highly efficient enzyme.

We have initially focused on the synthesis and characterization of four different 4OT analogues including: (*des*Pro^1)4OT, a truncated construct in which Pro^1 was deleted; (Cpc^1)4OT in which Pro^1 was replaced with cyclopentane carboxylate; a derivative [Met(O)^{45}]4OT in which the side chain sulfur of Met^{45} was oxidized to the sulfoxide; and an analogue (Nle^{45})4OT in which Met^{45} was replaced with norleucine. Our purpose in studying (Met(O)^{45}]4OT was to determine what effects the oxidation of the single Met residue in 4OT had on the enzyme's structure, because enzyme preparations containing partially oxidized material displayed reduced activity. The (Nle^{45})4OT construct was designed to eliminate the possibility of oxidizing the enzyme during sample handling. The (*des*Pro^1)4OT and (Cpc^1)4OT constructs were designed to investigate the catalytic significance of the N-terminal proline reside. It is noteworthy that the (Cpc^1)4OT analogue is only accessible by total chemical synthesis methods.

2.1. SYNTHESIS AND CHARACTERIZATION OF 4OT ANALOGUES

2.1.1. *Chemical Synthesis, Purification, and Folding of Proteins*

The 62 amino acid polypeptide chain of wild-type 4OT was synthesized from protected amino acids in stepwise fashion by highly optimized, solid phase peptide synthesis (SPPS) methods using *in situ* neutralization protocols for t-butoxycarbonyl (Boc) chemistry [9]. The crude polypeptide product was purified by reversed-phase HPLC. Typically, 50-100 mg of high purity 62mer could be obtained from a single 0.2 mmol synthesis (the smallest convenient scale of laboratory synthesis). Analytical reversed-phase HPLC and electrospray ionization mass spectrometry (ESI-MS) were used to confirm the identity and purity of the final polypeptide product. Results for the HPLC and ESI-MS analysis of purified, synthetic 4OT monomer are shown below in Figure 2. The purified synthetic peptide product eluted as a single peak using a shallow gradient (40-60% acetonitrile/water in 0.1% triflouroacetic acid over 30 minutes). The monomeric molecular weight of the 4OT sample was 6809.9 $\pm$ 0.7 (as determined by ESI-MS) and in good agreement with the expected mass, 6810.7 (average isotopic composition). The monomer polypeptide chains of 4OT were folded to form the homo-hexameric enzyme complex during a 2 hour, room temperature incubation in assay buffer (20 mM sodium phosphate, pH 7.4).

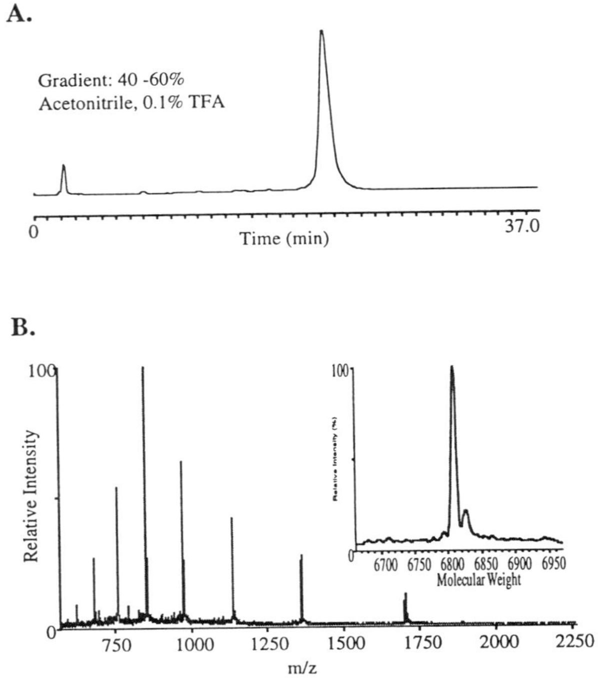

Figure 2. (A) Analytical reversed-phase HPLC analysis of purified 62 amino acid monomer of 4OT. (B) Electrospray mass spectra of HPLC purified 4OT.

In addition to the wild-type enzyme, we have also used total chemical synthesis to prepare the (*des*Pro^1)4OT, (Cpc^1)4OT, [Met(O)^{45}]4OT, and (Nle^{45})4OT analogues. The polypeptide chains of the (*des*Pro^1)4OT, (Nle^{45})4OT, and (Cpc^1)4OT constructs were each synthesized in stepwise fashion using SPPS methods. In the case of (Cpc^1)4OT, unprotected cyclopentanecarboxylic acid (Cpc) was used instead of Boc-proline in the last coupling reaction of the synthesis. The [Met(O)^{45}]4OT polypeptide chain was prepared by mild oxidation (0.3% v/v hydrogen peroxide in 0.1% triflouroacetic acid) of purified, synthetic 4OT. Analysis of the reaction by ESI-MS revealed that oxidation to the sulfoxide was complete and no sulfone was observed. Each 4OT analogue was folded in assay buffer as described above for the wild-type enzyme.

2.1.2. *Circular Dichroism Spectroscopy*

Circular dichroism (CD) spectroscopy was used to characterize the secondary structure of each folded 4OT analogue. The CD spectra recorded for the wild-type enzyme and for each analogue are shown in Figures 3A and B. The helical content of each analogue was estimated from the magnitude of its molar ellipticity at 222 nm [10]. The percent helicities calculated for 4OT, (Nle^{45})4OT, (Cpc^{1})4OT and [Met(O)^{45}]4OT were each

A.

B.

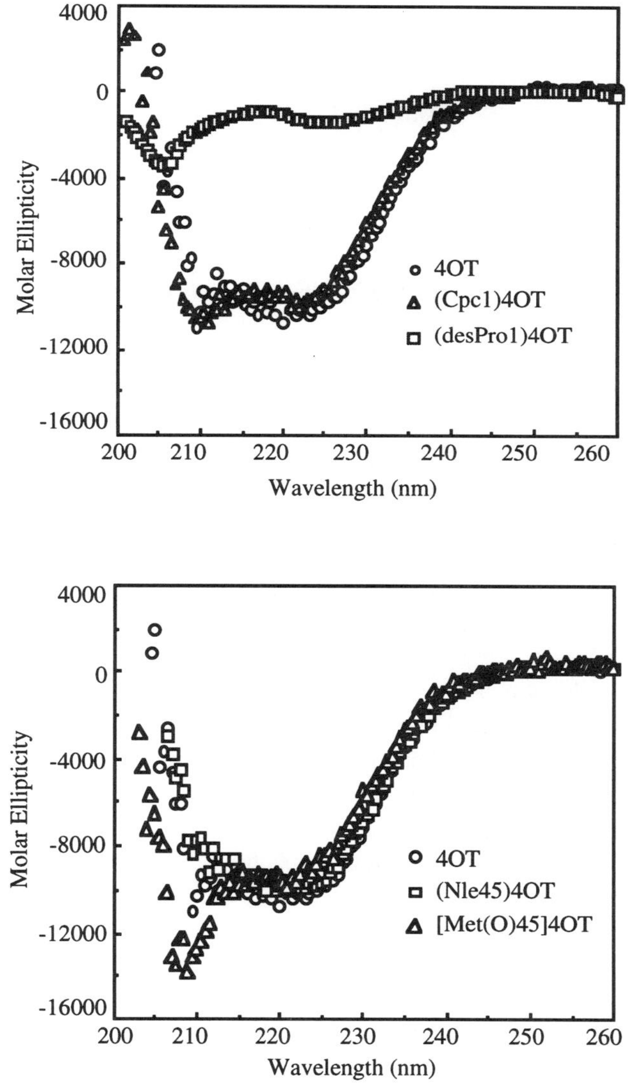

Figure 3. CD Spectra for 4OT and Analogues

approximately 26% and in relatively good agreement with that previously reported for recombinant 4OT (21%) [11]. Furthermore, the other features of the CD spectra recorded for (Cpc^1)4OT (see Figure 3A) and (Nle^{45})4OT (see Figure 3B) were essentially identical to the spectrum recorded for 4OT indicating that the secondary structures of these analogues are similar to those of the wild-type enzyme. In contrast, the unique minimum at 208 nm observed in the CD spectrum of [Met(O)^{45}]4OT (see Figure 3B) suggests that the secondary structure of this analogue differs in some aspects from that of the wild type enzyme. The essentially featureless spectrum of the (*des*Pro^1)4OT analogue (see Figure 3A) showed that this polypeptide has very little defined secondary structure.

2.1.3. *Electrospray Ionization Time-of-Flight (ESI-TOF) Mass Spectrometry*

ESI-TOF mass spectrometry was used to assess the oligomeric state of each 4OT analogue. The noncovalent hexamer of 4OT is readily detected by ESI-TOF at m/z values between 3000 and 3500, and the technique has proven very sensitive for examining the effects that subtle differences in covalent structure can have on an enzyme's higher order structure [12]. The ESI-TOF mass spectra recorded for the 4OT analogues in this study are shown in Figure 4 (a typical spectrum of wild-type 4OT is also included for comparison).

The spectra in Figure 4 were acquired under identical "native" electrospray conditions using the same buffer (5 mM ammonium bicarbonate, pH 7.5), needle position, and declustering potential. Strong hexamer signals were detected for the (Nle^{45})4OT and (Cpc^1)4OT samples indicating that 62 residue polypeptide chains of these analogues each fold in solution to form a hexameric enzyme complex. The absence of any multimeric species in the native mass spectrum of (*des*Pro^1)4OT (Figure 4E) suggests that this truncated analogue has no discernible quaternary structure. The presence (albeit weak) of multiply charged hexamer ions in the native mass spectrum of [Met(O)^{45}]4OT (Figure 4D) indicates that this construct is capable of forming a homo-hexameric complex. However, the ratio of hexamer to monomer in the [Met(O)^{45}]4OT spectrum was significantly smaller (~50-fold less) than the hexamer to monomer ratio in the 4OT, (Nle^{45})4OT, and (Cpc^1)4OT spectra. These results suggest that the noncovalent interactions defining the hexameric [Met(O)^{45}]4OT complex are significantly weaker than those in 4OT.

2.1.4. *Enzymatic Activity*

The catalytic efficiency of each 4OT analogue in this study was determined by studying the kinetics of the conversion of 2-hydroxymuconate (**2**) to 2-oxo-3-trans-hexenedioate (**3**) (see Figure 1). The K_M, and k_{cat} values determined for the wild-type enzyme and each analogue are given in Table 1.

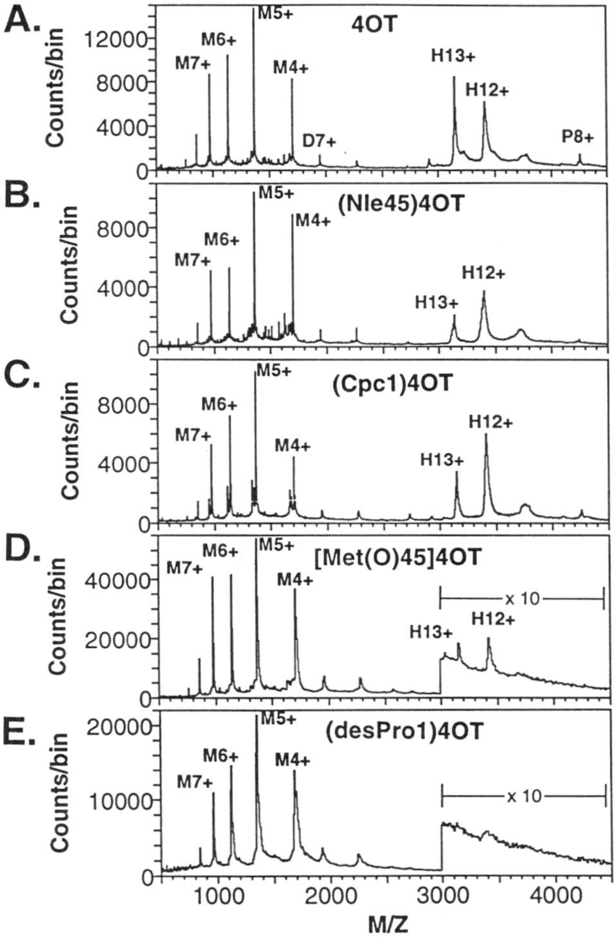

Figure 4. ESI-TOF Analysis of 4OT Analogues

The kinetic data show that the (Nle^{45})4OT and [$Met(O)^{45}$]4OT analogues displayed enzymatic activity comparable to that of wild-type 4OT (k_{cat} ~2900 s^{-1}). By contrast, for the (Cpc^1)4OT and (*des*Pro^1)4OT analogues no rate enhancement over the nonenzymatic chemical ketonization (uncatalyzed rate: 8.7 x10^{-4} s^{-1}) of 2-hydroxymuconate to 2-oxo-3-trans-hexenedioate was detected (detection limit: kcat ~1 s^{-1}). It is also noteworthy that the full enzymatic activity of [$Met(O)^{45}$]4OT reported in Table 1 was only obtained when enzyme stock solutions at concentrations greater than 40 μM (based on total amount of monomer). The [$Met(O)^{45}$]4OT analogue is apparently dissociated in more dilute solutions (<~40 μM) suggesting that the dissociation constant of this analogue is altered from that of the wild-type enzyme [12].

TABLE 1. Kinetic Parameters for 4OT and Analogues

Analogue	K_M, μM	k_{cat}, s^{-1}, x 10^{-3}
4OT	90 ± 20	2.9 ± 0.5
[Met(O)^{45}]4OT	290 ± 100	3.0 ± 1.0
(Nle^{45})4OT	50 ± 10	2.4 ± 0.2
(Cpc^{1})4OT	Not Active (k_{cat} <1 s^{-1})	
(*des*Pro^{1})4OT	Not Active (k_{cat} <1 s^{-1})	

2.2. STRUCTURAL AND MECHANISTIC CONCLUSIONS

Our physical and kinetic characterization of the synthetic 4OT analogues described above have allowed us to draw several conclusions about the structure and function of this small enzyme. Our studies with the (*des*Pro^{1})4OT and (Cpc^{1})4OT analogues have helped to establish both the structural and mechanistic significance of the Pro^{1} residue in 4OT. The CD data and ESI-TOF experiments both suggest that the (*des*Pro^{1})4OT analogue lacks any defined higher order structure and indicate that Pro^{1} is essential for the enzyme folding.

Our work with (Cpc^{1})4OT also made it possible to directly evaluate the mechanistic role of Pro^{1} in 4OT-catalysis. Replacement of the ring nitrogen of Pro^{1} by a methylene group in (Cpc^{1})4OT made it possible to effectively delete the N-terminal, secondary amine functionality in 4OT without significantly perturbing the enzyme's structure. The 2° and 4° structural features of the (Cpc^{1})4OT analogue were identical to those of the wild-type enzyme as determined by CD spectroscopy and ESI-TOF mass spectrometry. Thus, the lack of detectable activity in the (Cpc^{1})4OT could be unambiguously attributed to the absence of an N-terminal amine functionality in this construct, rather than a gross change in the protein's folded structure. These results therefore provide direct evidence that the ring nitrogen in Pro^{1} plays a critical role in the 4OT-catalyzed isomerization reaction.

Characterization of [Met(O)^{45}]4OT enabled us to understand the effects of oxidation of 4OT's only Met residue on the enzyme's structure and function. The kinetic data and ESI-TOF analysis of this construct revealed that this mutant folds to form a fully active hexameric enzyme complex resembling that of 4OT. However, it is apparent from the ESI-TOF results that Met^{45} oxidation weakens the noncovalent protein-protein interactions that define the hexameric, [Met(O)^{45}]4OT complex.

Work is in progress to further explore the general mechanistic and structural significance of 4OT's N-terminal residue. We have prepared an acyclic proline analog, (S)2(NMe)-n-butyric acid (MABA), which we have substituted for Pro[1]. Both the MABA and proline constructs of 4OT contain a secondary N-terminal amine functionality, but the nitrogen in the MABA analogue is not constrained in a ring structure. Preliminary kinetic results with the MABA 4OT analogue indicate that this construct is catalytically active and suggest that the N-terminal proline ring structure is not essential for function. We are also investigating whether or not there is any correlation between the enzyme's function and the degree of alkyl substitution at the N-terminal catalytic amino group.

3. A Combinatorial Chemistry Approach for Studying Protein Function

The 4OT analogues that we have studied to date contained modifications at only two of the 62 amino acid residues that define the primary structure of this small enzyme. In order to gain a more complete understanding of the factors that govern 4OT-catalysis, it would be useful to systematically introduce specific changes in chemical structure throughout the entire 62 amino acid sequence of the enzyme and examine their effect on function. For example, the synthesis and characterization of 4OT analogues where specific amino acid side chains have been systematically deleted throughout the sequence would be useful for determining which side chain functionalities were essential for the enzyme to function. Total chemical synthesis provides a means for doing this. However, the individual synthesis, purification, and subsequent characterization of multiple protein analogues can be extremely time consuming and the number of analogues that can be prepared is ultimately limited by this practical consideration. A better approach would be to carry out the synthesis and characterization of multiple protein analogues in a single experiment. In this section we describe such a combinatorial approach for studying protein function.

3.1. PROTEIN SIGNATURE ANALYSIS - OVERVIEW

This new technique which we have termed 'Protein Signature Analysis' involves a unique combination of chemical synthesis and mass spectrometry [13,14]. The approach, which is outlined in Figure 5, involves three steps: (1) chemical synthesis, (2) functional selection, and (3) chemical readout of a self-encoded array of protein analogues.

In the first step chemical synthesis is used to prepare an array of proteins analogues where each member of the array contains a specific mutation at a unique and defined position along the polypeptide chain. In a second step the protein array is subjected to a selection based on a specific functional property, so that the array is divided into two pools - one that is functional and one that is nonfunctional. Finally, the analogues present in each pool are read out using a built in chemistry. The readout technique that we have employed involves the use of matrix-assisted laser desorption/ionization (MALDI) mass spectrometry to unambiguously identify each of the modified proteins present in the functional and nonfunctional populations, as described below.

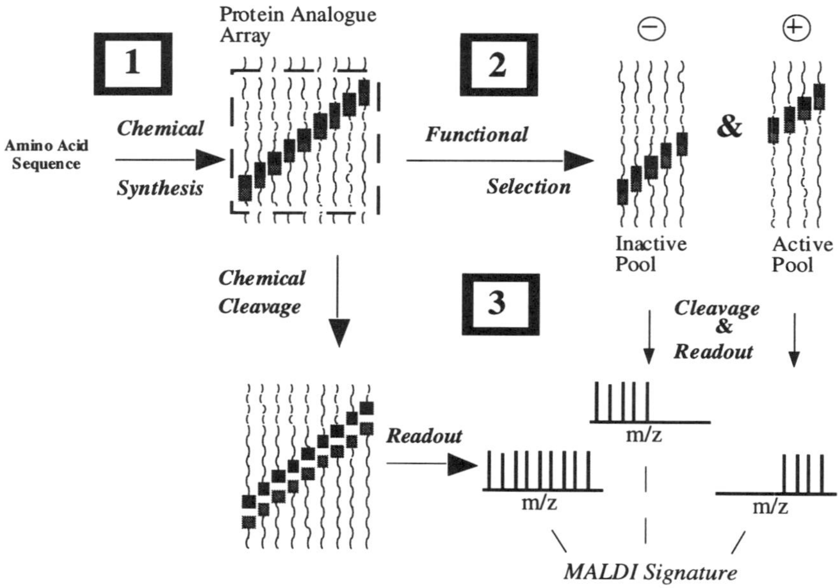

Figure 5. Protein Signature Analysis

3.1.1. *Chemical Synthesis of A Protein Analogue Array*

Standard stepwise SPPS methods have been used in a modified split resin procedure to prepare all members of a protein analogue array simultaneously in a single synthesis [13,14]. This procedure results in the production of multiple protein analogues that contain a specific chemical perturbation, or analogue unit, in their sequence. The protein analogue array is obtained as a single product mixture and most importantly, each protein analogue in the array contains only one copy of the analogue unit at a unique and defined position in the polypeptide chain.

The chemical structures of several analogue units that we have incorporated into peptides and proteins are shown in Figure 6A. In this case, the analogue units are dipeptide mimics that contain both side chain deletions and modifications to their peptide backbone. Another key feature that the analogue units also contain is a chemically cleavable bond. The ester bonds in each of the dipeptide analogue units are relatively stable to normal handling, but can be cleaved in an aminolysis reaction as shown in Figure 6B. This chemically cleavable bond is an important feature of the analogue unit as it serves to self-encode the members of a given protein array.

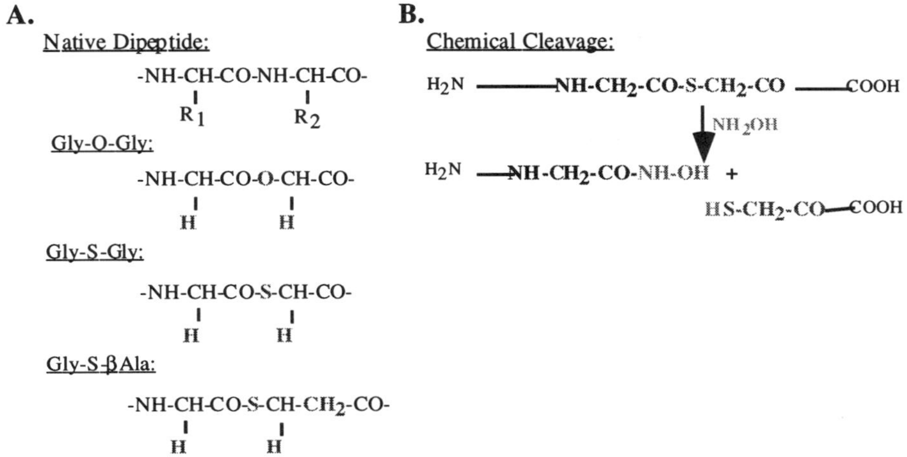

Figure 6. Analogue Units

3.1.2. *Readout of Protein Analogue Arrays*

The protein signature analysis technique relies on a unique chemical readout step to unambiguously identify the constituent protein analogues in a given array. Because a chemically cleavable bond is associated with each analogue unit, it is possible to "unzip" a protein array in a chemical cleavage step as shown in Figure 7.

The resulting N- and C-terminal sequence ladders can be analyzed by MALDI mass spectrometry to unambiguously identify each member of the original array [13-15]. Ultimately, the N- and C-terminal ladders that are generated in the MALDI readout of functional and nonfunctional pools create a signature that can be used to correlate changes in chemical structure with changes in function.

We have successfully used the protein signature analysis technique in an investigation of the folding/binding activity of an SH3 domain (a protein functional unit commonly found in signal transduction pathways) [14,16]. Currently, we are also using the technique in our studies of 4OT. Our preliminary results with this system are presented below.

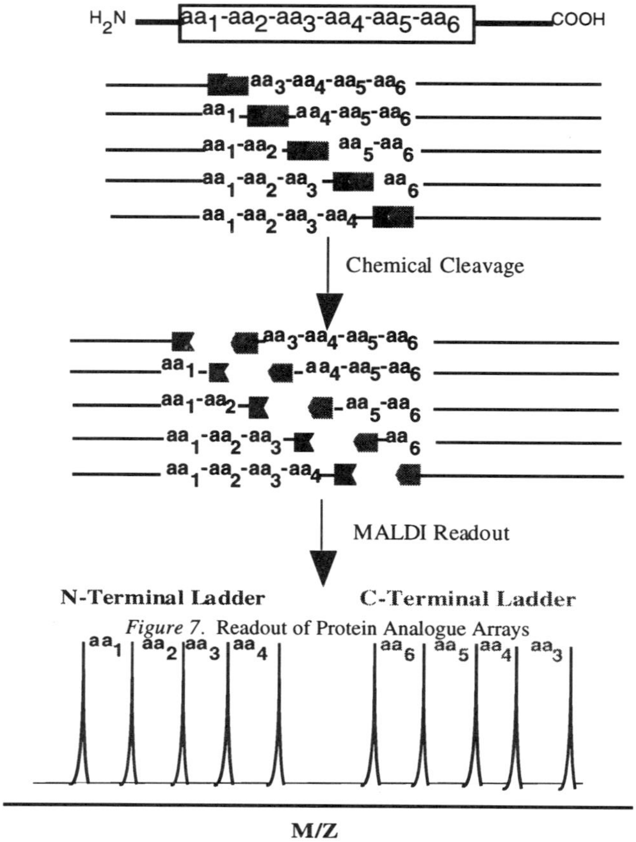

Figure 7. Readout of Protein Analogue Arrays

3.2. PROTEIN SIGNATURE ANALYSIS - APPLICATIONS TO 4OT

We have adapted the protein signature analysis technique to study the catalytic properties of 4OT, as shown in Figure 8. The approach is similar to the general protein signature analysis approach that is outlined in Figure 5 except that we employ two 'functional selections' - a separation based on folding, and a second covalent tagging based on catalysis. By subjecting 4OT protein analogue arrays to selections based on both folding and catalysis it is possible to correlate changes in the chemical structure of 4OT with their effect on each of these properties.

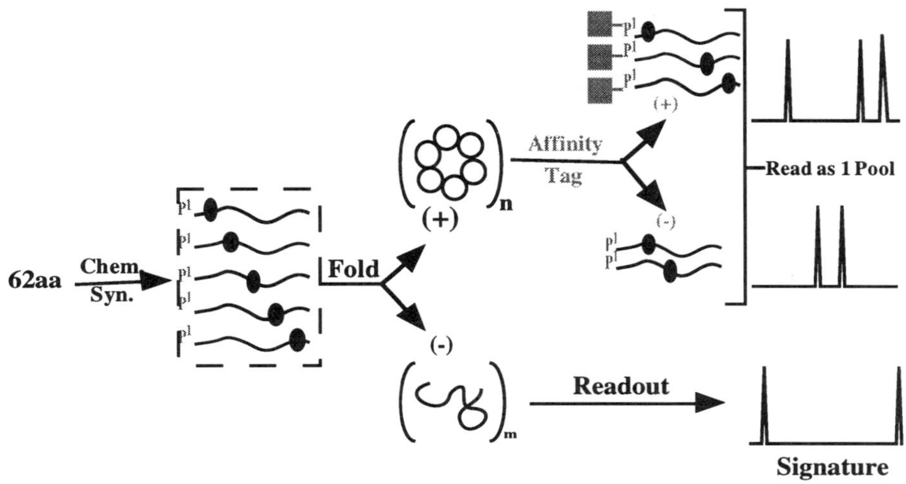

Figure 8. Probing the Chemical Basis of 4OT-Catalysis
using Protein Signature Analysis

In an initial study, we have incorporated two different analogue units into a 6 amino acid region that encompasses residues 8-13 of the 4OT polypeptide chain. It has been hypothesized that this portion of the sequence is involved in substrate binding [6,8]. In one case we prepared an array of five protein analogues placing the Gly-S-βAla dipeptide unit at each possible dipeptide position within the six amino acid stretch, 4OT(8-13) (see Figure 9). The MALDI analysis of this protein array before and after chemical cleavage is shown in Figure 9.

The expected masses for the five intact proteins in the array are 6697, 6696, 6726, 6737, and 6753 Da. The mixture is not easily resolved prior to chemical cleavage because several of the analogues in the mixture are very close in mass. However, after chemical cleavage each member of the array is easily identified from the C-terminal sequence ladder that is generated (Figure 9). Interestingly, this five component protein analogue array did not display any enzymatic activity.

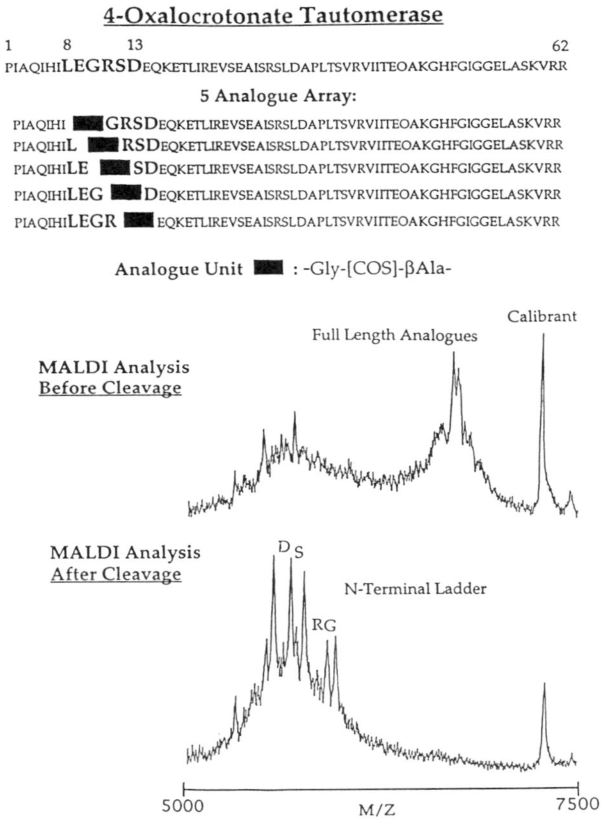

Figure 9. MALDI analysis of a 5 component, 4OT protein analogue array before and after chemical cleavage with hydroxylamine. The analogue unit was Gly-S-βAla.

We have also prepared a second array of 4OT protein analogues in which a Gly-O-Gly analogue unit was scanned through the same six amino acid sequence as above. The MALDI analysis of this protein analogue array before and after chemical cleavage is shown in Figure 10.

Apparently there was some cleavage of this array prior to it's treatment with hydrazine. However, four of the five analogues can be identified from the N-terminal ladder produced after treatment of the array with hydrazine. The fifth analogue apparently contains a deletion of the analogue unit as there is a prominent peak at m/z

6556 which is consistent with the mass expected for such a deletion in the first analogue (expected mass 6552 Da). Interestingly, this array of protein analogue was catalytically active, and we are currently the applying the selection methods outlined in Figure 8 to determine which analogues in the array are responsible for this activity.

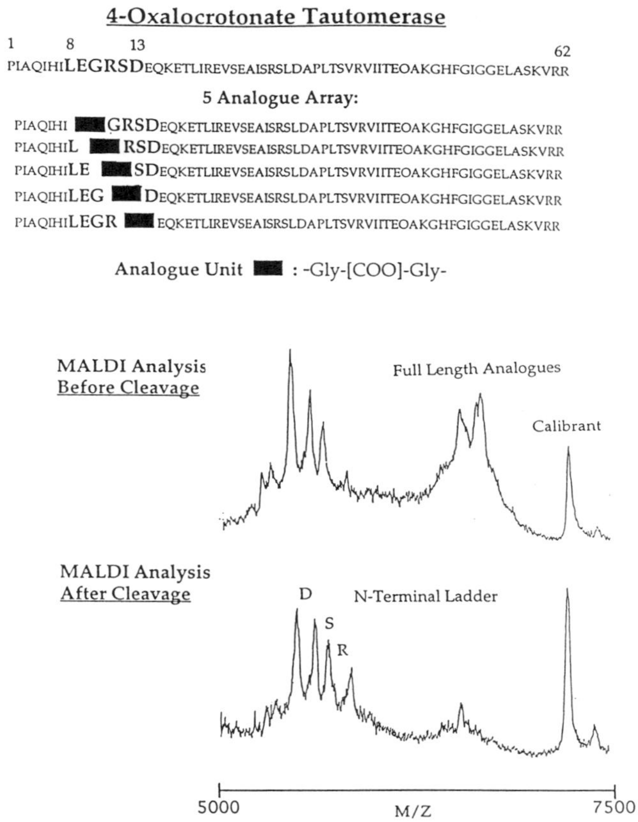

Figure 10. MALDI analysis of a 5 component, 4OT protein analogue array before and after chemical cleavage with hydrazine. The analogue unit was Gly-O-Gly.

4. Conclusion

Total chemical synthesis has proven to be an especially powerful approach for studying the catalytic function of 4OT. We have synthesized and characterized several individual 4OT analogues that have been instrumental in establishing the catalytic significance of several key residues in the enzyme. Our studies with the (*des*Pro^1)4OT and (Cpc^1)4OT analogues served to establish both the structural and mechanistic significance of the Pro[1] residue in 4OT. From our synthesis and characterization of [Met(O)[45]]4OT we discovered that Met[45] oxidation in this analogue apparently weakens the noncovalent protein-protein interactions that define the 4OT enzyme complex. Furthermore, the protein signature analysis technique that is described above promises to be useful in examining how the covalent structure of 4OT determines its function. We are optimistic that this new combinatorial chemistry approach for studying protein function will help dissect the molecular basis of 4OT-catalysis.

References

1. Muir, T. and Kent, S.B.H. (1993) The chemical synthesis of proteins, *Curr. Opin. Biotechnol.* **4**, 420-427.
2. Schnölzer, M., Alewood, P., Jones, A., Alewood, D., and Kent, S.B.H. (1992) In situ neutralization in Boc-chemistry solid phase peptide synthesis, *Int. J. Peptide Protein Res.* **40**, 180-193
3. Canne, L.E., Ferre-D'Armare, A.R., Burley, S.K., and Kent, S.B.H. (1995) Total chemical synthesis of a unique transcription factor-related protein: cMyc-Max, *J. Am. chem. Soc.* **117**, 2998-3007.
4. Harayama, S., Lehrbach, P.R., Timmis, K. (1984) Transposon mutagenesis analysis of *meta*-cleavage pathway operon genes of the TOL plasmid of *pseudomonas putida* mt-2, *J. Bacteriol.* **160**, 251-255.
5. Harayama, S., Rekik, M., Ngai, K.L., Ornston, L.N. (1989) Physically associated enzymes produce and metabolize 2-hydroxy-2,4-dienoate, a chemically unstable intermediate formed in catechol metabolism via *meta* cleavage in *pseudomonas putida*, *J. Bacteriol.* **171**, 6251-6258.
6. Subrammanya, H.C., Roper, D.I., Dauter, Z., Dodson, E.J., Davies, G.J., Wilson, K.S. and Wigley, D.B. (1996) Enzymatic ketonization of 2-hydroxymuconate: specificity and mechanism investigated by the crystal structures of two isomerases, *Biochemistry* **35**, 792-802.
7. Stivers, J.T., Abeygunawardana, C., Mildvan, A.S., Hajipour, G., Whitman, C.P., and Chen, L.H. (1996) Catalytic role of the amino-terminal proline in 4-oxalocrotonate tautomerase - affinity labeling and heteronuclear NMR studies, *Biochemistry* **35**, 803-813.
8. Stivers, J.T., Abeygunawardana, Whitman, C.P., and Mildvan, A.S. (1996) 4-Oxalocrotonate tautomerase, a 41-kDa homohexamer - backbone and side-chain resonance assignments, solution secondary structure, and location of active site residues by heteronuclear NMR spectroscopy, *Protein Science* **5**, 729-741.
9. Fitzgerald, M.C., Chernushevich, I., Standing, K.G., Kent, S.B.H., and Whitman, C.P. (1995) Total chemical synthesis and catalytic properties of the enzyme enantiomers L- and D-4-oxalocrotonate tautomerase, *J. Am. Chem. Soc.* **117**, 11075-11080.
10. Chen, Y.H., Yang, J.T., and Chau, K.H. (1974) Determination of the helix and β form of proteins in aqueous solution by circular dichroism, *Biochemistry* **13**, 3350-3359.
11. Stivers, J.T., Whitman, C.P., and Mildvan, A.S. (1994) NMR, CD, and Kinetic studies of 4-oxalocrotonate tautomerase (4-OTase), ACS Abstracts **208**, 261.
12. Fitzgerald, M.C., Chernushevich, I., Standing, K.G., Whitman, C.P., and Kent, S.B.H. (1996) Probing the oligomeric structure of an enzyme by electrospray ionization time-of-flight mass spectrometry, *Proc. Natl. Acad. Sci., USA* **93** (In Press).
13. Dawson, P.E., Fitzgerald, M.C., Muir, T.W., and Kent, S.B.H. (Submitted) A synthetic Approach to self-encoded arrays of peptide analogues, *J. Am. Chem. Soc.*
14. Dawson, P.E., Fitzgerald, M.C., Muir, T.W., and Kent, S.B.H. (Submitted) Chemical synthesis and functional readout of self-encoded arrays of protein analogues: the cCrk N-terminal SH3 domain, *J. Am. Chem. Soc.*
15. Chait, B.T., Wang, R., Beavis, R., and Kent, S.B.H. (1993) Protein ladder sequencing, *Science* **262**, 89-92.
16. Muir, T.W., Dawson, P.E., Fitzgerald, M.C., and Kent, S.B.H. (1996) Probing the chemical basis of binding activity in an SH3 domain by protein signature analysis, *Chemistry and Biology* **3** (10), 817-825.

APPROACHES TO THE CHARACTERISATION OF TERTIARY AND SUPRAMOLECULAR PROTEIN STRUCTURES BY COMBINATION OF PROTEIN CHEMISTRY AND MASS SPECTROMETRY

M. PRZYBYLSKI[1*], V. SCHNAIBLE[1], J. KAST[1], S. BÜHLER[1], J. MICHELS[1], A. WATTENBERG[1,3], T.A. FLIGGE[1], D. FORST[2], K. DIEDERICHS[2], U. NESTEL[2], K. ZETH[2], M.O. GLOCKER[1], AND W. WELTE[2]

[1]Faculty of Chemistry, and [2]Faculty of Biology
University of Konstanz, P.O. Box 5560 M731, 78457 Konstanz, Germany
[3]Present address:, Institute of Physical Chemistry
University of Frankfurt, Frankfurt, Germany

Abstract

Soft-ionisation methods, particularly electrospray (ESI) and matrix-assisted laser desorption (MALDI) have enabled a breakthrough in the mass spectrometric structure analysis of proteins. Whereas ESI-MS provides direct information about solution structures and non-covalent interactions, the combination of mass spectrometry with structure-specific protein chemistry is emerging presently as a powerful tool for characterising tertiary structures and structure-function relations. Recent developments of chemical modification reactions are summarised which are suitable to the mass spectrometric analysis of reactive sites, surface topology and antigenic determinants in protein-tertiary structures, and can be efficiently employed in x-ray crystallographic structure determinations. Applications to the structure elucidation, and functional characterisation of porin-channel proteins and to leucine zipper protein-nucleotide complexes illustrate their efficiency in the analysis of two most important topics of structural biology, molecular recognition structures and biomacromolecular interaction.

1. Introduction

The development of efficient soft-ionisation methods of mass spectrometry (MS) has led to a recent breakthrough in the direct, molecular characterisation of biopolymers,

* To whom correspondence should be addressed

W. Ens et al. (eds.), New Methods for the Study of Biomolecular Complexes, 17–43.

particularly polypeptides and proteins [1-3]. While fast atom bombardment (FAB-MS) and 252-Cf-plasma desorption (PD-MS) have enabled accurate molecular weight determinations of polypeptides and small proteins [3,4], biomacromolecules considerably beyond 100 kDa have become feasible by electrospray-ionisation (ESI-MS) and matrix-assisted laser desorption (MALDI-MS) [5-7]. In addition to the determination of molecular weights, desorption-ionisation MS methods already have a number of established applications to *primary structure* analysis of proteins, such as full or partial sequence determination, characterisation of intracellular proteolytic processing pathways and the identification of covalent post-translational structure modifications [1,8-10]. A combination of MS with specific chemical (e.g. enzymatic) degradation, or collision-induced fragmentation/tandem-MS has been mainly used in these applications; direct mass spectrometric fragmentation has been successful also, particularly by ESI-MS [3,7].

Beyond the applications to primary structure analysis, recent approaches to the characterisation of higher-order (tertiary) structures, to structure-function studies and even specific *non-covalent* interactions of proteins are presently emerging as exciting new areas of mass spectrometry [3,5,11]; the recent discovery of ESI-MS as a tool for the direct analysis of supramolecular complexes and molecular recognition processes representing a major theme for this NATO conference [12,13]. Furthermore, the feasibility of soft-ionisation-MS methods to the analysis of multicomponent proteolytic mixtures (peptide mapping) has been demonstrated in several applications, and has been applied successfully to the molecular characterisation of chemical modification sites in proteins [14-17]. Since the initial studies employing the combination of specific protein-chemical modification reactions and mass spectrometric peptide mapping [3,14], useful structural information has been obtained already, e.g. on surface topology, tertiary structure micro-environment and specific antigenic binding sites (epitopes) to antibodies [16-24]. Such applications have been further enhanced by the recent analytical development of ESI-MS showing that (i) specific biopolymer interactions can be characterised at near-physiological conditions in solution; and (ii) information about solution structures and conformation of proteins can be derived from multiply-charged ions in ESI spectra. [5,11].

In this article, analytical concept and hitherto successful protein-chemical modification reactions in the characterisation of protein tertiary structures and supramolecular association are summarised in a first paragraph. Model studies with proteins of known crystal structures show that a number of tertiary structure-selective modification reactions are available at present and further specific reaction types for single amino acids are likely to be developed in the near future. Recent applications to porin channel proteins are discussed in a second paragraph which demonstrate the efficiency of protein-chemical modification and mass spectrometry, in combination with x-ray crystallography for the structure determination of an important class of membrane proteins [25]. As described in several recent publications [5], the feasibility of ESI-MS for probing *non-covalent* biomacromolecular interactions has been well illustrated in studies on the structure elucidation of leucine-zipper polypeptide complexes with polynucleotides, encompassing two different types of hydrophobic and ionic interactions [26,27]. Recent studies with several leucine-zipper polypeptides as model structures for transcription activators are summarised in this article, demonstrating ESI-MS in combination with protein chemistry as a powerful methodology, and alternative to the classical electrophoretic analysis of protein-nucleotide interactions.

2. Characterisation of Tertiary and Supramolecular Protein Structures by Selective Chemical Modification and Mass Spectrometry

2.1 ANALYTICAL CONCEPT AND METHODOLOGY FOR THE CHARACTERISATION OF SITE-SPECIFIC MODIFICATION REACTIONS

Chemical modification reactions have long been employed in structure-function studies of proteins, e.g. for modifying enzymatic properties and immunological reactivity [28,29], and differential chemical modification has been used previously for probing specific protein-protein interactions such as in antigen-antibody complexes [30]. Although a number of reaction types for amino acid residues have been evaluated already until ca. 1980 (s. [28] for a review), their full development has been impeded by the lack of specific methods for characterising the multiple reaction products in proteins [3]. Conventional analytical methods generally require laborious procedures (such as radioactive labelling [29]); they do not enable a molecular homogeneity assessment of protein derivatives, and do not provide unambiguous identification of modification sites. Site-specific mutations in structure-function studies of proteins have become increasingly amenable in the last years through recombinant bacterial expression variants [31], yet this approach is restricted to proteinogenic amino acids. Thus, a number of recent studies have shown that there is a considerable need for complementary approaches that can provide information on chemical properties and reactivities in protein structures, in spite of the rapid development of x-ray crystallography and multidimensional NMR methods in the last years [32-36] (s. also cartoon in the preface of this volume).

In contrast to conventional methods based on electrophoretic or chromatographic procedures, mass spectrometric peptide mapping of proteolytic digest mixtures has been found highly suitable for the rapid and sensitive identification of multiple chemical modification sites in proteins [16,17]. Following initial studies showing the feasibility for the mass spectrometric analysis of *limited*, tertiary structure-selective chemical modifications [16,37], a general analytical scheme has been developed for the molecular characterisation of "native-like" chemically modified proteins and protein complexes, as shown schematically in Figure 1 [3]. In the first step of this procedure, the precise extent of modification in a protein, i.e. the number and distribution of partially modified residues are identified by direct mass spectrometric molecular weight determination. Both MALDI-MS and ESI-MS have been found suitable for this purpose, depending on the mass and/or component resolution required for the type of modification [3]. However, an additional dimension of information is provided by ESI-MS by the distribution of multiply charged ions of protein derivatives which have been termed "charge structures" [5] due to their correlation with (e.g. native or denatured) conformation states in solution [5,11,16]. This feature of ESI-MS provides a *relative* comparison of conformational states and hence a molecular assessment of the structural selectivity for a protein-chemical modification which has not been feasible hitherto by conventional methods. Thus, identical or very similar charge structures of ions for partially modified proteins indicate a native-like structure for modification reactions that do not cause charge alterations (*vd. infra*; Figure 2) [3]. Corresponding data from ESI mass spectra have been particularly useful for the crystallisation of selectively modified proteins [38], and in the functional characterisation of protein complexes, as shown below for ion-channel proteins.

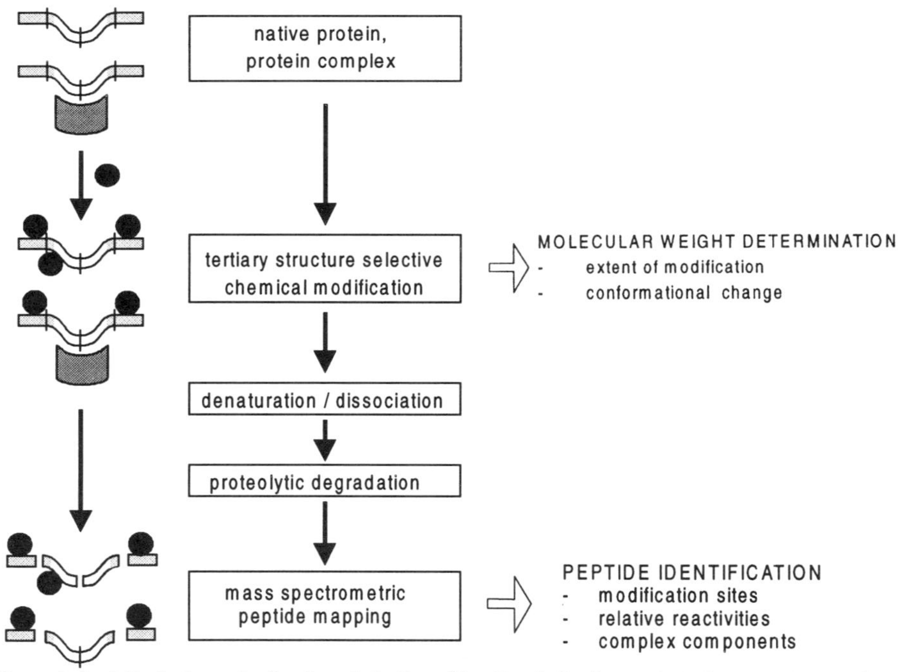

Figure 1. Analytical scheme for the characterisation of functional sites in protein tertiary structures and supramolecular complexes by selective chemical modification and mass spectrometric peptide mapping. Modification groups and proteolytic cleavage sites are indicated by filled circles and lines, respectively. The white area represents a shielded sequence in a protein complex, as adopted from the general mass spectrometric epitope mapping procedure for protein antigens in ref. [18].

Following the direct characterisation of protein derivatives, denaturation (complex dissociation) and proteolytic digestion is performed, and the sites of modifications are identified by mass spectrometric peptide mapping of peptide fragment mixtures [16]. Moreover, assignments of relative reactivities at specific amino acids can be derived from series of partially modified proteins or protein complexes. Such direct determinations of reactive sites in intact protein structures can be assessed by comparison of chemical modifications in proteins with model peptides, under identical reaction conditions, and have been found highly useful in structure-function studies [5,17,39].

Examples of modification reactions at specific amino acid residues evaluated by mass spectrometric peptide mapping, and hitherto successful applications in protein structural studies are given in Table 1. The extent of modifications (conversion curves) can vary from essentially quantitative modification of lysine-ε-amino groups in globular proteins according to their relative surface accessibilities [16], to high structural selectivity as has been observed in the modification of Arg residues by 1,2-diketone-derivatives [17,40]. Thus, amino acylations of Lys residues in aqueous solution (pH 6.5), e.g. by acetyl or succinyl groups have provided astounding, detailed structural information [3]; acylation sites are readily identified by peptide mapping using serine proteases, confining cleavage sites to Arg residues. Reactivities at lysine-

TABLE 1: Examples of modification reactions of specific amino acid residues employed in the mass spectrometric characterisation of protein tertiary structures and protein interactions.

amino acid	reagent	product	application	refs
lysine, N–Terminus	carbonic acid anhydride	acyl-Lys	surface topology, reactive sites, epitope mapping binding surfaces	[3, 16, 45]
arginine	1,2-diketone	N7,N8-(1,2-dihydroxyethylene)-Arg	microenvironment structures	[17, 22]
tyrosine	iodine	3-iodo- and 3,5-diiodo-Tyr	heavy atom insertion/x-ray crystallography surface topology	[36, 37]
	tetranitro-methane	3-nitro-Tyr	surface topology	[37]
cysteine(s)	iodoacetamide	s-carbamido-methylene-Cys	redox catalysis; protein folding	[10, 39]
	arsonous acid derivatives	arsonous acid dithioester-Cys		[23]
histidine	diethylpyro-carbonate	carbethoxy-His	reactive centres	[22]

amino groups were found to correlate well with relative surface accessibilities (e.g. at a 1.4 Å van der Waals sphere of solvent accessibility [41]) for several proteins with known structures, whereas pK_a-differences were only found significant for the rapid N-terminal acylation [17,42]. This feature has provided useful molecular information on surface topology and accessible reactive sites, as illustrated in structure-function studies of porin ion-channel proteins (see also ref [43], this volume). In hen egg-white lysozym (HEL), the most reactive Lys-97 residue has the highest surface accessibility [17]. The selectivity of the acylation has been established by crystal structure determination of its Lys-97-monosuccinyl derivative (s. 2.2; Figures 4,5). A new application is the recently explored fixation of heavy-atom derivatives in proteins by amino-acylation (s. 2.2).

Furthermore, Lys-amino-acylation has been successfully employed in the characterisation of protein interactions, such as antigenic determinant structures (epitopes) to monoclonal antibodies and binding surfaces in complexes of the elongation factor EF-Tu/Ts proteins [44]. Approaches to the identification of epitope structures, including conformationally assembled epitopes, by differential chemical modification and mass spectrometric peptide mapping of antigen-antibody complexes (epitope-excision), and applications to Troponin-T and HEL epitopes have been described in recent publications [20,21,45].

In contrast to the amino-acylation, high structure-dependent selectivity has been found for the bifunctional modification of cysteine-thiol groups by arsonous acid derivatives (phenyl-arsinoxides [23]) which have been employed in cysteine folding pathways [3, 10] and for the Arg-specific modification by 1,2-cyclohexanedione (CHD) [17]. The latter reaction is an example for the efficiency of mass spectrometric peptide mapping in probing microenvironment structures. Reactivity of Arg residues with 1,2-diketones and formation of dihydroethylene adducts is only found for suitable intramolecular proton acceptor residues and sufficient accessibility of the guanidino-nitrogen atoms, as illustrated by the reaction pathway in Figure 2. Partial structures around two reactive Arg residues (Arg-10, -33) in bovine RNAse A and HEL are shown in Figure 2. In HEL, selective modification was found for Arg-5, -12, -112, and -73 which all have suitable intramolecular hydrogen acceptor groups [17]. The only modification site in RNAse A was identified at the Arg-39 residue having suitable *asymmetric* accessibility for the guanidino group; the lack of reactivity at other Arg residues is explained either by effective shielding of both guanidino-nitrogen atoms (Arg-10) or the lack of deprotonating residues in the microenvironment (Figure 2) [46].

The present applications and recent studies utilising new types of protein-chemical reactions indicate broad analytical potential for structure-reactivity correlations by selective chemical modification and mass spectrometry. Although still in the stage of analytical development, the further evaluation of this protein-chemical approach as a complement to site-specific mutation using recombinant biotechnology will be interesting in the future.

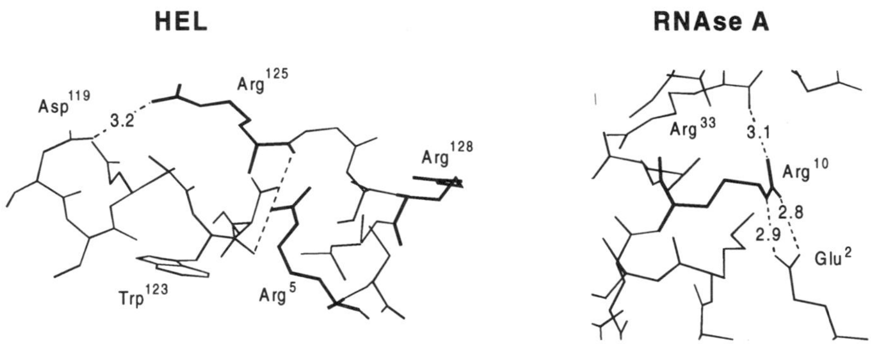

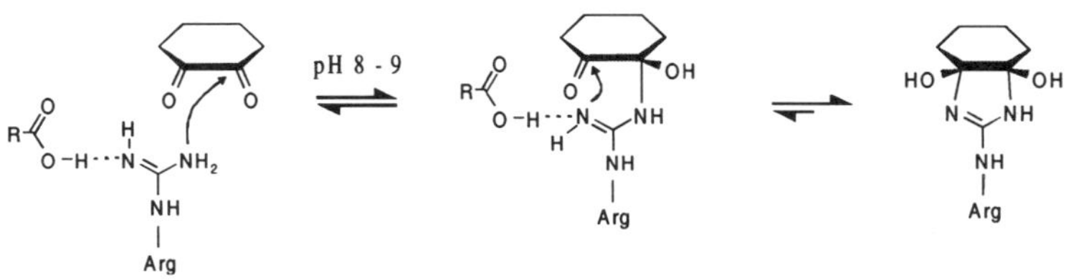

Figure 2. Tertiary structures around Arg-5 and Arg-125 residues in HEL and Arg-10, Arg-33 residues in RNAse A, and reaction scheme of the Arg-specific modification by 1,2-cyclohexanedione. Proteins were incubated for 2 hrs with a 50-fold excess of CHD in sodium borate buffer (pH 8.5), under identical conditions. Modification sites were determined by trypsin digestion and PD-MS and MALDI-MS peptide mapping. The reactive Arg-125 and Arg-5 residues have suitable hydrogen-bond distances to the Asp-119-carboxylate and the Trp-123 peptide bond, respectively [17]. The two nitrogen atoms of Arg-10 in RNAse A are at hydrogen-bond distance to Glu-2-carbonyl (2.8 Å) and the Arg-33-peptide bond (3.3 Å); no acceptor group is found in the environment of Arg-33. Structural data in the upper panel are taken from the Brookhaven National Laboratory, protein data bank.

2.2 PROTEIN-CHEMICAL MODIFICATION REACTIONS IN STRUCTURE DETERMINATION BY X-RAY CRYSTALLOGRAPHY

Specific covalent modification reactions have not yet been considered in detail and experimentally explored in X-ray crystallographic structure determination procedures of proteins. The lack of work in this area is in marked contrast to the successful use of recombinant protein mutations for crystallisation and structure elucidation [33,36]. Only few previous studies have been performed using chemically modified proteins in crystallisation trials; however, an illustrative example is the recent crystallisation of a chlorophyll membrane protein which was successful only upon a specific N-alkylation reaction [38]. Major obstacles, considered in the past to prevent the use of chemical modification have been (i), problems in the preparation/purification of protein derivatives with suitable homogeneity; and (ii), insufficient characterisation methods. As pointed out in the preceeding paragraph, mass spectrometry enables the molecular characterisation of chemically modified proteins and, hence, provides the basis for evaluating the homogeneity of proteins suitable for X-ray crystallography.

The well-known structure of HEL was utilised as a model system to study the effect of amino-acylation on the crystallisation and structure determination. A series of succinylated HEL derivatives were prepared by reaction with succinic-anhydride as previously described [16], and protein derivatives with different extent of modification (1-4 succinyl groups) purified by gel filtration and HPLC (Figure 3). Two succinylated proteins were characterised by MALDI-MS and ESI-MS having average contents of one and two succinyl groups, and yielded isomorphous crystals with the native protein (Figure 4). The succinylation site was identified at Lys-97 by mass spectrometric peptide mapping using Lys-C-protease, consistent with the previously described highest surface accessibility for this most reactive lysine residue [3].

The structure determination of the monosuccinylated-HEL, was performed by multiple isomorphous replacement with the crystals shown in Figure 4 diffracting to 2.6 Å, and provided unit cell constants and difference Fourier electron density maps nearly identical with that of the native protein [47]. In the structure (Figure 5) the

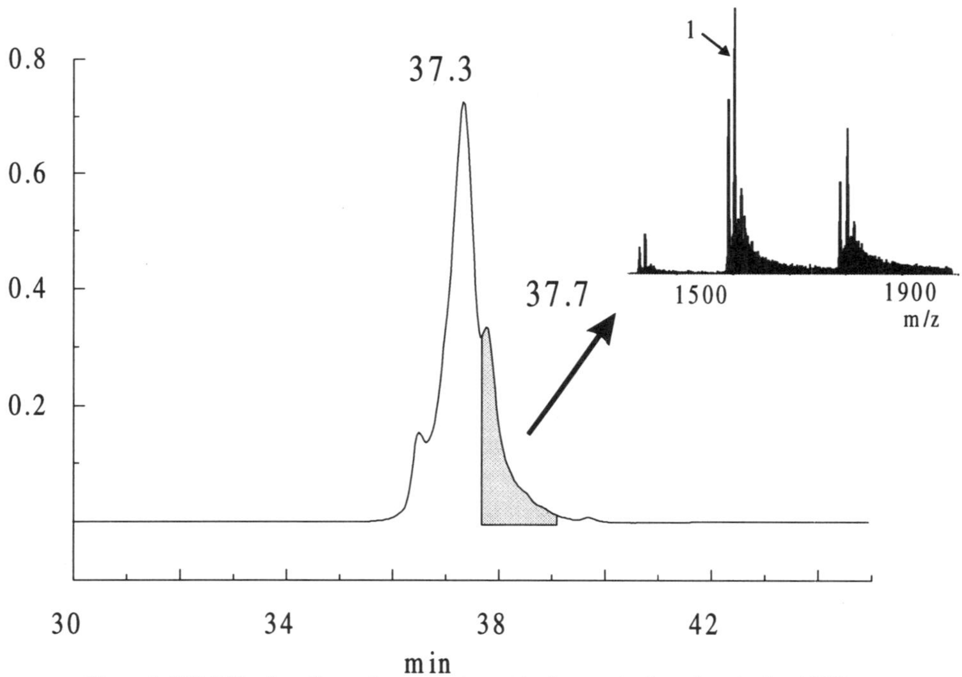

Figure 3: HPLC Fractionation and mass spectrometric characterisation of succinylated HEL, prepared with a 50-fold molar excess of succinic-anhydride. The insert shows the ESI spectrum of the separated component indicated by the shaded area, with the $[M+9H]^{9+}$ ion of monosuccinyl-HEL as the major product (M_r, 14440 Da); average extent of succinylation was 1.3. Reaction conditions for succinylation, and HPLC-separation conditions were as described in refs. [16,47].

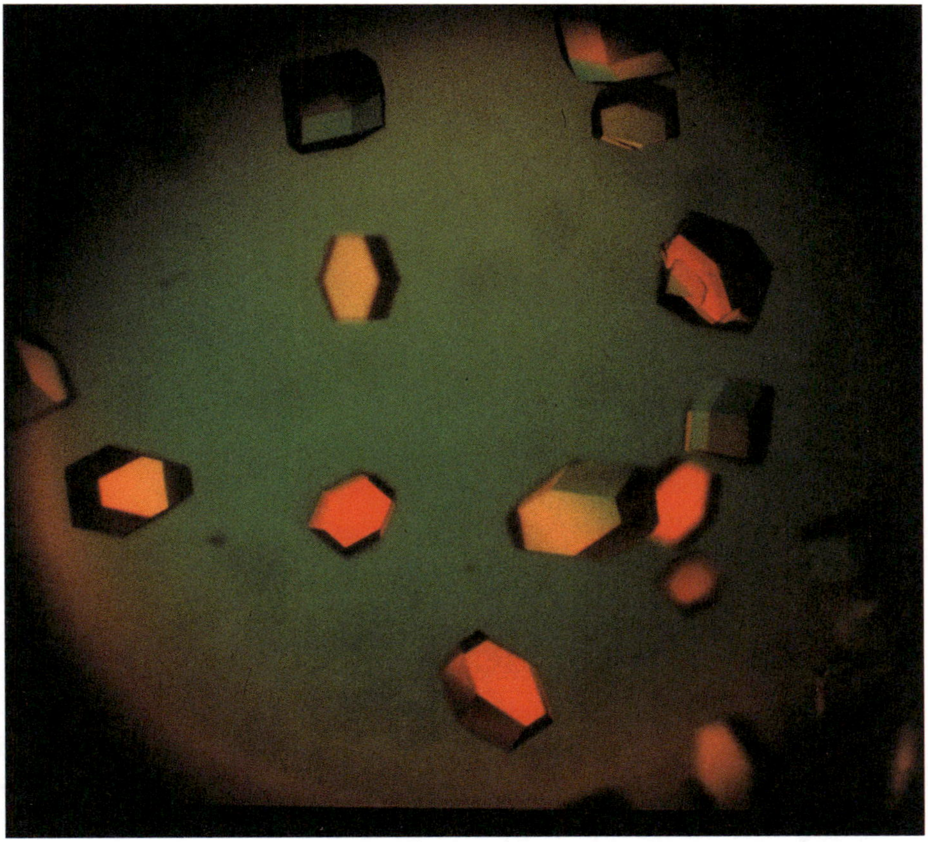

Figure 4. Photograph of hexagonal single crystals of succinylated HEL isolated by HPLC (s. Fig. 3).

electron density at Lys-97 was not sufficient to ascertain the location of the succinyl group, so that this group was added arbitrarily. The failure to identify the modified residue may be explained by a high flexibility of the succinyl group at the surface-exposed Lys-97 residue. This result is consistent with recent x-ray crystallographic studies of chemically modified porins in which modification sites were not detected from electron density maps, despite the preparation of suitably diffracting crystals (see below, 3.1).

Several effects may be responsible for the failure in crystal structure analysis to identify partial structures such as modification groups of high mobility which are difficult to predict with the present methodology of difference electron density maps [25,48]. Hence, the capability of MALDI-MS and ESI-MS to directly assess complete chemical structures and post-translational modifications in proteins, renders x-ray

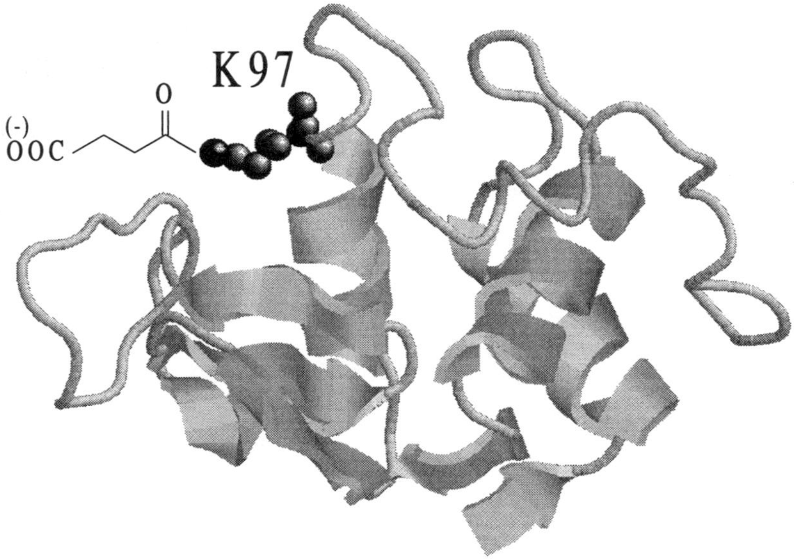

Figure 5. Structure model of lysine-97-succinylated HEL determined by x-ray crystallography (s. Fig. 3). The succinyl-group at K-97 was not determined from the electron density map.

crystallography and mass spectrometry particularly effective as *complementary* methods. An important application area may be the further development and refinement of isomorphous replacement methods by heavy-atom derivatives for phase assignment in crystal structure analysis. This methodology is largely based at present on trial- and error-procedures (e.g. soaking or co-crystallisation [36]). A new concept pursued in our laboratory is the development of defined procedures for the *covalent* fixation of heavy-atom derivatives at conditions that still maintain intact protein structures, e.g. by the amino-acylation described above.

First model studies have been performed with iodinated derivatives of lysozyme (cf. Table 1), and with HEL amino-acylated by [(succinimidooxy)carbonyl]methyl-mercuri chloride. The ESI mass spectrometric characterisation of the corresponding chloromercuri-acetyl-HEL provides the possibility of selecting suitable derivatives with exactly defined chemical structure, such as for the Lys-97-monosubstituted HEL derivatives (Figure 6). The direct characterisation of molecular stoichiometries for heavy-atom protein derivatives may lead to improved crystallisation methods, as well as procedures for the assignment of electron density maps.

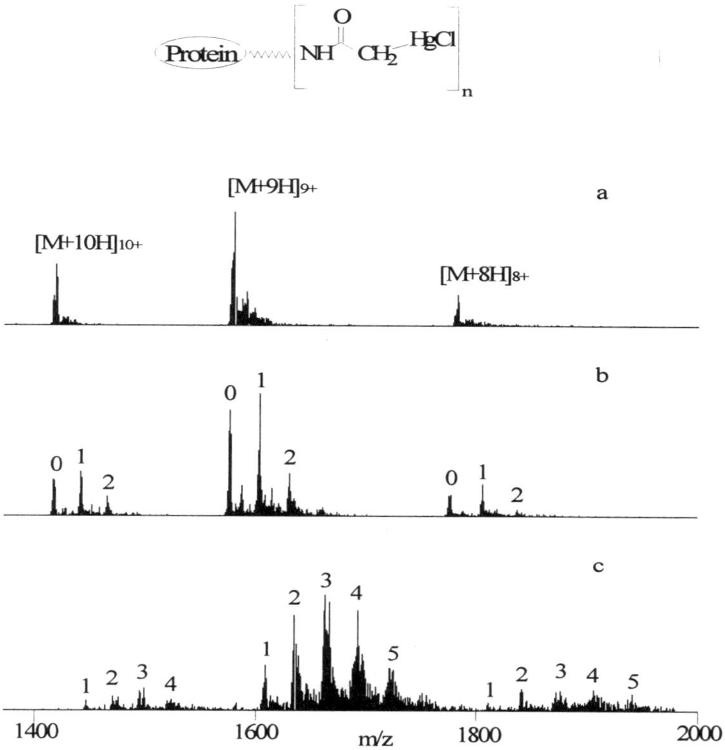

Figure 6. ESI-mass spectra of (a) native HEL; (b,c), chloromercuri-acetyl-HEL derivatives modified with a two-fold and five-fold molar excess of [(succinimidooxy)carbonyl]methyl-mercuri chloride, respectively. The numbers denote the number of chloromercuri-acetyl groups introduced.

3. Mass spectrometric and X-Ray Crystallographic Structure Determination of Porin-Channel Proteins

3.1 STRUCTURE DETERMINATION AND ION TRANSPORT PROPERTIES OF SUCCINYLATED PORIN FROM *RHODOBACTER CAPSULATUS*

The permeability barrier of gram-negative bacteria is overcome by water-filled channel-type porins which form the major protein component in the outer membrane and allows the permeation of ions and other small molecules up to a molecular weight of ca. 800 Da [25,49]. To date, the structures of only five bacterial porins have been completely determined by X-ray crystallography, such as the ion-channel porin from *Rhodobacter capsulatus* (R.c.-porin) [25]. Generally, they form trimeric complexes consisting of 16-18-stranded β-barrels of high stability, which have a similar periplasmic pore mouth but different loops in the medium-opposed mouths (s. Figures 7 and 8, and ref. [43], this volume). The structures, transport functions and selectivities, and membrane association of general diffusion channels as well as specific porins have been subject of detailed studies by several groups in the last years, both in the soluble and the solid (crystalline) state [25,43,50]. Methods for the direct mass spectrometric characterisation of porins in solubilised, detergent-containing preparations have been recently developed [51-53].

A characteristic structure element of R.c.-porin is a central constriction loop inside the β-barrel cylinder which constricts the free cross section to approximately 8x10Å and shows a highly asymmetric charge distribution of acidic and basic amino acids, opposite to each other [51]. Ion conductance studies with lipid membranes *in vitro* have strongly suggested this peptide loop as the central site determining cation/anion permeability and transport selectivity [43]. The functional role of charges near the pore mouth of R.c.-porin and other ion channels has been intensively discussed in theoretical work. In order to introduce new negative charges, the R.c.-porin was succinylated with succinic anhydride at pH 6.5, using conditions that where established previously to provide selective, native-like acylation of amino groups [16]. The structure of the major, purified succinylated porin derivative was identified by X-ray crystallography and mass spectrometry. An isomorphous crystallisation with space group and unit cell constants very similar to the native porin was found which enabled the structure elucidation at 2.4 Å resolution (s. Figure 7 and ref. [51]). However, the electron density maps did not permit an assignment of the succinylated groups except for Lys-46, one of the lysine residues at the constriction site of the inner channel surface. Ion transport experiments in black lipid membranes revealed substantial channel forming activity of the succinylated protein with a higher single channel conductance than the native porin, and a significant increase of the cation selectivity (Table 2).

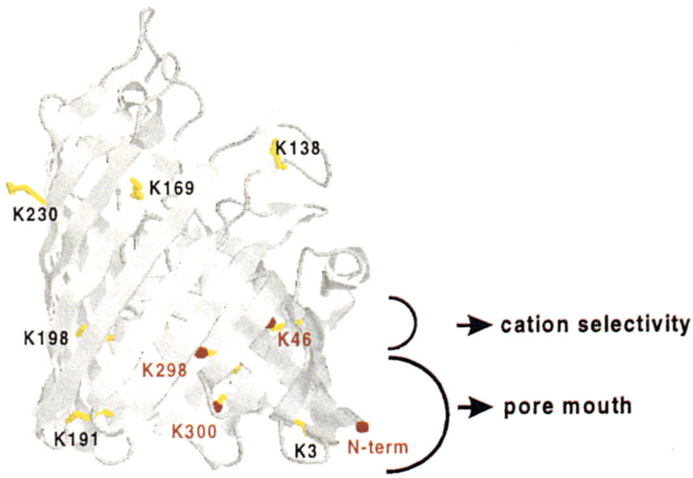

Figure 7. Point-charge structure model of tetrasuccinylated porin from *Rhodobacter capsulatus* [51].

The precise extent and the sites of succinylations were determined by MALDI-MS which revealed the presence of a mono- (MS-porin) and exactly tetra-succinylated protein (TS-porin), for the porin derivative with isomorphous crystal structure (M_w, 31541 Da for the native R.c.-porin; 31941 Da for TS-porin) [51]. The complete identification of the succinylation sites in TS-porin was obtained byMALDI-MS peptide mapping of proteolytic peptides, and by mass spectrometric and sequence analysis of HPLC-isolated peptide fragments. These results revealed the selective modification of three Lys-amino groups (K-46, K-298, K-300) and the N-terminus. All succinylation sites were found at the inner channel surface. Lys-46 is located at the channel constriction site, while the other succinylated residues are all near the periplasmic pore mouth. As shown by the structure model of TS-porin (Figure 7), this specific pattern is well compatible with a point charge model [50] that directly provides evidence to explain the concomitant increase of both the single-channel conductance and the cation selectivity.

TABLE 2: Modification sites of succinylated porins and ion transport properties through black-lipid membranes.

Protein	succinylation sites/ proteolytic peptides[a]	single-channel conductance[b]		membrane potential[c]	
		c/KCl	G(nS)	Vm (mV)	P
n-porin[d]	934 (1-9)	0.10	0.32	39	9.0
	4345 (27-69)	0.03	0.09		
	6824 (231-298)	0.01	0.03		
MS-porin	α-NH_2/1035 (1-9)*	0.10	0.5	44	13
	K-46/2199 (42-62)*	0.03	0.24		
		0.01	0.09		
TS-porin	α-NH_2/1035 (1-9)*	0.10	0.90	49	23
	K-46/2199 (42-62)*	0.03	0.53		
	K-298, K-300/3390 (270-301)**	0.01	0.30		

[a] Molecular weights and sequences determined by MALDI of tryptic peptides and Edman sequence analysis [51].

[b] Average single channel conductance (G) in diphytanoyl-phosphatidylcholine/n-decane membranes [50]. c, molar KCl concentration.

[c]Vm, zero-current membrane potential (s. ref. [25]); P, permeability ratio P_{cation}/P_{anion}

[d]n-porin, native porin; MS-porin, monosuccinylated porin; TS-porin, tetrasuccinylated porin.

The succinylation of native R.c.-porin indicated the combination of tertiary structure-selective chemical modification and mass spectrometry as a powerful approach for the molecular structure-function evaluation of channel proteins. These results have been already extended to other ion channels, using succinylation as a specific chemical probe, such as in the crystallisation and mass spectral characterisation of the Omp32 porin from *Comamonas acidovorans* [52]. Moreover, a *water-soluble*, tetrasuccinylated form of R.c.-porin has been recently prepared; this succinyl-derivate had still native-like ion transport properties, yet characteristic, different succinylation sites at the inner channel surface were found [54]. These results suggest possible applications of chemical modification and mass spectrometry for probing the conformation-dependent gating effects of ion channels [55].

3.2 STRUCTURE ELUCIDATION OF A SUCROSE-SPECIFIC PORIN FROM *SALMONELLA THYPHIMURIUM*

In comparison to ion channels, the structure elucidation and characterisation of transport properties of specific porins are only in an initial state. A first x-ray crystal structure has been obtained for the sucrose-specific porin from *Salmonella Thyphimurium* (scry-porin; s. Figure 8) [56]. This porin has recently found much interest due to the presence of both a β-barrel-type channel structure, and an additional N-terminal polypeptide chain containing a putative leucine-zipper motif [53] (s. Figure 9). The crystal structure of scry-porin is shown in Figure 8 which clearly delineates the β-barrel transmembrane topology, as well as periplasmic and external loops [57,58]. Surprisingly, no suitable electron density has been found for the N-terminal polypeptide chain to derive information about its structure and topology; hence a possible periplasmic association and function of this N-terminal partial sequence (1-63) is unknown at present (Figure 9).

The systematic analysis of porins, as well as other detergent-solubilised membrane proteins by MALDI-MS with high sensitivity has recently become amenable through the development of suitable matrix/sample preparation procedures [53]. A crucial step for optimal sample preparation has been the addition of detergent-containing protein solution to a preformed layer of matrix (such as α-cyano-4-hydroxy-cinnamic acid [6]) by treatment with micro-ultrasonication which results in homogeneous microcrystalline samples [52,53]. The importance of using suitable detergent types and concentrations is illustrated in Figure 10 by MALDI spectra of native scry-porin, isolated from *S. Thyphimurium.* The spectrum obtained from a sample solubilised in n-octylglucoside (n-OG) above the critical micelle concentration (cmc; 0.26 %) yielded abundant singly and doubly charged molecular ions providing an exact molecular weight determination for the intact protein ((1-483); M_W, 53164 Da). By contrast a sample rebuffered in n-dodecylmaltoside at low concentration showed considerable degradation with molecular ions (M_D, 46241 Da) due to removal of the N-terminal polypeptide (1-63).

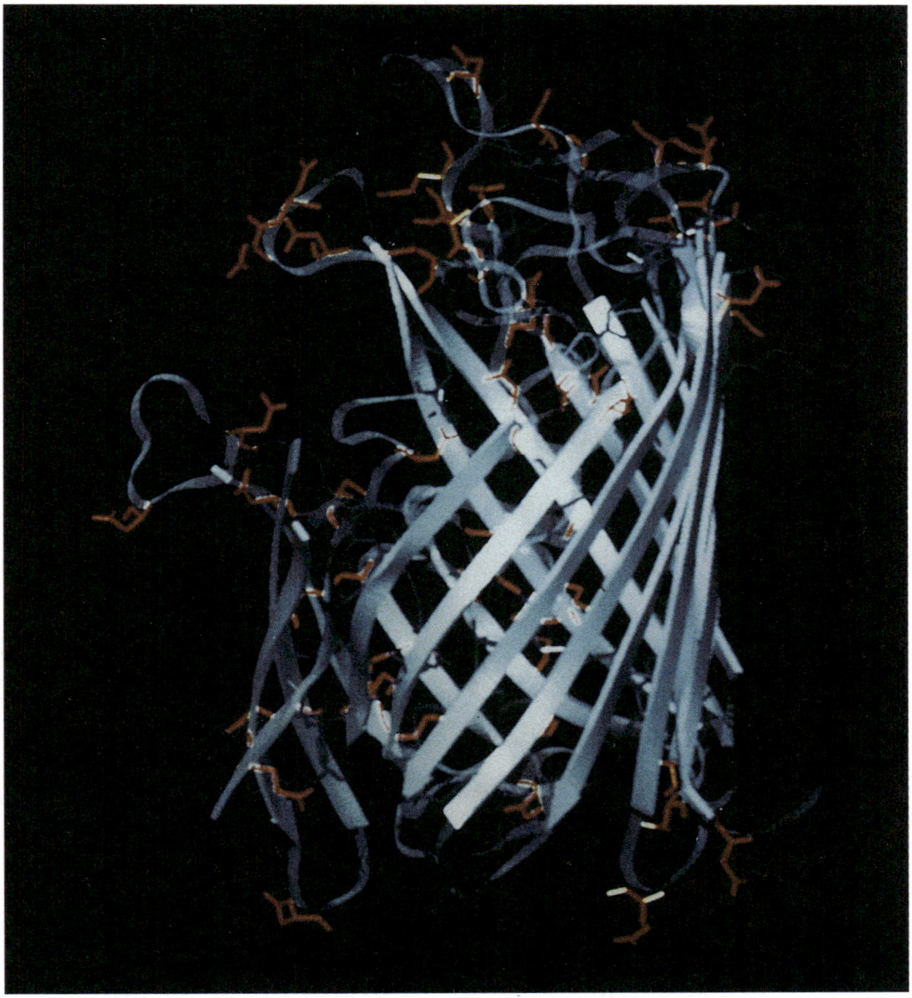

Figure 8. Structure of scry-porin from x-ray crystallography, showing the channel composed of β-barrel transmembrane domains in the membrane plane. The N-terminal seq (1-63) is not detected in the crystal structure [58].

The formation of a truncated form of scry-porin was further ascertained by ESI-MS which yielded molecular ions with charge distributions due to the presence of a native-like and a denatured conformation (M. Przybylski et al., manuscript in preparation). This instability of scry-porin as revealed by MALDI-MS may be well explained if shielding of the protein by the detergent is insufficient to prevent its degradation by traces of proteolytic enzymes remaining from the isolation.

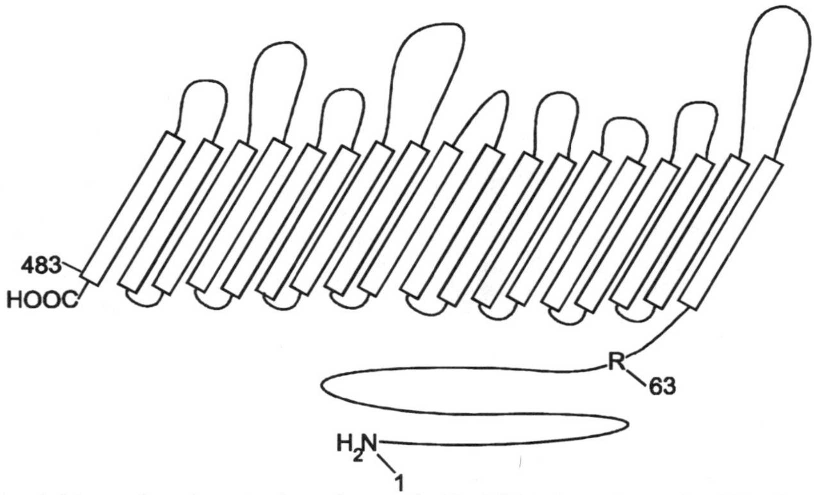

Figure 9. Scheme of membrane topology of scry-porin. The full-length protein contains 483 amino acid residues (M_W calculated: 53164 Da). The β-strand structure elements comprising the channel transmembrane domain are boxed. The cytoplasmic polypeptide sequence (1-63) is not detected in the crystal structure; cf. Figure 8.

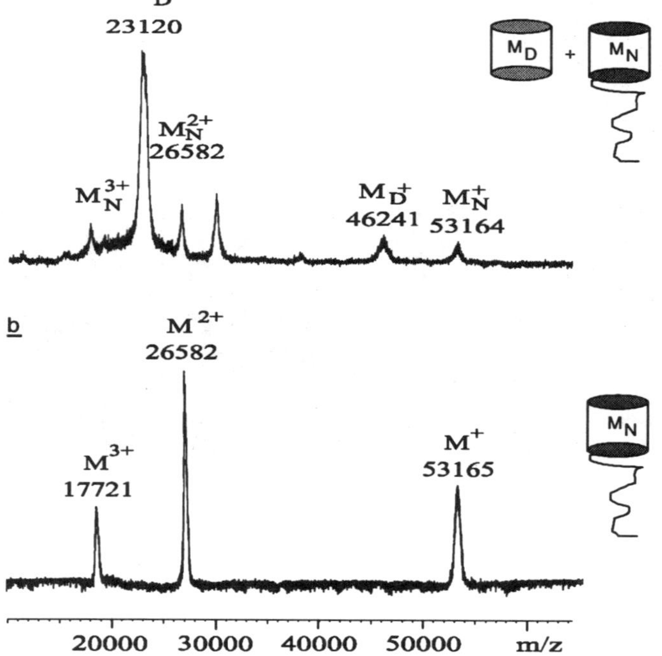

Figure 10: MALDI-mass spectrum of scry-porin (M_w, 53164 Da) prepared from a solution in 0.04% n-dodecylmaltoside (a) and 0.4% n-octylglucoside (b). Sample preparation was performed by ultrasonic treatment using HCCA as matrix as described in [53]; sample amount ca. 10 µg. M_N, native full-length protein (1-483); M_D, degraded form of scry-porin lacking the N-terminal sequence (1-63).

The current results on the characterisation of scry-porin further illustrate the efficiency and complementarity of x-ray crystallography and mass spectrometry in protein structure analysis. As suggested by the topology scheme (Figure 9), recent studies indeed indicate a high susceptibility to proteolytic degradation at the conjunction of the β-barrel core to the N-terminal polypeptide, whereas native porin channels have been found quite resistant against proteases [53,54]. For a separate structure-function analysis, the N-terminal polypeptide (1-63) has been recently prepared by chemical synthesis.

4. Mass Spectrometric Approaches to the Characterisation of Protein-Nucleotide Complexes

4.1 ELECTROSPRAY MASS SPECTROMETRY OF LEUCINE-ZIPPER POLYPEPTIDES

Successful analyses by ESI mass spectrometry have been obtained for a variety of non-covalent complexes of biomacromolecules already as shown in recent publications, including several contributions to this volume [3,5,11-13,59]. Examples to date include a wide range of biomacromolecular complexes such as proteins with low molecular weight ligands (e.g. substrates, inhibitors), specific cofactor and metal-ion binding to proteins, antigen/epitope:antibody complexes up to homo- and heteromeric multicomponent assembly [5,60-62]. Furthermore, they encompass different binding types such as ionic interaction, co-ordinative binding and hydrogen bonds with a wide range of binding/dissociation constants. (for complexes of known structures), thus indicating the broad potential of ESI-MS. The application to "coiled-coil" types of interactions, as exemplified by the leucine-zipper polypeptides, was considered to be a particular challenge for ESI-MS due to the predominant hydrophobic binding, and their structural specificity in the interaction with poly-nucleotides [63,64].

Leucine zipper classes of proteins have attracted much interest in the last years because of their functions in transcription regulation and their structure and specificity in DNA recognition [26,27,64]. Leucine zipper structural elements, consisting of repeating heptads of leucine residues and dominating hydrophobic and non-polar interactions, followed by a basic DNA-binding sequence have been identified in a number of transcription activator proteins, and recently in ribosomal RNA-binding proteins [64]. The hitherto best known leucine zipper is the C-terminal domain of the yeast transcription-activator GCN4, the structure of which has been characterised by x-ray crystallography as a complex with a specific palindromic nucleotide recognition sequence [26]. As shown by the crystal structure (s. Figure 11), GCN4 assembles to a dimeric coiled-coil complex in which hydrophobic interactions between the α-helical polypeptide chains are the dominant stabilizing forces. This structure thus appeared to be an appropriate model system for studying the corresponding protein-protein and protein-nucleotide interactions by ESI-MS. Initial ESI mass spectra of synthetic GCN4 peptides revealed [65,66] that (i) non-covalent dimers could be identified by the specific *odd-charged* ions of complexes; (ii) however, their abundances relative to the monomers were weak suggesting a low stability of hydrophobic interactions under ESI conditions [5].

Synthetic polypeptides, comprising the GCN4 dimerisation and DNA-binding regions (221-281), and the suitable oligonucleotide recognition sequences are shown ion Figure 11. In a systematic study, the coiled-coil complexes of several variant peptide sequences were studied in homo- and hetero-dimerisation experiments using different equilibria (pH, buffer, concentration) [67], and varying desolvation potential differences between capillary tip and skimmer (repeller) electrode (ΔCS [5], in an ESI source without a countercurrent gas [68]. The identification of a leucine zipper peptide, GCN4-p1, displaying a characteristic $[M+5H]^{5+}$ ion of the dimer (molecular weight,

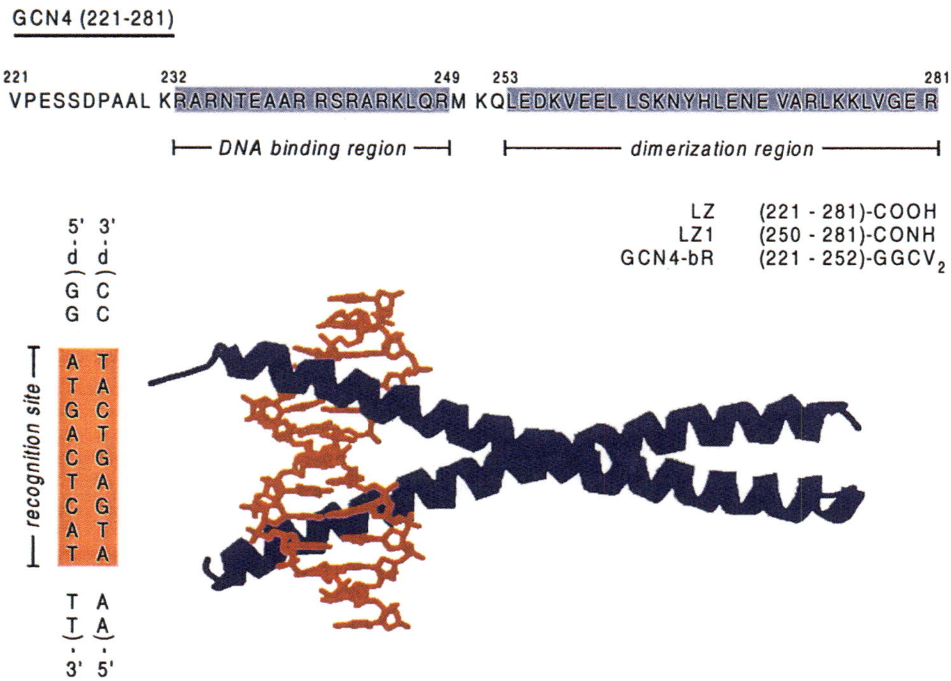

Figure 11. Structure the leucine zipper -oligonucleotide complex comprising the GCN4 polypeptide (221-281) determined by x-ray crystallography, and synthetic dimerisation and DNA binding peptides. The DNA binding and dimerisation regions are shaded in blue, the palindromic nucleotide recognition site is shaded in red. the crystal structure of the GCN4-nucleotide complex was depicted from the Brookhaven National Laboratory, protein structure bank.

8078 Da) at low desolvation potential of 10 V, and its complete dissociation at ΔCS = 60 V are illustrated in Figure 12. 5+ charged ions of complexes from several synthetic leucine zipper peptides are summarised in Table 3.

These ESI spectra enabled the identification of homo- as well as hetero-dimers under mild conditions, although no quantitative formation of the weak coiled-coil complexes was obtained (presumably by adsorption processes and other dissociative conditions [67,69]). Moreover, an important result of this study was the direct identification of trimeric complexes in addition to dimers at appropriate solution equilibrium conditions and with peptide sequences that were previously shown to favour the formation of triple-stranded coiled-coil complexes [63,69].

The successful identification of intact double-stranded oligonucleotides, as well as higher-order nucleotide complexes by negative-ion ESI-MS has been shown in several recent publications [60,70]. as a further pre-condition for the direct analysis of protein-DNA interactions. In most cases, remarkable stabilities of multiply charged duplex ions were observed, despite some limitations imposed by purification problems of polynucleotides from contaminating alkali salts [67] and by limitations of the molecular weight (mass/charge) range attainable [5]. First model studies of polypeptide complexes with small nucleotides such as ATP/GTP successfully revealed intact molecular ions using a conventional quadrupole mass analyser with limited m/z range [44,67]. However, they failed to provide ions of higher charge state for leucine zipper-nucleotide complexes. It should be pointed out that the effects on charge structures of proteins upon polar interaction with an oligonucleotide backbone are still unknown, as

Ac-RMKQLEDK VEELLSK NYHLENE VARLKKL VGER-$CONH_2$

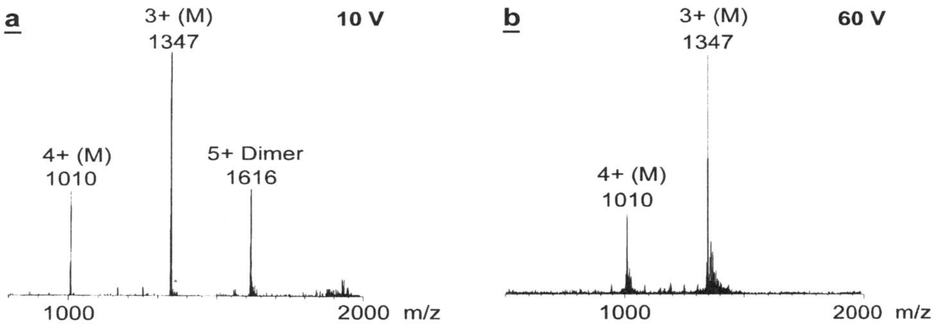

Figure 12. ESI Mass spectra of leucine zipper polypeptide GCN4-p1 (249-281) at 10 V (a) and 60 V (b) declustering voltage. The $[M+5H]^{5+}$ ion of the dimer is only observed at ΔCS = 10 V. Spectra were obtained by injection of 5 μl of a 50 mM peptide solution in amixture of 2 mM ammoniumacetate : methanol, 9:1 at pH 6.0 [67].

are the multiply charged ions detectable by ESI-MS. Recently, the use of ESI-time-of-flight instruments have clearly shown advantages for detecting high-mass complex ions [71], as illustrated by the ESI spectra of the intact complex of the GCN4 (221-281) leucine zipper with its recognition duplex-oligonucleotide. The molecular weight of the synthetic complex was determined with high accuracy despite its relatively low purity (Figure 13). These results give promise to the further evaluation of basic leucine zipper-DNA complexes by ESI-(TOF)-MS.

TABLE 3. ESI-Mass spectrometric identification of dimer and trimer coiled-coil complexes of synthetic peptides comprising dimerisation regions of leucine zippers (cf. Figure 11).

polypeptide	sequence/mw (Da)[a]	complex identified[b]		
		mw (Da)	charge	strands
GCN4-p1	Ac-(249-281)-$CONH_2$ / 4039	8079	5+	dimer
LZ1	H-(250-281)-$CONH_2$ / 3956	7941	5+	dimer
GCN4-35	H-(246-281)-$CONH_2$ / 4456	8911	5+	dimer
LZ[c]	Ac-(1-29)-$CONH_2$ / 3308	9925	5+	trimer
				dimer
LZ12A[c]	Ac-(1-29)-$CONH_2$-A12L / 3266	9697	5+	trimer
				dimer

[a] Sequence numbering of GCN4 peptides refers to numbers in Figure 11. All peptides were prepared by solid-phase synthesis procedures as described in [67].

[b] Molecular weights were derived from the most abundant charge states of complex ions.

[c] "Coil-Ser"leucine-zipper [63]; LZ, Ac-EYEALEKKLAALEAKLQALEKKLEALEHG-NH_2; LZ12A, exchange of L/A at position 12.

4.2 CHARACTERISATION OF LEUCINE ZIPPER-NUCLEOTIDE INTERACTIONS BY CHEMICAL MODIFICATION AND MASS SPECTROMETRY

The successful application of specific chemical modifications at basic residues in proteins [16,17] should render this method to be a promising tool for the molecular characterisation of interaction regions in leucine zipper proteins. In a recent model study, synthetic sequences of GCN4 encompassing the DNA recognition region were subjected to chemical modification reactions of Lys and Arg residues, aimed at modifiying the structure and charges interacting with the corresponding duplex nucleotide. An illustrative example is the modification of the peptide GCN4bR (221-252) which was converted into a covalent dimer by introducing an intermolecular cystinyl-disulfide, by lysine-succinylation [67]. The mass spectrometric analysis revealed specific succinylation at three Lys residues (Lys-232, -247, -252) and the N-terminus of each peptide chain (Figure 14). DNA binding studies with this peptide showed a dramatic effect of the charge reversal due to the succinylation, despite the

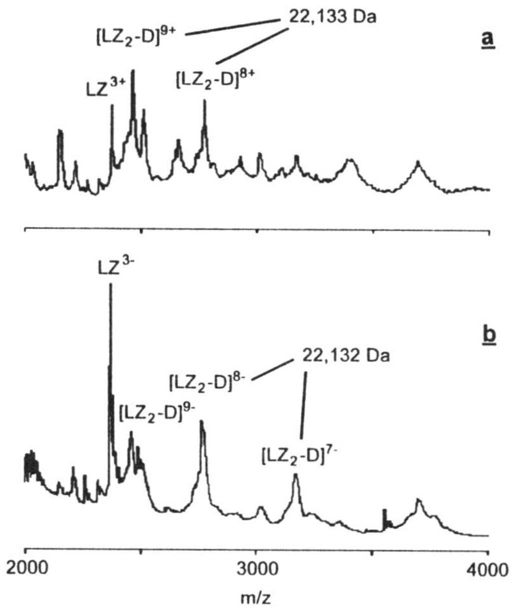

Figure 13. Positive ion (a) and negative ion (b) ESI-TOF mass spectrum of a GCN4 (221-281)-leucine zipper complex (molecular weigh, 22,131 Da). The synthetic oligonucleotide duplex used is shown in Fig. 11. Sample preparation conditions were as described in refs. [67,69] (spectra obtained at the University of Manitoba; courtesy of I. Chernushevitch).

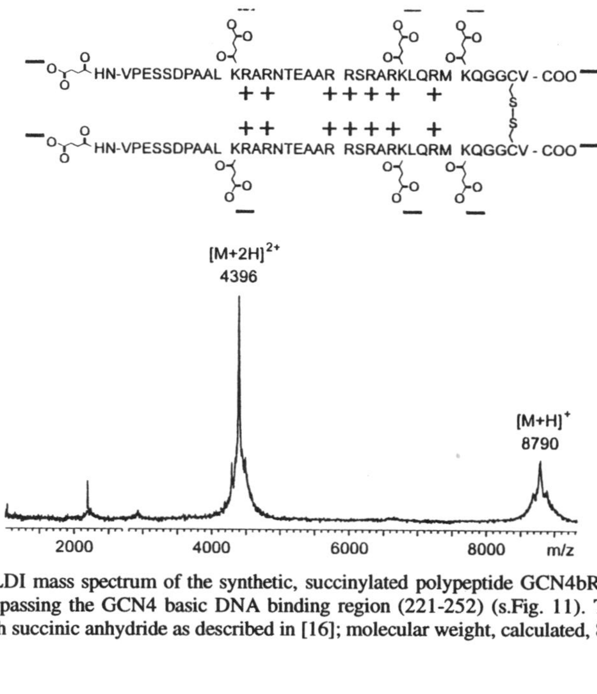

Figure 14. MALDI mass spectrum of the synthetic, succinylated polypeptide GCN4bR, intermolecular disulfide, encompassing the GCN4 basic DNA binding region (221-252) (s.Fig. 11). The peptide was succinylated with succinic anhydride as described in [16]; molecular weight, calculated, 8790 Da.

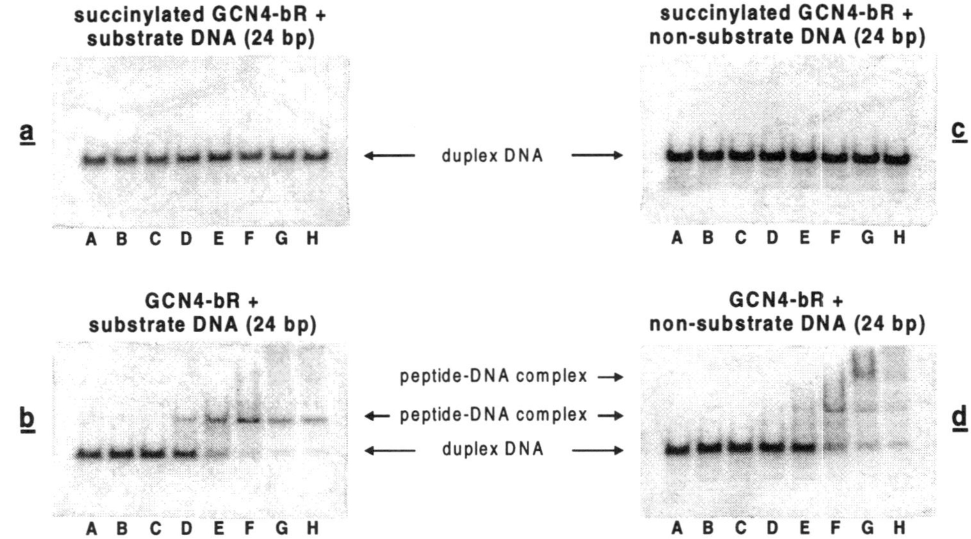

Figure 15. Polyacrylamide gel elctrophoresis band shift analysis upon binding of unmodified and octa-succinylated GCN4bR peptide to substrate and non-substrate oligonucleotide . The substrate 24 bp nucleotide contained the 9 bp palindromic GCN4 recognition site shown in Figure 11. The non-substrate synthetic nucleotide was 5'-ATCATCATCATCAGTAGTAGTAGT-3'. Binding studies were performed with equimolar mixtures of complementary oligonucleotides which were annealed for 30 min at 80°C in 0.1 M ammonium acetate. Different amounts of unmodified or succinylated peptides were then added, respectively, and each solution was incubated for 30 min at 25°C.

presence of multiple cationic sites at Arg residues. Using a suitable substrate 24-bp oligonucleotide (with the palindromic GCN4-specific site) and a non-substrate DNA (24-bp duplex without GCN4 recognition site), the unmodified GCN4-bR provided the specific peptide-nucleotide complex, and also bound non-substrate DNA to several unspecific duplex/peptide adducts. In the succinylated peptide, binding to the recognition-nucleotide was completely abolished as revealed by gel electrophoresis band shift analysis, and no binding was observed with non-substrate DNA (Figure 15). Corresponding modifications in synthetic partial sequences hence appear feasible to the detailed evaluation of the structural specificity and sites of DNA interaction. Such studies with peptide mixtures by ESI-MS are performed at present in our laboratory.

5. Conclusions and Perspectives

The analytical development of "soft"-ionisation methods of mass spectrometry (in particular, ESI-MS and MALDI-MS), and recent results by several laboratories both give promise to the perspective of protein-chemical modification in combination with mass spectrometry, as a powerful tool for characterising higher-order protein structures and supramolecular association [3,5]. A considerable part of these new possibilities has come from the development of ESI-MS for the analysis of proteins and, particularly, the feasibility to specific non-covalent complexes [5,7,72,73]; in the latter area, ESI-MS provides a new powerful dimension and complements classical techniques of lower specificity for probing biomacromolecular interaction such as gel electrophoresis and sedimentation equilibrium analysis [74-79]. It will be interesting to evaluate to which extent the initial attempts for probing non-covalent interactions by MALDI-MS from the solid state [80,81] can be developed to comparable selectivity by ESI-MS. Furthermore, the feasibility of both ESI-MS and MALDI-MS for the exact characterisation of chemically modified proteins, even with partial modifications and multicomponent mixtures, is a first key step in applications to structure-function studies. This feature not only provides the precise determination of modification degrees and sites (using suitable degradation and peptide mapping methods), and thereby the basis for the development of high chemical selectivity in amino acid modifying reactions [3,82]. The comparative analysis of multiple charge structures of ions by ESI-MS also provides a suggestion as to the presence of a still functional "native-like" protein derivative [5,11]. Hence, mass spectrometry appears to constitute the analytical basis for applications to two most important topics of structural biology which hitherto have not been fully amenable to protein chemistry, molecular recognition structures and biomacromolecular interaction.

6. Acknowledgements

We gratefully acknowledge the enthusiasm, ideas and contributions by several co-workers of our laboratories at the Faculty of Chemistry and Biology, particularly Christoph Borchers, Martin Blüggel, Winfried Fiedler, Martina Jetschke, Marcus Macht, and Petra Mayer. Furthermore, we wish to thank many colleagues for their helpful suggestions and advice, amongst them K.P. Schäfer, B. Penke, U. Krawinkel, G. Jung, K. Tomer, M. Sprinzl, and P. Roepstorff. We also thank I. Chernushevitch and K. Standing for their help with ESI-time-of-flight spectra. This work has been supported by the Deutsche Forschungsgemeinschaft; Bundesministerium für Wissenschaft und Forschung, Bonn, FRG; The European Union (Network: "Peptide and Protein Structure Elucidation by Mass Spectrometry"), and the Baden-Württemberg - Hungarian Science Collaboration Programme.

7. References

1. Burlingame, A.L. and Carr, S.A. (1996) *Mass Spectrometry in the Biological Sciences*. Humana Press, Totowa (N.J.).

2. Hillenkamp, F., Karas, M., Beavis, R.C., and Chait, B.T. (1991) Matrix-assisted laser desorption/ionisation mass spectrometry of biopolymers, *Anal. Chem.* **63**, 1193A-1203A.
3. Przybylski, M. (1995) Mass spectrometric approaches to the characterization of tertiary and supramolecular structures of biomacromolecules, *Adv. Mass Spectrom.*, **13**, 257-283.
4. Loo, J.A., Edmonds, C.G., Smith, R:D., Lacey, M.P., and Keough, T. (1990) Comparison of electrospray ionisation and plasma desorption mass spectra of peptides and proteins, *Biomed Env. Mass Spectrom.* **19**, 286-294.
5. Przybylski, M. and Glocker, M.O. (1996) Electrospray mass spectrometry of biomacromolecules complexes with non-covalent interactions - new analytical perspectives for supramolecular chemistry and molecular recognition processes, *Angew. Chem. Int. Ed. Engl.*, **35**, 806-826.
6. Overberg, A., Hassenbürger, A., and Hillenkamp, F. (1994) Laser desorption mass spectrometry. Part II: Performance and applications of matrix-assisted laser desorption/ionization of large biomolecules, in *Mass Spectrometry in the Biological Sciences: A Tutorial*, M. Gross, Ed., Kluwer Academic Publishers, Dordrecht, pp. 181-197.
7. Smith, R.D., Loo, J.A., Ogorzalek-Loo, R.R., Busman, M., and Udseth, H.R. (1991) Principles and practice of electrospray ionization-mass spectrometry in peptide and protein chemistry, *Mass Spectrom. Rev.*, **10**, 359-452.
8. Roepsdorff, P., Nielsen, P.F., Klarskov, K., and Højrup, P. (1988) Application of plasma desorption mass spectrometry in peptide and protein chemistry, *Biomed. Env. Mass Spectrom.*, **16**, 9-18.
9. Svoboda, M., Przybylski, M., Schreurs, J., Mijajima, A., Hogeland, K., and Deinzer, M. (1991) Mass spectrometric determination of glycosylation sites and oligosaccharide composition of insect-expressed mouse interleukin-3. *J. Chromatogr.*, **562**, 403-411.
10. Glocker, M., Arbogast, B., Milley, B., Cowgill, C., and Deinzer, M.L. (1994) Disulfide linkages in the *in vitro* refolded intermediates of recombinant human macrophage-colony-stimulating factor: Analysis of the sulfhydryl alkylation of free cysteine residues by fast atom bombardment mass spectrometry (FAB-MS), *Proc. Natl. Acad. Sci. USA*, **91**, 5868-5872.
11. Smith, R.D. and Light-Wahl, K.J. (1993) The observation of non-covalent interactions in solution by electrospray ionization mass spectrometry: promise, pitfalls and prognosis, *Biol. Mass Spectrom.*, **22**, 493-501.
12. Aplin, R.T. (1997) Probing molecular association by ESI-MS: From enzyme active site mimics to protein-protein interactions; presented at this meeting.
13. Van Dorsselaer, A. (1997) Use of ESI-MS for characterisation of supramolecular edifices: formation of intermediates and determination of stability constants; presented at this meeting.
14. Nielsen, P.F., Schneider, K., Suckau, D., Landis, B., and Przybylski, M. (1990) Applications of selective chemical reactions in combination with 252-californium plasma desorption mass spectrometry in protein structure analysis, E. Hilf, (ed.), *Mass Spectrometry of Large Non-Volatile Molecules* **3**, World Scientif., London, pp. 194-207.
15. Krell, T., Horsburgh, M.J., Cooper, A., Kelly, S.M., and Coggins, J.R. (1996) Localization of the active site of type II dehydroquinases, *J. Biol Chem.*, **271**, 24492-24497.
16. Glocker, M.O., Borchers, C., Fiedler, W., Suckau, D., and Przybylski, M. (1994) Molecular characterization of surface topology in protein tertiary structures by amino-acylation and mass spectrometric peptide mapping, *Bioconj. Chem.*, **5**, 583-590.
17. Suckau, D., Mák, M., and Przybylski, M. (1992) Protein surface topology probing by selective chemical modification and mass spectrometric peptide mapping, *Proc. Natl. Acad. Sci. USA*, **89**, 5630-5634.
18. Suckau, D., Köhl, J., Karwath, G., Schneider, K., Bitter-Suermann, D., and Przybylski, M. (1990) Molecular epitope identification by limited proteolysis of an immobilized antigen-antibody complex and mass spectrometric peptide mapping, *Proc. Natl. Acad. Sci. USA*, **87**, 9848-9852.
19. Steiner, R.F., Albaugh, S., Fenselau, C., Murphy, C., and Vestling, M. (1991) A mass spectrometry method for mapping the interface topography of interacting proteins, illustrated by the melittin-calmodulin system, *Anal. Biochem.*, **196**, 120-125.
20. Fiedler, W., Macht, M., Glocker, M.O., Kaufmann, I.G., Kürzinger, K., and Przybylski, M. (1995) Mass spectrometric characterization of epitope structures in protein antigens -affinity mass spectrometry, *Protein Sci.*, **4**, 122.

21. Macht, M., Fiedler, W., Kürzinger, K., and Przybylski, M. (1996) Mass spectrometric mapping of protein epitope structures of myocardial infarct markers myoglobin and troponin T, *Biochemistry*, **35**, 15633-15639.
22. Kalkum, M., Glocker, M.O., Jetschke, M.R., Borchers, C., Przybylski, M., Saletti, R., and Foti, S. (1995) Modification of histidine and arginine residues in peptides and proteins with amino acid selective reagents and mass spectrometric characterization of the reaction products, *Eur. J. Clin. Chem. Clin. Biochem.*, **33**, A26-A27.
23. Kussmann, M. and Przybylski, M. (1995) Tertiary structure-selective characterization of protein dithiol groups by phenylarsinoxide modification and mass spectrometric peptide mapping, L. Packer, (ed.), *Methods in Enzymology* , Academic Press, pp. 430-435.
24. Papac, D.I., Hoyes, J., and Tomer, K.B. (1994) Epitope mapping of the gastrin-releasing peptide/anti-bombesin monoclonal antibody complex by proteolysis followed by matrix-assisted laser desorption ionisation mass spectrometry, *Prot. Sci.*, **3**, 1485-1492.
25. Welte, W., Nestel, U., Wacker, T., and Diederichs, K. (1995) Structure and function of the porin channel, *Kidney Int.*, **48**, 930-940.
26. Ellenberger, T.E., Brandl, C.J., Struhl, K., and Harrison, S.C. (1992) The GCN4 basic region leucine zipper binds DNA as a dimer of uninterrupted alpha helices: crystal structure of the protein-DNA complex. *Cell*, **71**, 1223.
27. O'Shea, E.K., Klemm, J.D., Kim, P.S., and Alber, T. (1991) X-ray structure of the GCN4 leucine zipper, a two-stranded, parallel coiled coil, *Science*, **254**, 539-544.
28. Glazer, A.N. (1976) The chemical modification of proteins by group-specific and site-specific reagents, H. Neurath and R.L. Hill, (eds.), *The Proteins*, **2**, Academic Press, New York, pp. 1-103.
29. Kaplan, H., Stevenson, K.J., and Hartleys, B.S. (1971) Competitive Labelling, a method for determining the reactivity of individual groups in proteins. *Biochem. J.*, **124**, 289-299.
30. Burnens, A., Binz, H. Bosshard, H.R., Corradin, G. and Demotz, S. (1987) Epitope mapping by chemical modification of free and antibody-bound antigen. *Science*, **235**, 780-783
31. Berzofsky, J.A. (1985) Intrinsic and extrinsic factors in protein antigenic structure, *Science*, **229**, 932-940.
32. Wüthrich, K. (1989) The development of nuclear magnetic resonance spectroscopy as a technique for protein structure determination, *Acc. Chem. Res.*, **22**, 36-44.
33. McPherson, A. (1990) Current approaches to macromolecular crystallization, *Eur. J. Biochem.*, **189**, 1-23.
34. Miranker, A., Radford, S.E., Karplus, M., and Dobson, C.M. (1991) Demonstration by NMR of folding domains in lysozyme, *Nature*, **349**, 633-636.
35. Smith, L.J., Sutcliffe, M.J., Redfield, C., and Dobson, C.M. (1991) Analysis of Phi and Chi-1 torsion angles for hen lysozyme in solution from H-1-NMR spin spin coupling-constants, *Biochemistry*, **30**, 986-996.
36. Drenth, J. (1994) *Principles of protein x-ray crystallography*. Springer,New York.
37. Przybylski, M., Borchers, C., Sukau, D., Mak, M., and Jetschke, M. (1993) Selective chemical modification and mass spectrometric peptide mapping- a new approach for the molecular characterisation of surface topology and microenvironment in protein tertiary struktures. In: "*Peptides 1992*" (C.H. Schneider, A.N. Eberle, eds.), Escom Sci. Publ., Amsterdam, 83-85).
38. Rayment, I., Rypniewski, W.R., Schmidt-Base, K., Smith. R., Tomchick, D.R., Benning, M.M., Winkelmann, D.A., Wesenberg, G. and Holden, H.M. (1993) Three-dimensional structure of myosin subfragment-1: a molecular motor, *Science*, **261**, 50-58.
39. Happersberger, P., Cowgill, C., Deinzer, M.L., and Glocker, M.O. (1996) Characterization of protein-folding intermediates by chemical trapping and mass spectrometry, *44th ASMS Conference on Mass Spectrometry and Allied Topics*, 1050.
40. Patthy, L. and Smith, E.L. (1975) Reversible modification of arginine residues. Application to sequence studies by restriction of tryptic hydrolysis to lysine residues. *J. Biol. Chem.*, **250**, 557.
41. Lee, B. and Richards, F.M. (1971) The interpretation of protein structures: estimation of static accessibilities., *J. Mol. Bio.*, **55**, 379-400.
42. Spassov, V.Z., Karshikov, A.D., and Atamasov, B.P. (1989) Electrostatic interactions in proteins. A theoretical analysis of lysozyme ionization, *Biochim. Biophys. Acta*, **999**, 1-6.

43. Welte, W., Przybylski, M., Glocker, M.O., Diederichs, K., Benz, R., and Breed, J. (1997) X-ray crystallographic and mass spectrometric structure determination and functional characerisation of succinylated porin from *Rhodobacter capsulatus*: Implications for ion selectivity and single-channel conductance, pp. 239-276.
44. Glocker, M.O., Nock, S., Sprinzl, M., and Przybylski, M. (1997) Characterisation of surface topology and binding surfaces in complexes of the elongation factor proteins EF-Ts and EF-Tu from *Thermus thermophilus* by chemical modification and mass spectrometry, *Chemistry*, in press.
45. Fiedler,W., Borchers, C., Kürzinger, K., Macht, M., Deininger, S., and Przybylski, M. (1997) Molecular characterisation of conformational epitopes of hen eggwhite lysozyme by differential chemical modification of immune complexes and mass spectrometric peptide mapping, *Bioconj. Chem.*, submitted for publication.
46. Borchers, C. (1996), *PhD dissertation*, University of Konstanz.
47. Wattenberg A. (1996), *Dipl.-Thesis*, University of Konstanz.
48. Brünger, A.T. (1993) Assessment of phase accuracy by cross validation: the free R-value. Methods and applications, *Acta Crystallogr.*, **D49**, 24-36.
49. Nestel, U., Wacker, T., Woitzik, D., Weckesser, D., Kreutz, W., and Welte, W. (1989) Crystallisation and preliminary X-ray analysis of proin from *Rhodobacter capsulatus, FEBS Lett.*, **242**, 405.
50. Benz, R. (1988) Structure and function of porins from gram-negative bacteria, *Annv. Rev. Microbiol.*, **42**, 359-393.
51. Przybylski, M., Glocker, Nestel, U., Schnaible, V., Blüggel, M., Diederichs, K., Weckesser, J., Schad, M., Schmid, A., Welte, W., and Benz, R. (1996) X-ray crystallographic and mass spectrometric structure determination and functional characterisation of succinylated porin from *Rhodobacter capsulatus*: Implications for ion selectivity and single-channel conductance, *Protein Sci.*, **5**, 1477-1489.
52. Zeth, K., Schnaible, V., Przybylski, M., Welte, W., Diedrichs, K., and Engelhardt, H. (1997) Crystallisation and mass spectrometric analysis of native and chemically modified porin Omp32 from *Comamonas acidovorans*, *FEBS-Lett.*, in press.
53. Schnaible, V.,Michels, J., Zeth, K., Freigang, J., Welte, W., Bühler, S., Glocker, M.O., and Przybylski, M. (1997) Approaches to the characterisation of membrane channel proteins (porins) by UV-MALDI mass spectrometry, *Int. J. Mass Spectrom. Ion Proc.*, in press.
54. Bühler, S., Schnaible, V., Glocker, M.O. Michels, J., Zeth, K., Welte, W., and Przybylski, M. (1997) Conformational differentiation of two porin forms by selective chemical modification and MALDI-MS, *14th International Mass Spectrometry Conference*.
55. Jordon, P. (1987) How pore mouth charge distributions alter the permeability of transmembrane ionic channels, *Biophys. J.*, **51**, 297-311.
56. Forst, D., Schülein, K., Wacker, T., Diederichs, K., Kreutz, W., Benz, R., and Welte, W. (1993) Crystallisation and preliminary X-ray diffraction analysis of Scry, a specific bacterial outer membrane porin., *J. Mol. Biol.*, **229**, 228-262.
57. Schülein, K., Anderson, C., and R., B. (1995) The deletion of 70 amino acids near the N-terminal end of the sucrose-specific porin Scry causes its functional similarity to LamB in vivo and in vitro., *Mol. Microbiol.*, **17**, 757-767.
58. Forst, D. (1996), *PhD dissertation*, University of Konstanz.
59. Henion, J., Li, Y.T., Hsieh, Y.L., and Ganem, B. (1993) Mass spectrometric investigations of drug-receptor interactions, *Therap. Drug Monitor.*, **15**, 563.
60. Light-Wahl, K.J., Springer, D.L.,Winger, B.E., Edmonds, C.G., Camp, D.G., Thrall, B.D., and Smith, R.D. (1993) Observation of a small oligonucleotide duplex by electrospray ionisation mass spectrometry, *J. Am. Chem. Soc.* **226**, 803-804.
61. Baca, M. and Kent S.B.H. (1992) Direct observation of a ternary complex between the dimeric enzyme HIV-1 protease and a substrate-based inhibitor, *J. Am. Chem. Soc.*, **114**, 3992-3993.
62. Leize, E., van Dorsselar, A., Krämer, R. and Lehn, J.M. (1993) Electrospray mass spectrometry of the self assembly of a capped polymetallic complex, J. Chem. Soc. Chem. Commun., 990-993.
63. Lovejoy, B., Choe, S., D, C., McRorie, D.K., DeGrado, W.F., and Eisenberg, D. (1993) Crystal structure of a synthetic α-helical bundle, *Science*, **259**, 1288-1293.
64. Hurst, R.S. (1994) Transcription factors 1:bzip proteins, P. Sheterline, (ed.), *Protein Profile*, London, pp. 123-168.

65. Li, Y.-T., Hsieh, Y.-L., Henion, J.D., Senko, M.W., McLafferty, F.W., M., and Ganem, B. (1993) Mass spectrometric studies on noncovalent dimers of leucine zipper peptides, *J. Am. Chem. Soc.*, **115**, 8409-8413.
66. Przybylski, M., Glocker, M.O., Maier, C., Borchers, C., Dürr, E., Fiedler, W., Kast, J., Wendt, H., and Bosshard, H.R. (1995) Direct characterization of supramolecular complexes of polypeptides and proteins by electrospray mass spectrometry, H. Maia, (ed.), *Peptides 1994* , Escom, Leiden, pp. 43.
67. Przybylski, M., Kast, J., Glocker, M.O., Dürr, E., Bosshard, H.R., Nock, S., and Sprinzl, M. (1995) Mass spectrometric approaches to molecular characterization of protein-nucleic acid interactions, *Toxicology Letters*, **82/83**, 567-575.
68. Allen, M.H. and Vestal, M.L. (1992) Design and performance of a novel electrospray interface, *J. Am. Soc. Mass Spectrom.*, **3**, 18-26.
69. Wendt, H., Dürr, E., Thomas, R.M., Przybylski, M., and Bosshard, H.R. (1995) Characterization of leucine zipper complexes by electrospray ionization mass spectrometry, *Protein Sci.*, **4**, 1563-1570.
70. Goodlett, D.R., Hardin, C.C., Corregan, M., and Smith, R.D. (1993) Estimation of quadruplex $DNAT_M$ by electrospray-ionization-mass spectrometry, *Proc. 41st Conf. Am. Soc. Mass Spectrom.*, 258.
71. Chernushevitch, I., Ens, W., and Standing, K.G. (1997) Measurement of noncovalent biomacromolecular complexes with high m/z by ESI-time-of-flight mass spectrometry, this volume.
72. Kebarle, P. and Tang, L. (1993) From ions in solution to ions in the gas phase, *Anal. Chem.*, **65**, 972A-985A.
73. Miranker, A., Robinson, C.V., Radford, S.E., Aplin, R.T. and Dobson, C.M. (1993) Detection of transient proteins folding populations by mass spectrometry, *Science*, **262**, 896-900.
74. Phisicky, E.M. and Fields, S. (1995) Protein-protein interactions: methods for detection and analysis, *Microbiol. Rev.*, **59**, 94-123.
75. Carr, D.W., and Scott, J.D. (1992) Blotting and band-shifting: techniques for studying protein-protein interactions, *Trends Biochem. Sci.*, **17**, 246-249.
76. Darawshe, S., Rivas, G., and Minton, A.P. (1993) Sedimentation equilibrium-quantitave ployacrylamide gel electrophoresis (SE-QPAGE): a new technique for the detection of associants in multicomponent solutions, *Anal. Biochem.*, **215**, 236-242.
77. Fields, S., and Sternglanz, R. (1994) The two-hybrid system: An assay for protein-protein interactions, *Trends Genet.*, **10**, 286-292.
78. Schmidt-Dorr.T., Oertel-Buchheit, P., Pernelle, C., Bracco, L., Schnarr, M., and Granger-Schnarr, M. (1991) Construction, purification and characterisation of a hybrid protein comprising the DNA binding domain of LexA repressor and the jun leucine zipper: a circular dichroism and mutagenesis study, *Biochemistry*, **30**, 9657-9664.
79. Stevens, F.J. (1986), Analysis of protein-protein interaction by simulation of small-zone size-exclusive chromatography: application to an antibody-antigen association, *Biochemistry*, **25**, 981-993.
80. Glocker, M.O., Bauer, S.H.J., Kast, J., Volz, J., and Przybylski, M. (1996) Characterization of specific noncovalent protein complexes by UV-MALDI-mass spectrometry, *J. Mass Spectrom.*, **31**, 1221-1227.
81. Lecchi, P. and Pannell, L.K. (1995) The detection of intact double-stranded DNA by MALDI, *J. Am. Soc. Mass Spectrom.*, **6**, 972-975.
82. Imoto, T. and Tamada, H. (1989) Chemical modification. In: Protein function, a pratical approach (T.E. Creighton, ed.), Oxford University press, pp. 247-277.

PROTEIN-LIGAND BINDING STUDIED BY AMIDE HYDROGEN EXCHANGE AND MASS SPECTROMETRY

DAVID L. SMITH AND KURUPPU DHARMASIRI
DEPARTMENT OF CHEMISTRY
University of Nebraska-Lincoln
Lincoln, NE 68588-0304

1. Abstract

The rates at which hydrogens located at peptide amide linkages in proteins undergo isotopic exchange when proteins are incubated in D_2O may be sensitive to their access to the aqueous solvent. If ligands shield the binding sites from the solvent, isotopic exchange of hydrogens located in the vicinity of the binding sites might be the basis of a method for locating ligand binding sites on proteins. Although successful applications of this approach have been reported, results of more recent studies indicate that structural changes occurring distant from the binding epitope may also change upon ligand binding, thus confounding interpretation of results. The suggestion is made here that isotopic exchange rates of a subset of amide hydrogens whose exchange rates depend primarily on the access of the amide hydrogens to the aqueous solvent might be more diagnostic for ligand binding sites. Fundamental mechanisms for amide hydrogen exchange and the protein fragmentation/mass spectrometry method for determining hydrogen exchange rates are reviewed with emphasis on kinetic measurements for determining isotopic exchange rate constants for the most rapidly exchanging peptide amide hydrogens in large proteins.

2. Introduction

In 1990 Paterson et al. (1) reported that amide hydrogen exchange rates were significantly and selectively reduced in the epitope region of cytochrome c to which a monoclonal antibody was bound. This interesting observation was attributed to shielding of the binding region by the antibody so that it had reduced access to the aqueous solvent. The broader significance of this observation suggested that amide hydrogen exchange might be generally useful as a tool for identifying protein-ligand binding sites. Although the study by Paterson et al. may be the first report of this use of amide hydrogen exchange, exchange rates of amide hydrogens have been used for

W. Ens et al. (eds.), New Methods for the Study of Biomolecular Complexes, 45–58.

more than three decades to distinguish and characterize different conformational states of proteins. Differences in the conformational states of ligated and unligated proteins have also been studied by this approach. Early studies used the rate of tritium uptake to determine whether ligation loosened or tightened the structure (2-4). More recently, Anderegg and Wagner (5) have used electrospray ionization mass spectrometry to show that the hydrogen/deuterium exchange rate in the Src homology 2 domain is decreased when bound to a peptide substrate. In 1987, Brandt and Woodward introduced the use of proton NMR to monitor hydrogen/deuterium exchange rates when bovine pancreatic trypsin inhibitor was bound to various substrates (6). They reported a reduction in the hydrogen exchange rates of six amide hydrogens, indicating a tightening of structure upon ligation. Loh et al. (7) reported a general reduction in hydrogen exchange rates when nuclease H124L was bound in a ternary complex. That is, isotopic exchange rates of peptide amide hydrogens located both within and distant from the ligand binding sites exchanged slower in the bound form. The notion that binding has a global effect on the structure and dynamics of protein structure has also been reported for cytochrome c and several of its monoclonal antibodies (8,9). The purpose of this presentation is to describe how the protein fragmentation/mass spectrometry approach for determining amide hydrogen exchange rates (10) might be used to investigate conformational changes associated with protein binding.

3. Amide Hydrogen Exchange in Peptides and Proteins

To understand why amide hydrogen exchange rates are sensitive to ligand binding, as well as global structural changes, one must consider the fundamental mechanisms of amide hydrogen exchange. The rates at which hydrogens located on peptide amide linkages in peptides undergo isotopic exchange is critically dependent on the pH of the aqueous solution, and can be described accurately as the sum of two terms, representing acid and base catalysis. The rate constants describing acid and base catalyzed isotopic

$$k_{ex} = k_H [H^+] + k_{OH} [OH^-] \tag{1}$$

exchange, k_H and k_{OH}, have been determined for model peptides and provide the basis for calculating the isotopic exchange rate of any peptide (11,12). For most peptides, acid and base catalyzed isotopic exchange rates are approximately equal when the pH is between 2 and 3. At higher pH, base catalysis dominates and the isotopic exchange rate increases 10-fold for each pH unit. As a point of reference, hydrogens located on peptide amide linkages are completely replaced with deuterium within 0.2-2 s when a peptide is incubated in D_2O at pH 7. This 10-fold range in exchange rates is due to different inductive effects of the various side-chains of the common amino acids.

Intramolecular hydrogen bonding and solvent shielding common to folded proteins lower isotopic exchange rates in folded proteins by as much as 10^8. The reduction in isotopic exchange rates in proteins depends, often critically, on protein structure, facilitating amide hydrogen exchange as a sensitive probe for conformational changes. One mechanism for hydrogen exchange is envisioned as a two-step process in which a folded protein momentarily unfolds, thereby enabling isotopic exchange (2,13,14). The unfolding process may involve global unfolding of the entire protein, facilitating isotopic exchange at all peptide linkages, or it may involve unfolding of short segments of a protein, leading to the possibility for exchange of all amide hydrogens in this segment. After an amide hydrogen has become enabled to undergo isotopic exchange, the rate of exchange may be similar to the rate expected for an unstructured peptide (Eq. 1). This two-step process for isotopic exchange in folded proteins is illustrated in Eq. 2 where *F* and *U* refer to the folded and unfolded protein.

$$\mathrm{f(H)} \underset{k_{-1}}{\overset{k_1}{\rightleftharpoons}} \mathrm{u(H)} \xrightarrow{k_2} \mathrm{u(D)} \underset{k_1}{\overset{k_{-1}}{\rightleftharpoons}} \mathrm{f(D)} \qquad (2)$$

Rate constants k_1 and k_{-1} describe the unfolding or exchange enabling event, while k_2 describes isotopic exchange from the unfolded form. The equilibrium concentrations of the folded and unfolded forms are described by an equilibrium constant, K_{unf} where $K_{unf} = k_1 / k_{-1}$. If $k_{-1} >> k_2$, the structural change facilitating hydrogen exchange occurs many times before isotopic exchange is complete, and the isotopic exchange rate constant k_{ex} is equal to $K_{unf}\, k_2$ (13,15,16). This mechanism prevails for conditions in which the folded protein is stable. For some conditions, $k_{-1} << k_2$ and all amide hydrogens that have become exchange competent in a given protein molecule will undergo isotopic exchange. The isotopic exchange rate constant k_{ex} for these conditions is equal to the protein unfolding rate constant, k_1. Amide hydrogens may also undergo isotopic exchange directly from the folded form of a protein (2,17). At any pH and temperature, some amide hydrogens may exchange by the unfolding mechanism, while others in the same protein exchange directly from the folded protein. Furthermore, both mechanism may be operational for any given amide hydrogen. For stable proteins, K_{unf} is small and isotopic exchange via the unfolding mechanism is relatively slow, while exchange directly from the folded protein may be fast. Since the exchange rates of the rapidly exchanging amide hydrogens are determined primarily by their access to the aqueous solvent, reductions in their exchange rates upon ligation may be highly diagnostic of the binding site.

4. Methods for Measuring Amide Hydrogen Exchange Rates

Prior to the decade of the 80's, most amide hydrogen exchange measurements were performed using tritium to determine global isotopic exchange rates in proteins. Similar information has been obtained when FTIR (18,19) and mass spectrometry(20-23) were used to determine hydrogen/deuterium exchange rates. Isotopic exchange in partially folded intermediates of proteins has also been determined from the molecular weights of intact proteins (24,25). Although these relatively simple experiments clearly demonstrate that exchange rates span a range of 10^8 and that specific exchange rates may be characteristic of specific structural forms of a protein, they do not lead to identification of regions or peptide amide linkages undergoing structural change. With the advent of multidimensional NMR, it has become possible to determine isotopic exchange rates of amide hydrogens located at individual peptide linkages in small proteins ($M_r < 30,000$) (7-9,15,26,27). The ability to detect and locate minor changes in protein structure and dynamics through the consequent changes in amide isotopic exchange rates has renewed interest in amide hydrogen exchange as an analytical tool.

Despite the extraordinary success of the combination of NMR and amide hydrogen exchange, this approach is limited by requirements for large quantities of sample, high protein solubility, and relatively low molecular weight proteins. Another method for determining localized hydrogen exchange rates, originally described by Rosa and Richards (28) and improved by Englander et al. (29), uses acid proteases to fragment proteins labeled with tritium. This approach has been adapted for measurements of hydrogen/deuterium exchange rates by mass spectrometry (10). The current version of this approach, which is termed the protein fragmentation/mass spectrometry method for measuring amide hydrogen exchange rates, has been used in several studies of protein structure (30-35). The remainder of this presentation will focus on use of the protein fragmentation/MS approach for determining amide hydrogen exchange rates, with emphasis on applications involving amide hydrogens whose exchange rates depend primarily on their access to the aqueous solvent.

The general procedure used to exchange deuterium into folded proteins and to determine the deuterium levels in short segments of the partially deuterated protein is illustrated in Figure 1. Isotope exchange occurs when the folded protein is incubated for a specified time in D_2O buffered to the desired pD. Following the deuterium exchange-in step, isotopic exchange at peptide amide linkages is quenched by decreasing the pD to 2.4 and the temperature to 0°C. If deuterium exchange-in were performed using typical conditions of neutral pH and 20°C, the isotopic exchange rates at the peptide amide linkages (k_2) would be decreased by approximately 10^5. Under quench conditions, the half-life for isotopic exchange at peptide amide linkages is 30-60 min (11). To determine isotopic exchange rates in small, defined regions of the protein, pepsin is added to the quench solution to fragment the protein into peptides. A large amount of enzyme (1:1 enzyme:substrate ratio) is required to produce sufficient fragments for analysis within 4-5 min. The peptic digest is analyzed by directly coupled HPLC MS to determine the molecular weights of the peptic fragments, from which the levels of deuterium present in the corresponding segments of the protein are determined. It is important to note that the HPLC separation, which is performed using Perfusion chromatography, is complete within 4-7 min. During this time, the peptides are separated from the buffer salts and fractionated by their hydrophobicity. In addition, since a protiated mobile phase is used for HPLC, deuterium located at rapidly exchanging sites, such as on the side-chains or on the N- and C-termini, are replaced with protium. Hence, the increment in molecular weight is attributable to deuterium located at peptide amide linkages.

The initial protein fragmentation/MS studies were performed using continuous-flow fast atom bombardment ionization on a Kratos MS-50 high resolution sector mass spectrometer (10,30,33,34). Similar results have also been obtained using electrospray ionization on a Sciex API III triple quadrupole mass filter (31,32) and a Micromass Autospec high resolution sector mass spectrometer equipped with array detection (35,36). Typical mass spectra obtained by the Autospec instrument, when used to analyze partially deuterated rabbit muscle aldolase, are illustrated in Figure 2. Rabbit muscle aldolase is a homotetramer of subunits, each containing 363 residues.

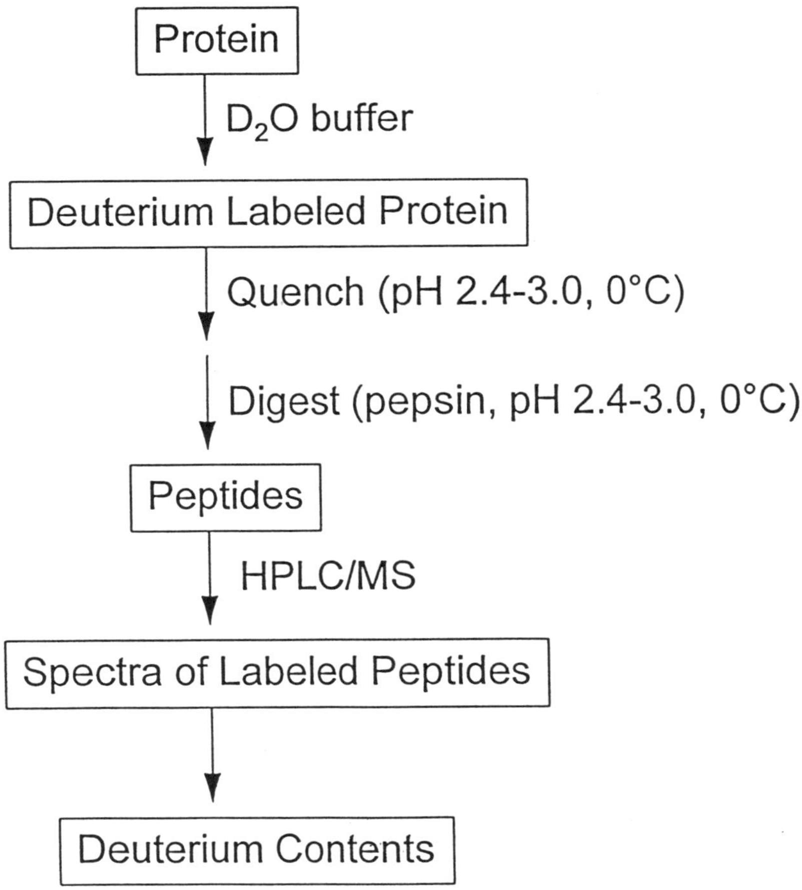

Figure 1 Diagram of the general procedure used to exchange deuterium into the intact protein, to fragment the protein into peptides, and to determine the deuterium contents of the peptides by HPLC MS.

The upper part of Figure 2 shows selected ion plots for the m/z intervals of 1054-1056 and 1133-1135, which indicate elution of peptides 4.2 and 4.4 min after injection. These m/z ranges include the singly-charged molecular ions of peptides derived from segments including residues 152-161 and 327-336 of aldolase. The composite mass spectrum obtained by summing all spectra taken during these elution intervals is given in the lower part of Figure 2. For these measurements, the m/z range from 400 to 1700 was scanned. Expansion of the m/z range including m/z 1040 to 1145 clearly shows the molecular ions of peptides eluting in this window of elution time and m/z space. The average m/z of these ions, when compared to the average m/z of the same ions derived from a non-deuterated standard, indicates that approximately 3.1 and 0.4, peptide amide hydrogens have been replaced with deuterium in the 152-161 and 327-336 segments, respectively.

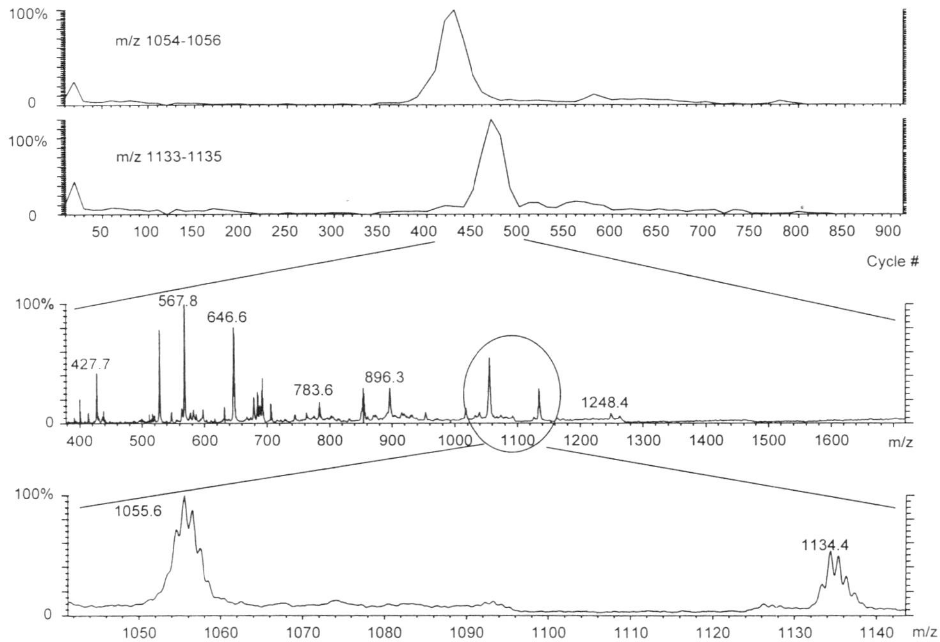

Figure 2 Typical mass spectral results obtained using a Micromass Autospec electrospray ionization mass spectrometer to analyze rabbit muscle aldolase that had been incubated in D_2O for 3 min and processed according to the procedure illustrated in Fig. 1

The fundamental information obtained by the protein fragmentation/MS hydrogen exchange method is the number of deuteriums located at peptide amide linkages in a specific segment of a protein following incubation of the intact protein in D_2O for given times and conditions. This information is taken from the centroid of the envelope of isotopic peaks for each peptide, as illustrated in Fig. 2, and adjusted for a small amount of exchange that occurs during digestion and HPLC MS analysis (10). Measurements are often made for a range of incubation times so that amide hydrogens with different exchange rates can be used to detect structural changes. For example, the deuterium level in the N-terminal segment of rabbit muscle aldolase containing residues 1-7 obtained following incubation of the intact protein in D_2O at pD 6.53 for 2.5 min to 44 h (37) is presented in Figure 3. This segment has 7 amino acid residues, one of which is proline. Therefore this segment has 5 hydrogens located on peptide amide linkages. Results presented in Figure 3 show that less than 0.2 deuteriums were present for the shortest incubation time, 2.5 min, while 2.5 deuteriums were present after incubation in D_2O for 44 h. It is significant to note that, for the present incubation

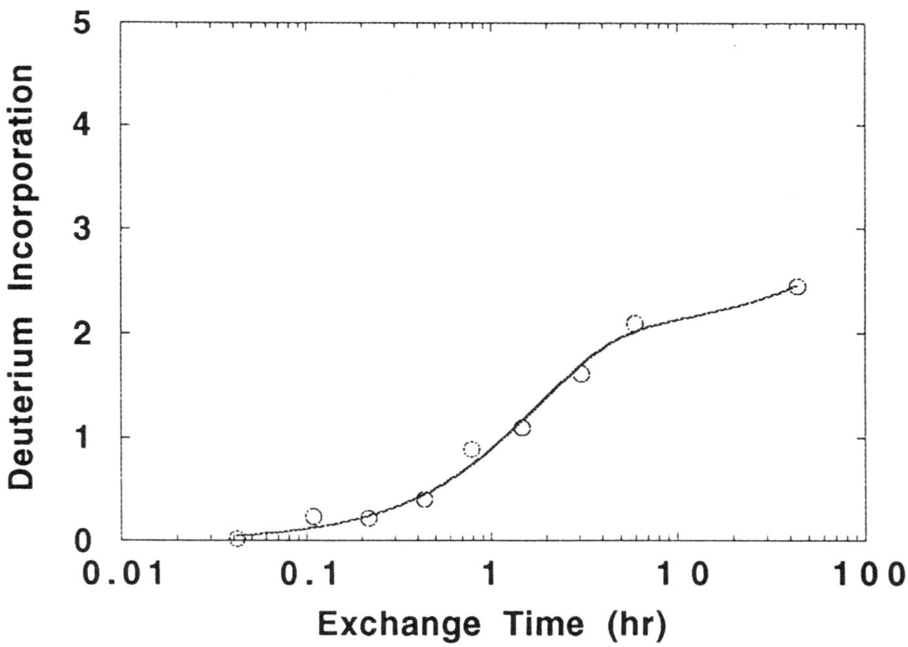

Figure 3 Level of deuterium located at peptide amide linkages in the segment containing residues 1-7 of rabbit muscle aldolase following incubation of the intact protein in D_2O for 2.5 min to 44 h (37).

conditions, all five amide hydrogens in this segment would have been replaced with deuterium within approximately 3 s if the protein were unfolded. It follows that the high-order structure of folded aldolase reduces the isotopic exchange rate in this segment by several orders of magnitude. To quantitatively assess the isotopic exchange rate in this segment, the deuterium exchange-in results presented in Figure 3 were fitted to the expression given in Eq. 3, where D is the deuterium content of this peptide,

$$D = N - \sum_{i=1}^{N} exp(-k_i t) \tag{3}$$

k_i is the hydrogen exchange rate constant for each amide hydrogen in this segment, and t is the time allowed for isotopic exchange. This analysis leads to determination of the distribution of isotopic exchange rate constants of amide hydrogens located on peptide linkages in this segment. Application of Eq. 2 to the data presented in Fig. 3 indicates that, among the 5 amide hydrogens located in the 1-7 segment of aldolase, 2 have rate constants of 0.60 h^{-1}, 1 has a rate constant of 0.014 h^{-1}, and 2 have rate constants less than 0.002 h^{-1}.

5. Isotopic exchange rates of surface amide hydrogens

As discussed above, the potential for using amide hydrogen exchange rates to determine ligand binding sites on protein surfaces, as well as to detect global changes accompanying binding, has been explored using NMR to detect isotopic exchange rates (1,7-9). These studies have used only the slowly exchanging amide hydrogens as probes because they are the only amide hydrogens whose isotopic exchange rates can be readily detected by high resolution NMR. Since the exchange rates of slowly exchanging amide hydrogens are likely to be sensitive to both intramolecular hydrogen bonding and their access to the aqueous solvent, interpretation of changes in their isotopic exchange rates upon binding may be ambiguous. However, isotopic exchange rates of rapidly exchanging amide hydrogens are regulated primarily by their access to the aqueous solvent. Hence, it is likely that the most rapidly exchanging amide hydrogens located on peptide linkages in proteins might be more useful for identifying ligand binding sites.

Amide hydrogen exchange rate constants in folded proteins at neutral pH have rate constants from as fast as 10^3 h^{-1} to as slow as 10^{-5} h^{-1}. For NMR measurements, isotopic exchange is quenched by placing the protein in a solution of the lowest pH compatible with retention of folded structure. Only those hydrogens with hydrogen exchange times greater than the time required for NMR analysis, which is usually a few hours, are detected. For the protein fragmentation/MS method, isotopic exchange is quenched by lowering the pH to 2.5 and the temperature to 0°C. For these conditions, the half-life for isotopic exchange of all peptide amide hydrogens is much greater than the analysis time. As a result, isotopic exchange rates at all peptide amide linkages can, in principle, be detected. Of particular importance is the possibility of detecting exchange at peptides linkages where isotopic exchange is relatively fast in the folded protein. For the incubation times used to obtain data presented in Figure 3, the maximum and minimum hydrogen exchange rate constants are approximately 50 h^{-1} and 0.002 h^{-1}, respectively. Thus, use of amide hydrogen exchange to detect structural changes by the protein fragmentation/MS method is usually limited to a subset of amide hydrogens, but only because a limited range of incubation times is used. Isotopic exchange rate constants greater than 50 h^{-1} could be determined if shorter incubation times were used. Likewise, amide hydrogens with exchange rate constants less than 0.002 h^{-1} could be used to sense structural changes if longer incubation times were used.

To measure the isotopic exchange rates of the most rapidly exchanging peptide amide hydrogens in proteins at pH 7, incubation times as short as 0.2 s are required because this is the approximate time needed for the most rapidly exchanging amide hydrogens in unfolded polypeptides to undergo isotopic exchange. To achieve short incubation times, Dharmasiri and Smith (35) have reported construction of a flow-quench apparatus that provides a semi-automated means of performing the procedures indicated in Figure 1. General features of the flow quench apparatus, which was constructed from PEEK tubing and low-volume tees, are illustrated in Figure 4. A protic solution of the protein is diluted 20-fold with D_2O at the first tee. Isotopic exchange occurs in the tube joining the two tees for a time determined by the volume of this tube and the total flow rate. At the second tee, acid and pepsin are added to quench isotopic exchange and to fragment the protein into peptides whose deuterium levels are determined by LCMS. In a study of hydrogen exchange in cytochrome c for incubation times from 0.2 s to 120 s, it was found that the isotopic exchange rates of 33 nonhydrogen bonded amide hydrogens were substantially less than expected for unfolded polypeptides. The ability to measure exchange rates of these rapidly exchanging amide hydrogens in folded proteins opens the possibility of using these hydrogens to investigate the extent to which binding alters accessibility of some regions of the backbone to the aqueous solvent.

Isotopic exchange rate constants of amide hydrogens whose exchange rates are determined primarily by the access of the amide hydrogens to the aqueous solvent have now been determined for several segments of a much larger protein, rabbit muscle aldolase. The general approach used to detect changes in the solvent accessibility of specific segments of aldolase can be illustrated using hydrogen exchange results for the segment encompassing residues 299-307. This segment, which has eight peptide amide hydrogens, includes the C-terminal residues of a short β-strand (comprising residues 296-302) and a loop joining this β-strand to a short α-helix (comprising residues 306-313) (33,38). When Zhang et al. (33) used relatively long isotope exchange times (2.5 min to 44 h), they found that 4 of the amide hydrogens had exchange rates greater than

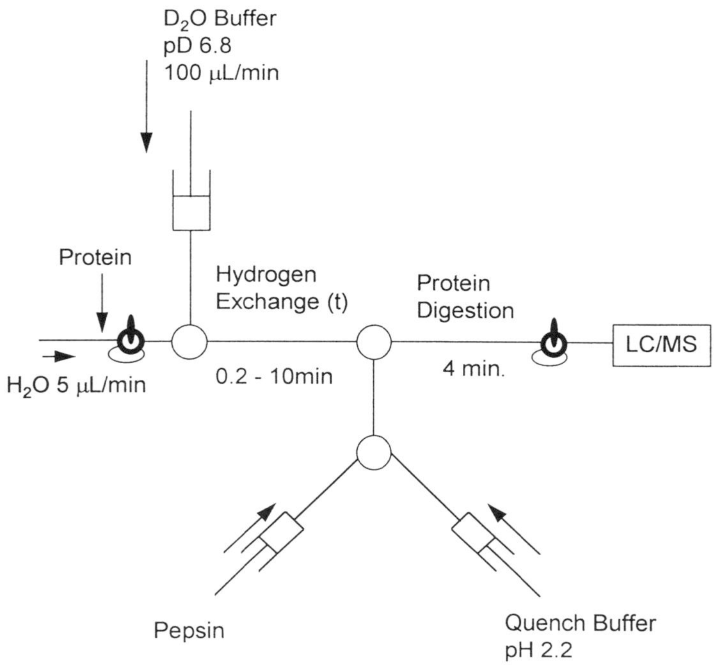

Figure 4 Flow-quench system constructed to determine hydrogen exchange rate constants of the most rapidly exchanging peptide amide hydrogens (Adapted from ref 35).

53 h^{-1} (0.01 s^{-1}). When the flow quench apparatus illustrated in Figure 4 was used to effect isotopic exchange in aldolase over a relatively short time scale (0.2 s to 180 s), isotopic exchange rate constants for these rapidly exchanging amide hydrogens could be determined. The increase in the deuterium level in the 299-307 segment over this time range is illustrated in Figure 5. Fitting these data to Eq. 2 shows that two amide hydrogens in this segment have rate constants of 7 s^{-1}, one has a rate constant of 0.1 s^{-1} and one has a rate constant of 0.01 s^{-1}. Similar measurements of deuterium uptake by short segments taken from throughout the aldolase backbone provide a base for estimating isotope exchange rate constants for amide hydrogens whose exchange rates are determined primarily by the access of the particular amide hydrogens to the aqueous solvent.

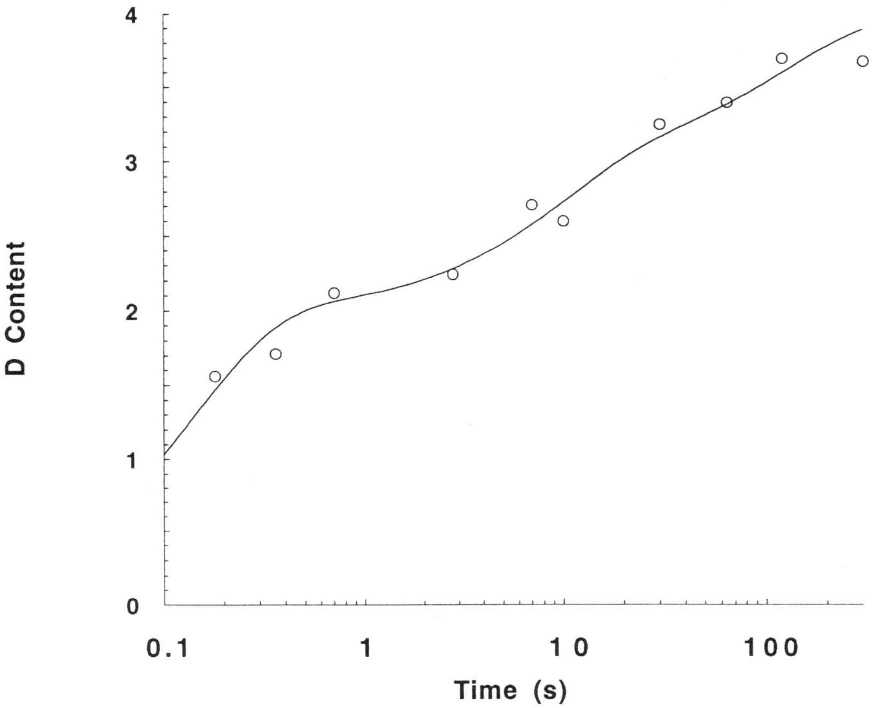

Figure 5 Deuterium level found in the segment including residues 299-307 of rabbit muscle aldolase when the intact, folded protein was incubated in D_2O and analyzed using the flow quench apparatus illustrated in Figure 4.

6. Summary

The binding of various ligands to proteins likely decreases access of the binding epitope to the aqueous medium. If this occurs, isotope exchange rates of amide hydrogens located in or near to the binding epitope may be decreased. The magnitude of the decrease is expected to depend on several factors, including the size of the ligand and its contact surface. Previous attempts to use diminished rates of isotopic exchange to identify epitope binding sites have not led to definitive identification of the binding surface because global changes distant from the binding site may also alter isotopic exchange rates. We have proposed using the most rapidly exchanging amide hydrogens, those whose isotopic exchange rates are determined primarily by their access to the aqueous solvent, as probes for identifying binding sites. To measure isotopic exchange rates of the rapidly exchanging amide hydrogens, a flow quench apparatus has been developed which facilitates exchange-in times as short as 0.1 s.

7. Acknowledgments

The authors are pleased to acknowledge the National Institutes of Health (GM RO1 40384) and the Nebraska Center for Mass Spectrometry for their support of this research, and to NATO for supporting this workshop.

8. References

[1] Paterson, Y., Englander, S. W., and Roder, H. (1990) An antibody binding site on cytochrome c defined by hydrogen exchange and two-dimensional NMR, *Science* **249**, 755-759.

[2] Woodward, C., Simon, I., and Tüchsen, E. (1982) Hydrogen exchange and the dynamic structure of proteins, *Mol. Cell. Biochem.* **48**, 135-160.

[3] Englander, S. W., Downer, N. W., and Teitelbaum, H. (1972) Hydrogen exchange, *Annu. Rev. Biochem.* **41**, 903-924.

[4] Barksdale, A. D., and Rosenberg, A. (1982) Acquisition and interpretation of hydrogen exchange data from peptides, polymers, and proteins, *Methods Biochem. Anal.* **28**, 1-113.

[5] Anderegg, R. J., and Wagner, D. S. (1995) Mass spectrometric characterization of a protein ligand interaction, *J. Am. Chem. Soc.* **117**, 1374-1377.

[6] Brandt, P., and Woodward, C. (1987) Hydrogen exchange kinetics of bovine pancreatic trypsin inhibitor β-sheet protons in trypsin-bovine pancreatic trypsin inhibitor, trypsinogen-bovine pancreatic trypsin inhibitor, and trypsinogen-isoleucylvaline-bovine pancreatic trypsin inhibitor, *Biochemistry* **26**, 3156-3162.

[7] Loh, S. N., Prehoda, K. E., Wang, J., and Markley, J. L. (1993) Hydrogen exchange in unligated and ligated staphylococcal nuclease, *Biochemistry* **32**, 11022-11028.

[8] Benjamin, D. C., Williams, D. C., Smith-Gill, S. J., and Rule, G. S. (1992) Long-range changes in a protein antigen due to antigen-antibody interaction, *Biochemistry* **31**, 9539-9545.

[9] Benjamin, D. C., Williams, D. C., Poljak, R. J., and Rule, G. S. (1996) Global changes in amide hydrogen exchange rates for a protein antigen in complex with three different antibodies, *J. Mol. Biol.* **257**, 866-876.

[10] Zhang, Z., and Smith, D. L. (1993) Determination of amide hydrogen exchange by mass spectrometry: A new tool for protein structure elucidation, *Protein Sci.* **2**, 522-531.

[11] Bai, Y., Milne, J. S., Mayne, L., and Englander, S. W. (1993) Primary structure effects on peptide group hydrogen exchange, *Proteins: Struct. Funct. Genet.* **17**, 75-86.

[12] Molday, R. S., Englander, S. W., and Kallen, R. G. (1972) Primary structure effects on peptide group hydrogen exchange, *Biochemistry* **11**, 150-158.

[13] Hvidt, A., and Nielsen, S. O. (1966) Hydrogen exchange in proteins, *Adv. Protein Chem.* **21**, 287-385.

[14] Englander, S. W., and Kallenbach, N. R. (1984) Hydrogen exchange and structural dynamics of proteins and nucleic acids, *Q. Rev. Biophys.* **16**, 521-655.

[15] Roder, H., Wagner, G., and Wüthrich, K. (1985) Amide proton exchange in proteins by EX1 kinetics: Studies of the basic pancreatic trypsin inhibitor at variable p^{2}H and temperature, *Biochemistry* **24**, 7396-7407.

[16] Bai, Y. W., Milne, J. S., Mayne, L., and Englander, S. W. (1994) Protein stability parameters measured by hydrogen exchange, *Proteins: Struct. Funct. Genet.* **20**, 4-14.

[17] Woodward, C. K., and Hilton, B. D. (1980) Hydrogen isotope exchange kinetics of single protons in bovine pancreatic trypsin inhibitor, *Biophys. J.* **32**, 561-575.

[18] Zhang, Y. P., Lewis, R. N. A. H., Hodges, R. S., and McElhaney, R. N. (1992) FTIR spectroscopic studies of the conformation and amide hydrogen exchange of a peptide model of the hydrophobic transmembrane α-helixes of membrane proteins , *Biochemistry* **31**, 11572-11578 .

[19] Heimburg, T., and Marsh, D. (1993) Investigation of secondary and tertiary structural changes of cytochrome c in complexes with anionic lipids using amide hydrogen exchange measurements: an FTIR study , *Biophys. J.* **65**, 2408-2417 .

[20] Katta, V., and Chait, B. T. (1991) Conformational changes in proteins probed by hydrogen-exchange electrospray-ionization mass spectrometry , *Rapid Commun. Mass Spectrom.* **5**, 214-217 .

[21] Thévenon-Emeric, G., Kozlowski, J., Zhang, Z., and Smith, D. L. (1992) Determination of amide hydrogen exchange rates in peptides by mass spectrometry, *Anal. Chem.* **64**, 2456-2458 .

[22] Wagner, D. S., Melton, L. G., Yan, Y. B., Erickson, B. W., and Anderegg, R. J. (1994) Deuterium exchange of alpha-helices and beta-sheets as monitored by electrospray ionization mass spectrometry, *Protein Science* **3**, 1305-1314.

[23] Wang, F., and Tang, X.-j. (1996) Conformational heterogeneity and stability of apomyoglobin studied by hydrogen-deuterium exchange and electrospray ionization mass spectrometry, *Biochemistry* **35**, 4069-4078.

[24] Miranker, A., Robinson, C. V., Radford, S. E., Aplin, R. T., and Dobson, C. M. (1993) Detection of transient protein folding populations by mass spectrometry, *Science* **262**, 896-900.

[25] Miranker, A., Robinson, C. V., Radford, S. E., and Dobson, C. M. (1996) Investigation of protein folding by mass spectrometry, *FASEB J.* **10**, 93-101.

[26] Tüchsen, E., and Woodward, C. (1987) Hydrogen exchange kinetics of surface peptide amides in bovine pancreatic trypsin inhibitor, *J. Mol. Biol.* **193**, 793-802.

[27] Robertson, A. D., and Baldwin, R. L. (1991) Hydrogen exchange in thermally denatured ribonuclease A, *Biochemistry* **30**, 9907-9914.

[28] Rosa, J. J., and Richards, F. M. (1979) An experimental procedure for increasing the structural resolution of chemical hydrogen-exchange measurements on proteins: application to ribonuclease S peptide, *J. Mol. Biol.* **133**, 399-416.

[29] Englander, J. J., Rogero, J. R., and Englander, S. W. (1985) Protein hydrogen exchange studied by the fragment separation method, *Anal. Biochem.* **147**, 234-244.

[30] Liu, Y., and Smith, D. L. (1994) Probing high order structure of proteins by fast-atom bombardment mass spectrometry, *J. Am. Soc. Mass Spectrom.* **5**, 19-28.

[31] Johnson, R. S., and Walsh, K. A. (1994) Mass spectrometric measurement of protein amide hydrogen exchange rates of apo- and holo-myoglobin, *Protein Science* **3**, 2411-2418.

[32] Ohguro, H., Palczewski, K., Walsh, K. A., and Johnson, R. S. (1994) Topographic study of arrestin using differential chemical modifications and hydrogen/deuterium exchange, *Protein Science* **3**, 2428-2434.

[33] Zhang, Z., Post, C. B., Smith, D. L. (1996) Amide hydrogen exchange determined by mass spectrometry: Application to rabbit muscle aldolase., *Biochemistry* **35**, 779-791.

[34] Zhang, Z., Smith, D. L. (1996) Detection of thermal-induced unfolding domains in aldolase by amide hydrogen exchange and mass spectrometry, *Protein Science* **5**, 1282-1289.

[35] Dharmasiri, K., and Smith, D. L. (1996) Mass spectrometric determination of isotopic exchange rates of amide hydrogens located on the surfaces of proteins, *Anal. Chem.* **68**, 2340-2344.

[36] Zhang, Z., and Smith, D. L. (1996) Detection of acid-induced unfolding of proteins by amide hydrogen exchange and mass spectrometry, (Unpublished Results) .

[37] Zhang, Z. (Unpublished Results) .

[38] Gamblin, S. J., Cooper, B., Millar, J. R., Davies, G. J., Littlechild, J. A., and Watson, H. C. (1990) *FEBS Lett.* **262**, 282.

STRUCTURAL STUDIES OF PROTEIN-PROTEIN INTERACTIONS USING PROTEOLYTIC FOOTPRINTING AND MALDI/MS

KENNETH B. TOMER, SONJA J. TROJAK AND CAROL E. PARKER
Mass Spectrometry Group, Laboratory of Structural Biology
National Institute of Environmental Health Sciences
P.O. Box 12233, Research Triangle Park, NC 27709

1. Introduction

With the advent of modern ionization techniques such as electrospray ionization (ESI) and matrix-assisted laser desorption/ionization (MALDI), mass spectrometry has become increasingly used in the study of molecular complexes of biomolecules. Applications have included observation of non-covalent complexes by ESI [1], information about protein folding from components of complexes that, by themselves, were beyond the mass range of the MS [2], the use of affinity techniques with MS [3], and the use of MS with proteolytic footprinting to define the region of contact between molecules in a non-covalent complex [4]. Mass spectrometry is especially well suited for use with proteolytic footprinting [5] because, with a suitable choice of reagents and enzymes, the area of contact between members of the complex can be narrowly defined.

Two approaches to proteolytic footprinting combined with MS have been reported. Most reports in this area have been on antibody-antigen complexes, so the approaches will be described assuming an antibody-antigen complex. In one method, proteolytic extraction or immunoprecipitation, the antigen is subjected to proteolysis. The resulting reaction mixture is allowed to react with the antibody to form a complex between the antibody and the proteolysis product containing the epitope. The immunoprecipitated complex can then be analyzed by MS, usually MALDI [6]. If an immobilized antibody is used, the antigen is eluted from the immobilized antibody prior to analysis [4]. This methodology is limited to continuous epitopes. In the second approach, epitope excision, the antigen is bound to an immobilized antibody prior to proteolysis. After proteolysis, the non-bound fragments are first washed from the column, then the fragment(s) still affinity bound are eluted from the column and analyzed by MS [7]. This approach should be useful with proteins in their native conformations and may prove useful for discontinuous as well as continuous epitopes. Most reports of epitope excision

W. Ens et al. (eds.), New Methods for the Study of Biomolecular Complexes, 59–65.

experiments, however, have determined epitopes on relatively small polypeptides [4,7].

We have combined epitope excision with direct analysis of affinity-bound analytes by MALDI [8]. In this procedure, instead of eluting the affinity-bound analytes from the immobilized antibody, an aliquot of the immobilized affinity column is placed on the MALDI target and analyzed without any additional steps. The MALDI matrix releases the analytes *in situ.*

In addition to simplifying the workup of the sample, the fine structure of the site of interaction (the epitope in this case) can be probed by consecutive reactions, permitting finer definition of the site of interaction. We have applied this concept to proteolytic footprinting of the epitope on gastrin releasing peptide recognized by an anti-bombesin monoclonal antibody [8] and the epitope on recombinant HIV-1_{IIIB} p26 recombinant protein recognized by a monoclonal antibody [9]. In this paper we illustrate the capability for incorporating consecutive reactions with proteolytic footprinting and MALDI/MS analysis for epitope determination on proteins in the native conformation by probing the epitope on recombinant HIV-1_{SF2} p24 recognized by the same monoclonal antibody used for the p26 study.

2. Materials and Methods

Enzymes were obtained from Boehringer Mannheim, Indianapolis, IN, USA (Carboxypeptidase Y and Trypsin-TCPK). The mouse monoclonal anti-p24 antibody (clone 13-102-100) was obtained from Advanced Biotechnologies, Inc, Columbia, MD, USA. Recombinant HIV-1_{IIIB} was purchased from Intracel Corp., Cambridge, MD. Affinity columns were prepared from CNBr-activated Sepharose columns (Pharmacia Biotech, Piscataway, NJ, USA) according to the procedure supplied by the company. α-Cyano-4-hydroxycinnamic acid was purchased from Aldrich Chemical Co., Milwaukee, WI.

2.1 AFFINITY-BINDING OF P24 TO THE IMMOBILIZED ANTIBODIES

The prepared antibody beads were incubated with 20 micrograms of truncated p24 for 3 hours at room temperature with slow rotation. The columns were then washed three times with 0.3 mL pH 7.2 phosphate buffered saline (PBS) buffer to remove any unbound protein and a 1 μL aliquot was removed for MALDI analysis.

2.2 PROTEOLYTIC DIGESTION PROCEDURES

2.2.1 *Trypsin*

On-column digestion of affinity-bound p24 was done with trypsin sequencing grade enzyme. A 100 μg aliquot of the enzyme was dissolved in 100 μL 1 mM HCL. Approximately 300 μL of the digestion buffer, Tris-HCl pH 8.4, and 3 μL of the enzyme solution (3 μg enzyme) were added to the affinity-bound p24 column. The column was then rotated at 37° C for 3 h. The beads were rinsed three times with the digestion buffer and 1 μL was spotted on the MALDI target for analysis.

2.2.2 *Carboxypeptidase Y*

On-column digestions of affinity-bound p24 tryptic fragments were done with carboxypeptidase Y. Approximately 300 μL of the digestion buffer (150 mM NaCl, 150 mM sodium citrate, pH 6.5 and 3 μL of the enzyme solution, 3 micrograms enzyme) were added to the affinity-bound p24 column. The column was then rotated at 37° C for 0.25, 1.25, and 2 hour time increments. The beads were rinsed three times with the digestion buffer and a 1 μL aliquot was spotted on the MALDI target for analysis.

2.3 MASS SPECTROMETRY

A Voyager-RP MALDI mass spectrometer (PerSeptive Biosystems, Framingham, MA) was used to obtain all spectra. The instrument was equipped with a nitrogen laser (λ=337nm). An accelerating voltage of 30 kV was used. Mass calibration was performed using an external calibration spot that contained standards with two masses bracketing the mass range of interest.

All samples were analyzed using a modified Chait matrix [10]. A saturated solution of recrystallized α-cyano-4-hydroxycinnamic acid in 45:45:10 ethanol:water: concentrated formic acid (v:v:v) was used as the matrix solution. A 1 μL aliquot of the affinity bead slurry was placed on the MALDI target, followed by a 1 μL aliquot of the matrix solution, and the sample was allowed to dry at room temperature.

3. Results and Discussion

The protein p24 (M_r 24k) is a core structural protein of the HIV virus and is derived from the *gag* genome. This protein is a major target for antibody response and antibodies to p24 are some of the first antibodies to appear during HIV infection. Blood supplies are also screened for antibodies to p24. Knowledge of the epitopes actually being recognized by the antibodies is vital because sequence variations in the antigens being measured could lead to false negative results. Recombinant truncated p24 is an 190 residue protein expressed in E. *coli* which differs from p24 in that 22 amino acids at the C-terminus are missing. The sequence is shown in Figure 1. It differs slightly from the catalog sequence in that there is one additional amino acid at the N-terminus arising from a leader sequence used in purification [11].

SDTGHSNQVSQNYPIVQNIQGQMVHQAISPRTLNAWVKVV
EEKAFSPEVIPMFSALSEGATPQDLNTMLNTVGGHQAAMQ
MLKETINEEAAEWDRVHPVHAGPIAPGQMREPRGSDIAGTT
STLQEQIGWMTNNPPIPVGEIYKRWIILGLNKIVRMYSPTSIL
DIRQGPKEPFRDYVDRFYKTLRAEQA

Figure 1. Amino acid sequence of truncated p24, Mr 21247.

The MALDI spectrum of truncated p24 affinity-bound to the immobilized MAb 13-102-100, obtained by adding matrix solution to an aliquot of the column bed, is shown in Figure 2. The affinity-bound p24 is released into the matrix solution upon crystallization, and can be readily desorbed directly. Because the antibody is immobilized on the agarose beads, no ions are observed arising from the antibody.

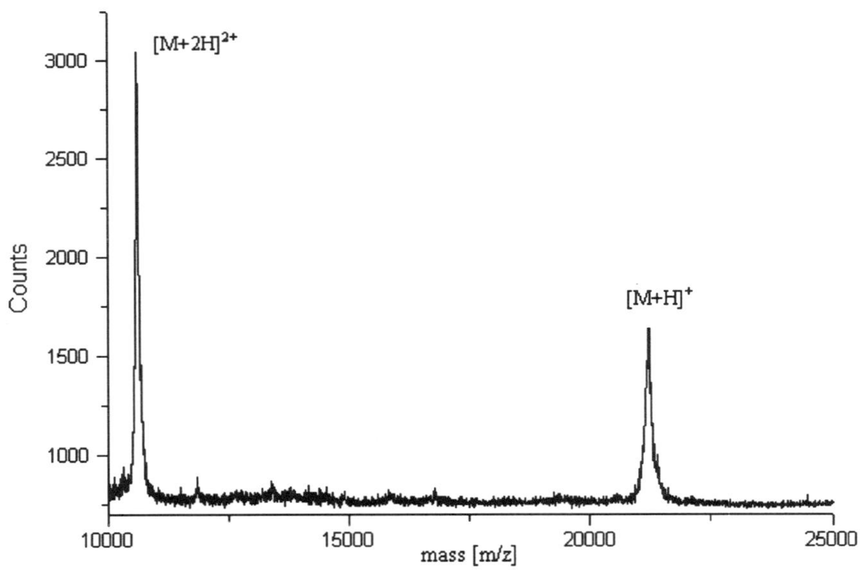

Figure 2. MALDI spectrum of truncated p24 bound to immobilized monoclonal antibody 13-102-100

The affinity-bound protein was then subjected to a trypsin digestion for four hours. At the end of the digestion, the column was drained and then washed with three column volumes of phosphate-buffered saline. This procedure removed any non-affinity-bound peptides from the column. An aliquot of the column bed was then directly analyzed by MALDI (Figure 3). Only peptides that remained affinity-bound to the monoclonal antibody are detected at this point. This spectrum shows ions due to tryptic peptides T_6 and $T_{5,6}$, and $T_{5,6,7}$. The presence of these three tryptic peptides indicates that the epitope lies within the T_6 residues and that the cleavage site between T_5 and T_6, and between T_6 and T_7 are somewhat protected in that, while cleavage does occur, they are not cleaved as readily as other tryptic cleavage sites. It should be noted that the peptides tend to become oxidized while affinity bound to the beads in solution leading to ions 16, 32, and 48 Da above the non-oxidized forms. Also, relatively little background is observed coming from the antibody, because the antibody tends to be protected from proteolysis due to its high degree of glycosylation.

T_5 T_6 T_7
ETINEEAAEWDR VHPVHAGPIAPGQMR EPR

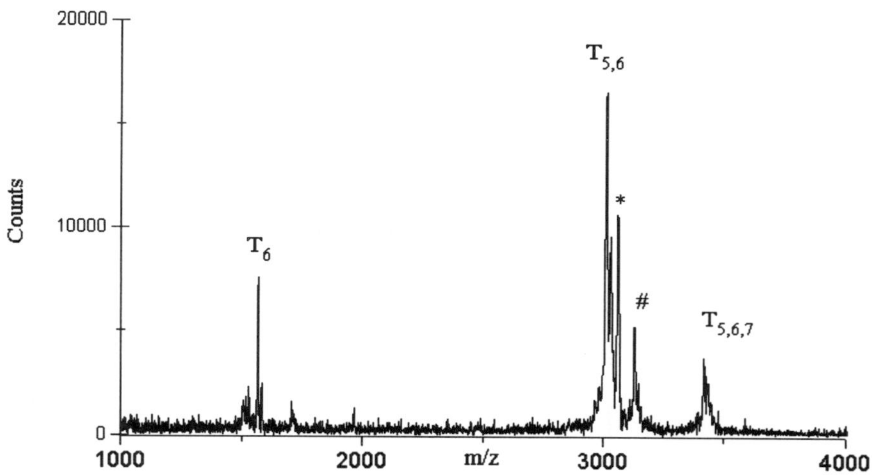

Figure 3. MALDI spectrum of the tryptic peptides still affinity-bound to Mab 13-102-100 after digestion of affinity-bound truncated p24. * denotes oxidized species. # denotes background peak.

After digestion with trypsin, the remaining affinity-bound analytes were subjected to a second digestion step, this time with carboxypeptidase Y. After the unbound fragments have been washed off, the MALDI spectrum of an aliquot of the column bead was obtained. The spectrum of the 1.25 hr timepoint is shown in Figure 4. The loss of arginine from both the T_5 and $T_{5,6}$ fragments is readily observed. Further loss of methionine from these two fragments can also be identified. Loss of both arginine and proline from $T_{5,6,7}$ can also be noted in an expanded plot of the spectrum. These results narrow the epitope to within the residues VHPVHAGPIAPGQ, in agreement with the epitope determined on recombinant p26 that is recognized by this antibody [9].

The relative abundance of the oxidized peaks in Figure 4 is greatly reduced once the methionine is lost, indicating that most of the oxidation has occurred at the methionine.

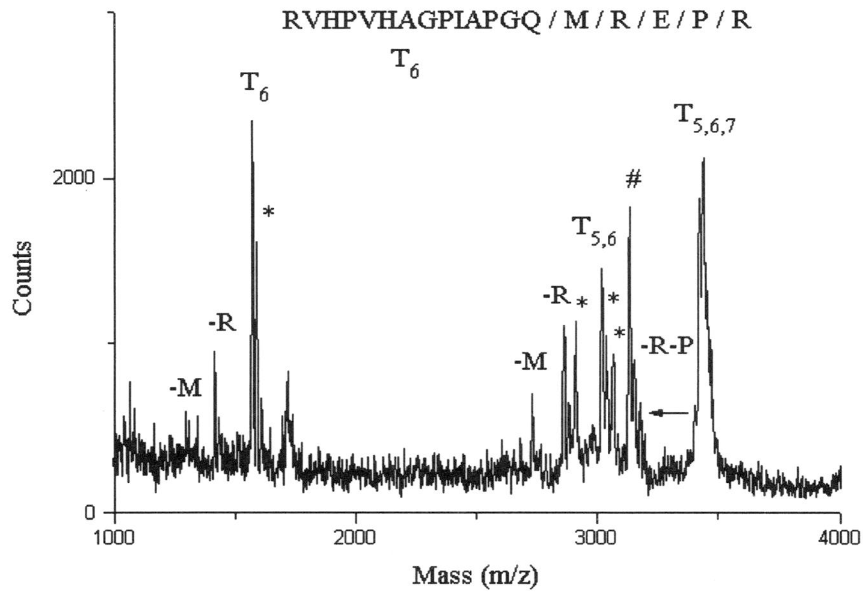

Figure 4. MALDI spectrum of the fragments remaining affinity-bound after consecutive trypsin and carboxypeptidase Y cleavages. # denotes a background ion. * denotes oxidized ions.

4. Conclusion

The results presented here demonstrate that tertiary structural features of large proteins can be probed using a combination of proteolytic footprinting and MALDI. With the use of immobilized antibodies, consecutive reactions can be successfully carried out on affinity-bound analytes, without the need of a separate elution step prior to detection of the antigen. There is also no need for rebinding the antigen to the antibody prior to the next digestion. Another advantage of the technique is that the protein is mapped in its native folded state, so that the possibility of detecting discontinuous epitopes is not precluded.

5. References

1. Przybylski, M. And Glocker, M.O. (1996) Electrospray mass spectrometry of biomacromolecular complexes with noncovalent interactions - New analytical perspectives for supramolecular chemistry and molecular recognition processes, *Angew. Chem. Int. Ed. Engl.* **35**, 807-826.
2. Miranker, A., Robinson, C.V., Radford, S.E., and Dobson, C.M. (1996) Investigation of protein folding by mass spectrometry, *FASEB J.* **10**, 93-101.
3. Papac, D.I., Hoyes, J., and Tomer, K.B. (1994) Direct analysis of affinity bound analytes by MALDI/TOF MS, *Anal. Chem.* **66,** 2609-2613.
4. Przybylski, M. (1995) Mass spectrometric approaches to the characterization of tertiary and supramolecular structures of biomolecules, in I. Cornides, G. Horváth and K. Vekey (eds.), *Adv. Mass Spectrom.* **13**, 257-283.
5. Jemmersen, R., and Paterson, Y. (1986) Mapping epitopes on a protein antigen by the proteolysis of antigen-antibody complexes, *Science*, **232**, 1001-1004.
6. Zhao. Y., and Chait, B.T. (1994) Protein epitope mapping by mass spectrometry, *Anal. Chem.* **66**, 3723-3726.
7. Suckau, D., Köhl, J., Karwath, G., Schneider, K., Casaretto, M., Bitter-Suermann, D., and Przybylski, M. (1990) Molecular epitope identification by limited proteolysis of an immobilized antigen-antibody complex and mass spectrometric peptide mapping, *Proc. Natl. Acad. Sci. USA,* **87**, 9848-9852.
8. Papac, D.I., Hoyes, J., and Tomer, K.B.(1994) Epitope mapping of the gastrin releasing peptide/anti-bombesin monoclonal antibody complex by proteolysis followed by matrix-assisted laser desorption mass spectrometry, *Protein Science*, **3**, 1488-1492.
9. Parker, C.E., Papac, D.I., Trojak, S.K., and Tomer, K.B. (1996) Epitope mapping by mass spectrometry: determination of an epitope on HIV-1_{IIIB} p26 recognized by a monoclonal antibody. *J. Immunol.*, **15**, 198-206.
10. De Llano, J.J.M., Jones, W., Schneider, K., Chait, B.T., Manning, J.M., Rodgers, G., Benjamin, L.J., and B. Weksler (1993) Biochemical and functional properties of recombinant human sickle hemoglobin expressed in yeast, *J. Biol. Chem.,* **268**, 27004-27011.
11. Qian, X.-H., and Tomer, K.B. manuscript in preparation.

Investigation of the Effect of Protein Oxidation Upon the Lipid-Binding Activity of the Cerebroside-Sulfate Activator Protein (Saposin B).

JULIAN P. WHITELEGGE, K.F. FAULL, B. PENN, T. TO, A. WARING, C. FLUHARTY AND A.L. FLUHARTY

Center for Molecular and Medical Sciences Mass Spectrometry, Departments of Chemistry & Biochemistry and Psychiatry & Biobehavioral Sciences, University of California, Los Angeles, California 90095, USA.

Abstract

The cerebroside-sulfate activator (CSact or Saposin B) protein is a multifunctional lipid-binding protein that is required for efficient metabolism of some brain membrane lipids. Deficiency of the protein leads to a lethal form of metachromatic leukodystrophy due to a disruption in sulfatide metabolism and demyelination. The porcine protein, whose primary structure has been solved [Stevens R.L., Faull K.F., Conklin K.A., Green B.N. and Fluharty A.L. (1993), *Biochemistry* **32**, 4051-4059], contains seven methionine residues which are apparently susceptible to oxidation as revealed by electrospray mass spectrometry of isolated protein.

Different species of the oxidized protein can be separated from each other and from non-oxidized protein by using C8 reverse-phase chromatography; the 'oxoforms' are in all cases less retained consistent with an increase in polarity accompanying oxidation of methionine residues to their sulfoxides. Full reversibility of oxidation with dithiothreitol treatment further confirms presence of the sulfoxide. The chromatogram revealed a collection of peaks that were partially resolved and since mass analyses confirmed that all the material carried only a single oxygen (+16 mu) it was concluded that these peaks represent different populations of protein each singly oxidized on a specific methionine residue. We are currently identifying which methionines of the primary sequence are modified in each peak by using a combination of cyanogen bromide mapping and MS/MS methodology. Since this protein has yet to yield to either genetic manipulation or successful chemical synthesis, the resolution of these 'oxoforms' provides a unique and novel structure/function probe.

W. Ens et al. (eds.), New Methods for the Study of Biomolecular Complexes, 67–73.

The effect of individual oxidations upon the 3-sulfo-galactosylceramide (sulfatide) binding parameters of the protein is under investigation. One way to assess lipid binding activity of CSact is to use the CSact assay which tests protein for its ability to activate the hydrolysis of sulfate from sulfatide by arylsulfatase A in vitro. The specific activity of material that was oxidized on only a single methionine was lowered by 20 to 60 % compared to non-oxidized material when assayed in this way. We conclude that the interaction of sulfatide and possibly other lipids with CSact may be sensitive to oxidation of single methionine residues.

1. Introduction

Protein oxidation is an unavoidable consequence of life in an aerobic environment. Cells limit oxidative modification of proteins primarily to the formation of essential disulfides by maintaining a reducing environment and actively scavenging the various 'active' oxygen species. However, under certain conditions oxidative stress may build, leading to potentially damaging levels of reactive species. These events may be associated with brief physiological periods such as the oxidative burst of the immune response or may arise as a more chronic condition associated with aging. The established correlation between increasing levels of protein oxidation with age [1] leads to two conclusions. In younger people cells are successful in minimizing protein oxidation but the necessary mechanisms become less efficient during aging. Elevated levels of protein oxidation become detectable in older subjects however, but are apparently tolerated. One possible consequence is that elevated levels of protein oxidation contribute to the triggering and progression of late onset diseases associated with mid and old age.

The CSact protein is missing in a rare but lethal variety of metachromatic leukodystrophy (MLD) which is a lipid-storage disease resulting from disruption of sulfatide metabolism and demyelination [2; reviewed in 3]. The protein is concluded to play a role in sulfatide metabolism and indeed can activate sulfate hydrolysis when sulfatide and arylsulfatase A (araA) are incubated together in vitro [4]. We are investigating the role of CSact in membrane assembly, modification and degradation with particular emphasis on the myelin membrane. As part of our effort to understand the function of CSact in vivo we are currently directing an intensive effort toward solving its tertiary and quaternary structure. During the course of our experiments it became apparent that we could resolve populations of minimally oxidized deglycosylated CSact, raising the potential of investigating this rarely considered aspect of protein structure and function. We believe this a timely study given the current surge of interest in the possible link between oxidative stress and the onset of neurodegenerative diseases.

2. Materials and Methods

2.1. ISOLATION AND DEGLYCOSYLATION OF CSACT

CSact protein was isolated from porcine kidney [5] and treated with PNGase F in order to remove the glycosyl moiety which we have shown to be attached to N21 [6]. Hydrolytic removal of the glycosyl group leaves an aspartic acid residue at position 21.

2.2. C8 REVERSE-PHASE CHROMATOGRAPHY

A sample of the deglycosylated/oxidized CSact was clarified (10 k x g ; 1 min) loaded onto an analytical C8 reverse-phase HPLC column previously equilibrated in 10 column volumes of 0.1 % aqueous TFA containing 5 % acetonitrile (v/v) at 1 ml/min. An extended linear gradient was used to separate the differentially oxidized forms of CSact protein; the percentage of B (0.1 % TFA in acetonitrile) was raised to 40 % in 10 minutes and slowly raised to 60 % over a further 200 minutes prior to a final column wash at 100% B. The different forms of CSact protein eluted in the 20 to 110 minute range (see Figure 1). The eluting fractions were dried under vacuum (SpeedVac < 100 mTorr) until acetonitrile was removed and then frozen at - 80 °C before lyophilization to dryness.

2.3. ELECTROSPRAY IONIZATION MASS SPECTROMETRY

CSact protein samples were dissolved in acetonitrile:water:formic acid (50:50:0.1; v/v) and subjected to flow injection analysis (FIA) using an appropriately configured Perkin Elmer Sciex API III operating in positive ion mode. Flow rate was 10 μl/min and orifice voltage was 90 V.

2.4. CSACT ACTIVITY ASSAY

Column fractions were redissolved in Tris buffer (20 μl 25mM Tris-HCl, pH 7.5). Protein assays (Lowry) were performed on aliquots (5 μl) of each redissolved sample. Additional aliquots of each fraction containing 1/20th of the protein were diluted to 30 μl with Tris buffer and mixed with Na-acetate (10 μl; 1 M; pH 4.2) and ^{35}S-sulfatide preparation (10 μl; 10000 cpm; 10 nmol of mixed rat and bovine brain sulfatides). The reaction was started by adding arylsulfatse A (10 μl; 250 U/ml; isolated from human liver and dissolved in Tris buffer containing 1 mg/ml BSA) and incubated (1 hour; 37 °C). The reaction was quenched by addition of chloroform/methanol (1 ml; 2:1) and Na_2SO_4 (200 μl 4 mM in 0.73 % NaCl). After vortexing and centrifugation (16 k x g; 2 min) 400 μl of the upper aqueous phase was removed and mixed with 400 μl of equilibrated lower phase. After a second vortex/centrifugation step, an aliquot of the upper phase (200 μl) was counted by liquid scintillation (2 ml Biosafe 2). Specific activity was calculated as nmol sulfatide hydrolysed/hr/mg CSact protein.

3. Results and Discussion

CSact contains seven methionine residues in its primary sequence with four of these clustered in a QMMMHMQ motif characteristic of the protein. The content of alpha-helix is very high and the protein has been modeled as a polyhelix bundle consisting of two pairs of alpha-helices each interconnected by disulfide bonds and hinged at one end [2] allowing the opening of a non-polar cleft that is thought to provide the binding site for a single lipid molecule. The binding specificity is moderately broad and evidence suggests that binding affinity and turnover are highly sensitive to lipid class as well as fine structure such as fatty-acyl chain length [7; reviewed in 3]. One of the most significant steps in the purification of CSact is chromatography on Octyl-Sepharose. It seems reasonable that this step constitutes a partial affinity chromatography involving an interaction between the octyl chains of the support and the lipid-binding pocket of the protein [5]. The CSact protein is eluted with high concentrations of octyl-glucoside and little resolution of sub-populations is observed. However, reverse-phase chromatography under conditions that do not denature CSact provides substantial resolution of different forms of the protein and we are currently using a C8 support to optimize our separation as we isolate quantities of deglycosylated CSact for structural analysis. The deglycosylated protein is most highly retained with the remaining glycosylated populations eluting some tens of minutes earlier. A small amount of deglycosylated but oxidized protein was identified in the region of the chromatogram immediately preceding the non-oxidized/deglycosylated material and was pooled for later analysis. Reverse-phase chromatography on an analytical C8 column was used to further investigate the oxidized material after adding some non-oxidized protein into the mixture to provide control samples (Figure 1). Significant resolution of components in the oxidized mixture was achieved. Most of the material eluted in five close peaks (1A-1E/F/G) and ESI mass-spectrometric analysis (data not shown) revealed that all material in this region of the chromatogram had a mass corresponding to the addition of a single oxygen (+16 mu). Since the oxidation can be reversed by extended treatment with 1 M dithiothreitol (data not shown) we conclude that the observed oxidation is due to conversion of single methionine residues to their sulfoxides. Furthermore, we have confirmed the presence of methionine sulfoxide at position 10 of a tryptic peptide derived from pooled oxidized material by MS/MS (data not shown) although it was not clear from this study what proportion of the total oxidation was accounted for by this specific modification. The most reasonable explanation for the resolution of the five peaks shown in Figure 1 (1A-1E/F/G) is that the mobility of the singly oxidized material is dependent upon which of the seven methionines of the original protein is existing as its sulfoxide. We propose to call the different oxidized forms of the protein 'oxoforms'.

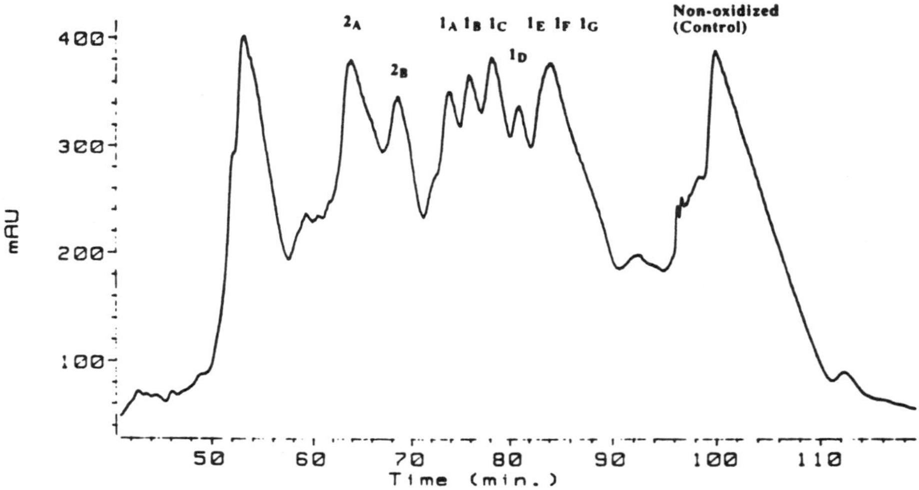

***Figure 1.* Oxidized CSact species are less retained during C8 reverse-phase chromatography.**

Pooled and dried oxidized deglycosylated CSact protein (5 mg) was mixed with non-oxidized deglycosylated material (200 μg) , dissolved in 250 μl of 0.1 % TFA in water and loaded onto a C8 analytical column (Keystone 4.6 x 250 mm Deltabond Octyl 5 μm particle size; 300 Å pore size) after clarification. The different forms of the deglycosylated protein eluted in the 56 - 114 minute region of the chromatogram as shown ($A_{214\ nm}$). The protein in 1 ml fractions was frozen at - 80 °C and dried (SpeedVac, < 100 mTorr) prior to assay of CSact activity and mass analysis. CSact modified by the addition of a single oxygen (1A-1G) elutes earlier than non-oxidized protein (see Table I).

We are using a mapping technique, which relies upon the documented resistance of methionine sulfoxide to cyanogen bromide cleavage [8], as well as MS/MS methodology to fully identify oxidation sites. Figure 1 also shows peaks corresponding to the addition of two oxygens (+32 mu) and some of the glycosylated component with a single sugar residue (least retained).

The CSact activity assay provides a useful assessment of the performance of this lipid binding protein in presenting sulfatide to arylsulfatase A in vitro. The total amount of sulfatide hydrolysed during the assay period however depends upon several interrelated factors including, 1. efficiency of presentation of the lipid to the sulfatase and dependence upon lipid structure; 2. turnover of cerebroside and sulfatide on the CSact protein; 3. relative concentrations of cerebroside and sulfatide; 4. CSact protein concentration and protein assay accuracy. Clearly, interpretation of assay results is complex. The observation that the different oxoforms of the CSact protein demonstrate impaired performance suggests some change in the properties of the protein upon single oxidations. The increased polarity implied by the lowered retention of the oxoforms in

C8 chromatography hints that the lipid-binding domain of the protein, which is expected to be the same non-polar region that interacts with the octyl groups on the column, is being modified in a way that makes it less favorable to lipid binding. If sulfatide spends less time associated with the protein it will be less likely to get desulfated by the sulfatase. However, the observed inhibition of the CSact activity assay could equally result from change in shape of the lipid/protein complex or inhibition of lipid turnover.

Table I. Suppression of CSact Activity Accompanying Protein Oxidation.

Label (Fig.1)	CSact Protein Species[A]	CSact Activity (% Control)[B]
Control	Non-oxidized	100
1A	One Met Sulfoxide	66
1B		73
1C		42
1D		61
1E		84
1F		57
1G		77
2A	Two Met Sulfoxide	50
2B		75

A. The identity of the CSact protein species was assigned based upon the increase in mass measured; +16 = one sulfoxide, +32 = two sulfoxide. The modifications were concluded to be methionyl sulfoxides because they could be fully reversed by prolonged treatment of the protein with 1 M dithiothreitol. The mass of non-oxidized deglycosylated CSact was measured to be 8845.12 mu compared to the 8845.20 mu predicted from the primary structure.

B. The CSact assay which tests protein for its ability to activate the hydrolysis of sulfate from sulfatide by arylsulfatase A (araA) in vitro was used to assess CSact lipid-binding activity. Results from individual 1 ml fractions (Fig. 1) were averaged, Control 100-102 min, 1A 73-4 min, 1B 75-6 min, 1C 78-9 min, 1D 80-1 min, 1E 82-3 min, 1F 84-6 min, 1G 87-9 min, 2A 64-5 min, 2B 68 min. Standard deviation on these measurements was in the range 0.6 - 18 % control. The fifth peak of the singly oxidized material (82-89 min) was subdivided into three because we have observed resolution of three peaks in this region when smaller amounts of material are loaded.

The assay results can still provide potentially significant information as part of a structure/function analysis. The problem here is obtaining homogenous oxidized material. It is quite clear that while the different peaks in Figure 1 represent separation of individual oxoforms, all of the fractions contain somewhat mixed populations since resolution is not to baseline. We are currently investigating whether a better separation can be achieved by manipulation of the gradient and loading conditions. It is hoped that future measurements may reveal whether or not some of the oxoforms are more highly inhibited than others - a conclusion that would be hard to draw from data presented here. A structure/function study using defined oxidation of methionine residues represents a novel approach to the analysis of lipid binding by proteins. Our results to date indicate that oxidation of the CSact protein provides a useful model with which to examine the consequences of oxidative stress upon protein function. Long term goals include the development of mass-spectrometric assays for the analysis of the lipid-protein complex to allow us to monitor changes in lipid-binding specificity that may arise due to protein oxidation.

4. Acknowledgements

Supported in part by NS31271 to A.L. Fluharty and by a grant from the W.M. Keck Foundation.

5. References

1. Stadtman, E.R. (1992) Protein Oxidation and Aging, *Science* **257**, 1220-1224.

2. Stevens, R.L., Fluharty, A.L., Kihara, H., Kaback, M.M., Shapiro, L.J., Marsh, B., Sandhoff, K. and Fischer, G. (1981) Cerebroside Sulfatase Activator Deficiency Induced Metachromatic Leukodystrophy, *Am. J. Hum. Genet.* **33**, 900.

3. Fluharty, A.L. (1995) The Cerebroside Sulfate Activator (Saposin B, SAP-1): A Many faceted Protein, *Trends in Glycoscience and Glycotechnology* **7**, 167-189.

4. Fisher, G. and Jatzkewitz, H. (1975) The Activator of Cerebroside Sulfatase: Purification From Human Liver and Identification as a Protein, *Hoppe-Seyler's Z. Physiol. Chem.* **356**, 605-613.

5. Fluharty, A.L., Katona, Z., Meek, W.E., Frei, K. and Fowler, A.V. (1992) The Cerebroside Sulfate Activator From Pig Kidney - Purification and Molecular Structure, *Biochem. Med. Met. Biol.* **47**, 66-85.

6. Stevens, R.L., Faull, K.F., Conklin, K.A., Green, B.N. and Fluharty, A.L. (1993) Porcine Cerebroside Sulfate Activator - Further Structural Characterization, *Biochemistry* **32**, 4051-4059.

7. Vogel, A., Schwarzmann, G. and Sandhoff, K. (1991) Glycosphingolipid specificity of the human sulfatide activator protein, *Eur. J. Biochem.* **200**, 591-597.

8. Maier, K.L., Lenz, A.-G., Beck-Speier, I. and Costabel, U. (1995) Analysis of Methionine Sulfoxide In Proteins, *Methods In Enzymology* **251**, 455-461.

THE USE OF CHEMICAL LIGATION FOR INVESTIGATION OF THE STRUCTURE OF DUPLEXES FORMED BY LINEAR AND CIRCULAR OLIGONUCLEOTIDES

A.V. MAKSIMENKO, O.A. FEDOROVA,
M.B. GOTTICH, Z.A. SHABAROVA
*A.N. Belozersky Institute of Physical and Chemical Biology and Chemical Department,
Moscow State University, Moscow 119899, Russia*

Oligonucleotide base-paired duplexes have provided the major source of detailed structural and dynamic information about double helical structure. Polycircular compounds (catenanes) are formed during DNA replication and recombination [1, 2]. Here we report investigations of the possibility of using chemical ligation for the study of feasible types of linear oligonucleotide hybridization on linear and circular templates as well as for the design of polycircular oligonucleotides. Using chemical ligation it has been shown that the way linear oligonucleotide hybridization occurs on circular and linear templates depends on such parameters as: oligonucleotide and template sequences, the length of the duplex formed, and the ligation method. The structure of catenanes is confirmed by *Mva*I restriction endonuclease specific hydrolysis.

Peculiarities of Linear Oligonucleotide Ligation on Linear and Circular Templates

The study of duplex formation by the chemical ligation method is of particular interest in a system with several possible ways of hybridization. For this purpose the unusual DNA-duplexes presented in Figure 1 were designed.

It was previously shown that ligation in the system formed by linear oligonucleotides may result in the formation of a circular oligonucleotide (route A, Figure 2), a linear dimer of the linear precursor-oligonucleotide (route B), or a circular dimer of the linear precursor-oligonucleotide (route C). In choosing the sequence of the oligonucleotide-precursor for the formation of the circular oligonucleotide, one should take into account the possible formation of secondary structure elements (hairpins, etc.). The precursor secondary structure was proved to affect dramatically the pathways of the linear precursor hybridization with the template [3].

W. Ens et al. (eds.), New Methods for the Study of Biomolecular Complexes, 75–81.

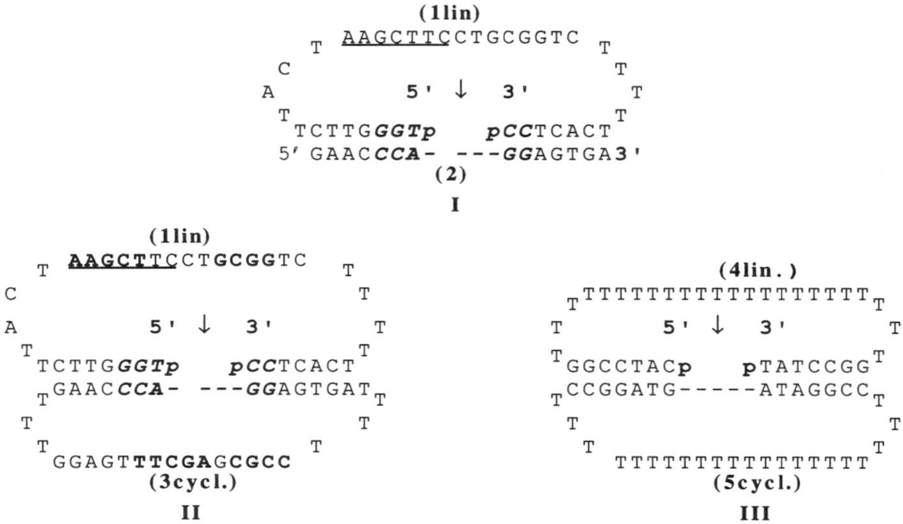

Figure 1. DNA-duplexes for investigation of the hybridization pathways. The site of internucleotide bond formation is indicated by an arrow. The *Mva*I recognition site is typed in bold italic letters. The *Hind*III recognition site is underlined. The 4-mer and 5-mer fragments of the 38-mer oligonucleotide ($1_{lin.}$) and the circular template ($3_{cycl.}$), which are able to hybridize in addition to the main 14-b.p. duplex, are typed in bold letters.

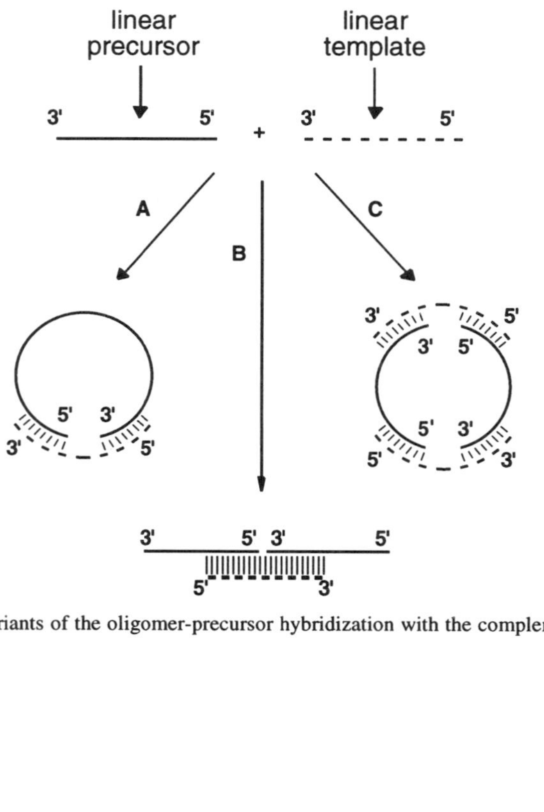

Figure 2. Feasible variants of the oligomer-precursor hybridization with the complementary template.

We are interested in studying the influence of the ligation methods: enzymatic ligation by T4 DNA ligase, BrCN-induced chemical ligation and chemical ligation using water-soluble 1-ethyl-3(3'-dimethylaminopropyl)-carbodiimide (CDI) on the reaction products. It was shown that DNA ligase catalyses formation of both the circular oligonucleotide and the linear dimer (Table 1). The same reaction course was observed for the CDI-induced reaction between 5'-hydroxyl and 3'-phosphate groups. In all other cases the circular oligonucleotide was the only ligation product. At the same time, the BrCN-chemical ligation of the 36-mer with the other structure (5'-**TTTGAGGTTTGAAC CCAGGAGTGATTTTCCGCGAGC**-3' ($3_{lin.}$)) on a 14-mer template (5'-**CCTCAA AGCTCGCG**-3' (6)) results in the formation of all three products: the circle 36-mer (yield 35-40%), the linear dimer (10-15%) and the circular dimer (20-25%).

TABLE 1. The influence of the ligation method on the reaction pathway and product yield[#].

Ligation method	Ligation site -GGT**x 5' 3'y**CCT- -CCA--------GGA- (I) X	Y	Product yield, % circle	linear dimmer
T4 DNA ligase	phosphate	-OH	60-65	10-15
BrCN	phosphate	-OH	45-50	0
	phosphate	phosphate	50-55	0
	-OH	phosphate	75-80	0
carbodiimide	phosphate	-OH	35-40	0
	phosphate	phosphate	75-80	0
	-OH	phosphate	35-40	25-30

[#]Oligonucleotide concentration (per monomer) is 10^{-4}M. Buffer for BrCN-induced ligation contained 1 M N-methylmorfolyne, 0.02 M $MgCl_2$, pH 7.5. Concentration of BrCN was 0.5 M. Buffer for CDI-induced ligation contained 0.05 M MES, 0.02 M $MgCl_2$, pH 5.0. CDI concentration is 0.2 M. Buffer for enzymatic ligation contained 0.05 M Tris-HCl, 0.01 M $MgCl_2$, 0.05 M dithiothreitol, 1 mM spermidine, pH 7.6. Activity of T4 DNA ligase was 30 E.

The peculiarities of linear oligonucleotide circularization on a circular template were studied on the systems (II) - (III) (Figure 1). The different pathways of the 38-mer linear oligonucleotide hybridization with the 36-mer circular template are presented in Figure 3. Ligation in the system formed by linear and circular oligonucleotides may result in the formation of the circular oligonucleotide (route A), the catenane (route B), the linear dimer of the 38-mer linear oligonucleotide (1_{lin}.) (route C) or the circular dimer of the 38-mer linear oligonucleotide (1_{lin}.) (route D).

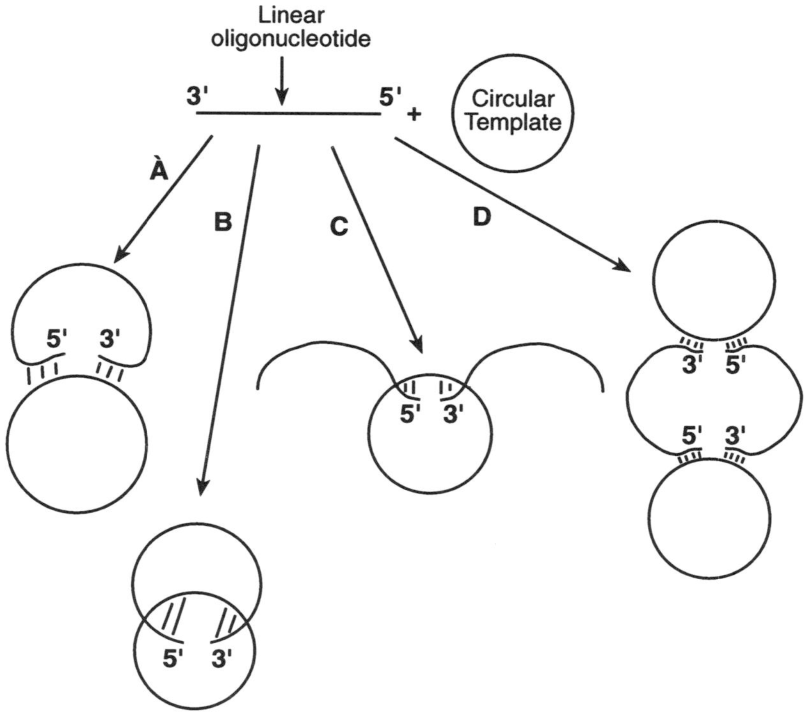

Figure 3. Feasible variants of the 38-mer linear oligonucleotide hybridization with the 36-mer circular template.

As shown in Table 2, ligation in system (II) gives rise to the formation of five product. The structure of the ligation products was confirmed by their hydrolysis using *Mva*I and *Hind*III restriction enzymes followed by Maxam-Gilbert sequencing. It was shown that the ($1_{cycl.}$) product is a 38-mer circular oligonucleotide, the ($1_{lin.dim.}$) product is a linear dimer of oligonucleotide ($1_{lin.}$), and the ($1_{cycl.dim.}$) product is a circular dimer of oligonucleotide ($1_{lin.}$). It should be noted that two products possessing the catenane structure were produced.

The formation of two catenanes in the case of the system (II) could be explained as follows. The single-stranded fragments of linear oligo ($1_{lin.}$) and circular template ($2_{cycl.}$) that do not participate in 14-b.p. duplex formation are able to hybridize forming an additional duplex. If the additional duplex is formed prior to the main 14-b.p. duplex, the catenane with two intertwinings between rings will be the ligation product (Figure 4, a). In the other case, ligation will lead to the classic catenane with only one intertwining between rings (Figure 4, b). This catenane is less compact than the former one and migrates more slowly on PAGE separation.

TABLE 2. The influence of the ligation method on the reaction pathway in the duplexes formed by circular and linear oligonucleotides and product yield.

Products	Product yield, %					
	T4 DNA ligase*		BrCN		CDI	
	ligation systems (see Fig. 1)					
	II	III	II	III	II	III
Circular dimer	0-1	0	1-3	0	0-1	0
Catenane 2	1-2	5-7	4-6	1-3	4-6	1-2
Catenane 1	0-1	0	6-8	0	8-10	0
linear dimer	12-14	7-9	3-5	0	6-8	0
38-mer circular oligonucleotide	32-34	76-78	7-9	30-32	26-28	86-88

*Internucleotide bond formation was occurred between 5'- phosphate and 3`-phosphate groups in the case of chemical ligation and between 5'- phosphate and 3'-hydroxyl groups in the case of enzymatic ligation.

In the cases of CDI-induced and enzymatic reactions, the circular oligonucleotide was the main ligation product (Table 2). The BrCN-chemical ligation results in the formation of all the products in equal quantities. At the same time, in the case of the BrCN-induced reaction in system (I), the chief product is circular oligonucleotide (5cycl.) (Table 1). This difference is due to the high speed of the chemical ligation induced by BrCN as shown in [4]. The ratio between reaction products accounts for the ratio between duplexes formed as a result of the linear oligonucleotide hybridisation on the circular template.

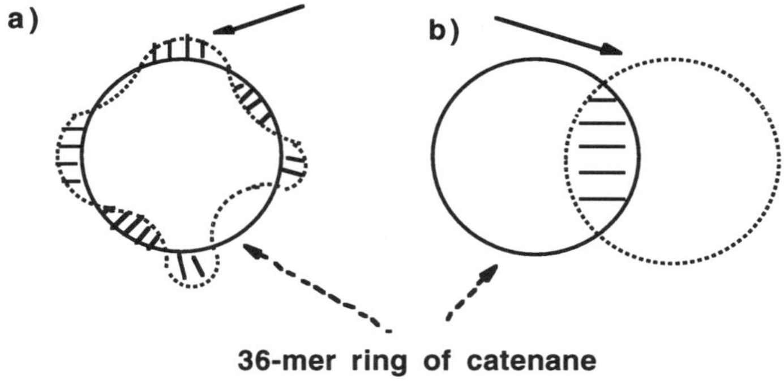

Figure 4. a) The structure of catenane with two intertwinings between rings ($1_{cat.1}$). b) The structure of catenane with only one intertwining between rings ($1_{cat.2}$).

The CDI-induced reaction proceeds during 40-48 hours. Thus reaction products are formed in the most thermodynamically stable duplexes. Consequently the duplex arising by linear oligonucleotide hybridization on the circular template (route A, Figure 3) is the most thermodynamically stable.

Ligation in system (III) results in the formation of three products (Table 2). In this case, ligation can result only in the classic catenane with one intertwining between rings. The formation of a more complicated catenane in system (III) is impossible. Enzymatic ligation gives rise to the circular oligonucleotide and linear dimer formation in system (II). In system (III) catenane formation takes place as well.

The influence on pathway reaction of the localization of the GC-tracts in the duplex structure was studied. The reaction pathway is shifted to formation of the 38-mer circular oligonucleotide, when the GC-tracts are displaced to the ends of the duplex (systems (II) and (III), Table 2). Consequently , catenane formation is favored when GC-tracts are localized on the duplex ends.

We also studied the influence of the duplex length on the pathway reaction and the product yield. The same products were observed in systems with duplex length from 10 to 14 b.p. However the highest efficiency of catenane production was observed in the case of a 14 b.p. duplex.

Conclusion

The present study establishes that chemical ligation can be used for the investigation of systems in which several pathways of oligonucleotide hybridization may take place. Duplexes formed from a linear oligonucleotide, both ends of which are complementary to the linear template or to the fragment of the circular template, were chosen as models of such systems. In this case, the reaction pathway depends on the chosen method and on the sequences of single-stranded fragments of linear and circular oligonucleotides.

The chemical ligation method may also be of interest for obtaining DNA fragments possessing an unusual spatial structure (catenanes). Displacement of the GC-tracts to the ligation site of the duplex was found to result in catenane formation. The length of the duplex must be no less than 14 b.p.

Polycircular oligonucleotides and complementary DNA fragments are able to form stable duplexes which are sufficiently good substrates for some restriction endonucleases. This approach may be fruitful for the design of multifunctional biosensors or of therapeutics directed to multiple medically important targets.

Acknowledgments

We thank T.S. Oretskaya and E.A. Romanova for the oligonucleotide synthesis.

References

1. Sundin, O., Varshavsky A. (1987) Arrest of segregation leads to accumulation of highly intertwined catenated dimers: dissection of the final stages of SV40 DNA replication, *Cell.* **25,** 659-669.
2. Bliska, J. B., Cozzarelli N. R. (1987) Use of site-specific recombination as a probe of DNA structure and metabolism in vivo, *J.Mol.Biol.* **194**, 205-218.
3. Dolinnaya, N.G., Blumenfeld, M., Merenkova, I.N., Oretskaya, T.S., Krynetskaya, N.F., Ivanovskaya, M.G., Vasseur, M., Shabarova, Z.A. (1993) Oligonucleotide circularization by template-directed chemical ligation, *Nucl. Acids. Res.* **21**, 5403-5407.
4. Fedorova, O.A., Gottich M.B., Maksimenko A.V., Oretskaya T.C., Shabarova Z.A. (1995) The investigation of BrCN-induced chemical ligation mechanism. *Bioorgan. Khimiya.* **21,** 868-873. (Russ).

STUDYING NONCOVALENT PROTEIN-RNA INTERACTIONS AND DRUG BINDING BY ELECTROSPRAY IONIZATION MASS SPECTROMETRY

JOSEPH A. LOO, KRISTIN A. SANNES-LOWERY, PEIFENG HU, DAVID P. MACK, AND HOUNG-YAU MEI

Parke-Davis Pharmaceutical Research, Division of Warner-Lambert Company, 2800 Plymouth Road, Ann Arbor, MI 48105

1. Introduction

The field of biochemical analysis by mass spectrometry is maturing at an ever-increasing pace. The race to develop so-called "soft" desorption/ionization methods to vaporize intact, large biomolecules consumed many research laboratories prior to 1990. As important milestones along the molecular weight path were reached, from small peptides, to insulin (5.7 kDa), to trypsin (23 kDa), new applications were developed. However, the explosive growth in the field did not occur until electrospray ionization (ESI) [1] and matrix-assisted laser desorption/ionization (MALDI) [2] in the late 1980's-to-early-1990's came onto the scene. Biological compounds with M_r greater than 150 kDa can be measured with sub-picomole sensitivity by ESI-MS and MALDI-MS. Now, much of mass spectrometry has been put into the hands of the non-mass spectrometrist.

From the initial results demonstrated by the Fenn laboratory [1], it was evident that electrospray ionization mass spectrometry was truly a gentle technique. Molecular ions (albeit multiply charged) for intact peptide and protein species are generated by ESI, with little-to-no hint of fragmentation. Through collisional activation methods induced in the vacuum regions of the ESI-mass spectrometer (e.g., ESI interface, collision cell) [3], dissociation of the multiply charged molecules can occur for even large proteins [4, 5], thereby generating structurally-informative product ions. It is this ability to control the softness of the technique that has lead practitioners of ESI-MS to investigate its applicability for studying weakly-bound biochemical complexes. Water-bound adducts on protein ions can be observed by lowering the activation input as the highly charged droplets traverse the ESI interface [6, 7]. This suggested that noncovalently-bound complexes between peptides and other proteins [8], protein-cofactor [9, 10], enzyme-receptor [11], and other biochemical complexes [12] could remain intact in the

W. Ens et al. (eds.), New Methods for the Study of Biomolecular Complexes, 83–99.

gas phase for a mass spectrometer to measure. The question that remains to be resolved before ESI-MS can be a useful tool to study such systems is whether the complexes suggested in an ESI mass spectrum is faithful to the situation found in the solution phase.

This report will discuss some of our observations during the course of studying a variety of noncovalent complexes by ESI-MS. The efficiency for removing residual solvent molecules prior to detection of the gas phase complex is an important consideration. In addition, we will provide a few applications from our laboratory regarding the study of protein-ribonucleic acid complexes and inhibitor binding using electrospray ionization-MS.

2. Experimental Methods

ESI-mass spectrometry was performed with a forward geometry double focussing mass spectrometer (Finnigan MAT 900Q, Bremen, Germany) with a mass-to-charge (*m/z*) range of 10 000 at full acceleration potential (5 kV) [13]. The instrument is equipped with a position-and-time-resolved-ion-counting (PATRIC) scanning focal plane detector [14]. An 8% *m/z* range of the *m/z* centered on the array detector was used. The resolving power with the scanning array detector for this study was greater than 1000. An ESI interface based on a heated glass capillary inlet was used. Warm countercurrent nitrogen gas (~ 60 °C) with adjustable control of the flow (typically at 3 L min^{-1}) was used to aid the desolvation process. In addition, gas phase collisions in the ESI interface, controlled by adjustment of the voltage difference between the tube lens at the metallized exit of the glass capillary and the first skimmer element (ΔV_{TS}), was used to augment the desolvation of the ESI-produced droplets and ions. A stream of SF_6 coaxial to the spray suppressed corona discharges, especially important for ESI of aqueous solutions.

In most cases, protein and RNA solutions were buffered with 10-25 mM ammonium acetate, pH 6.9. The synthetic RNA samples were gel-purified; residual salts were removed by cold ethanol precipitation three times exchanged with ammonium acetate [15-17]. For the zinc-binding studies [18] and the protein-RNA experiments, methanol to the 10% (*v/v*) level was added to the aqueous ammonium acetate ESI-MS solution. This enhanced the stability of ESI-MS signal without altering the resulting spectra to a significant extent for these systems. Solution flow rates delivered to the ESI source were typically in the 0.5-1.0 μL min^{-1} range.

3. Desolvation of Noncovalent Complex Ions

The gentleness of electrospray ionization has allowed mass spectrometry to detect several examples of noncovalently-bound complexes. To preserve the proper protein conformation and/or complexation upon transfer from solution to the gas phase, ESI of aqueous solutions is usually necessary. Although the ESI process depends on efficient desolvation of the highly charged droplets prior to detection, it is sometimes difficult to completely remove *all* solvent molecules from the protein complex ions without affecting the gas phase complex (i.e., dissociation of the complex may occur). The resulting mass spectra can show variable amounts of solvation, depending on several experimental factors, such as collisional activation in the atmosphere/vacuum interface, drying gas flow rate and temperature, pressure in the high vacuum region, and ESI needle position [7, 19, 20].

3.1. EFFECT OF NITROGEN COUNTERCURRENT FLOWRATE

For many ESI interfaces, the use of a gas flow (e.g., N_2) countercurrent to the electrospray axis is necessary for efficient droplet and ion desolvation [1]. The N_2 flowrate can influence not only ESI sensitivity, but also the amount of residual solvation observed for the multiply charged ions, especially for ESI from aqueous solutions. Multimeric proteins presents an opportunity for ESI-MS to study very large molecular weight complexes. Many enzyme systems are composed of identical and nonidentical subunits that associate together to form the fully active species. Studying the nature of these interactions that maintain the quaternary structure of enzymes is essential for the understanding of cellular functions at the molecular level. Mass spectrometry experiments with protein oligomers can yield information on both the stoichiometry and molecular nature of subunit interactions [13, 20-25]. For avidin, a 64 kDa tetrameric protein, peaks for the protein tetramer ions are broad and shifted to higher *m/z* than expected with the N_2 flowrate of 2.5 L min^{-1} (Figure 1). Increasing the N_2 flowrate to 10 L min^{-1} produces sharper peaks and shifts the peaks to lower *m/z*, consistent for more desolvated ions.

3.2. EFFECT OF ENERGETIC COLLISIONS AND PRESSURE IN THE ESI INTERFACE

Optimizing the voltage difference between the various lens elements in the vacuum region of the ESI interface is critical for removing remaining solvent molecules from the ions. Shifting of ions to lower *m/z* with increasing ΔV_{TS} is a result of the ions sheading residual solvent molecules. In most cases studied, the shift to lower *m/z* is gradual. However, the spectra for lysozyme in Figure 2 represent a more unusual example. At low ΔV_{TS}, lysozyme peaks are broad and at relatively high *m/z*. Slightly increasing ΔV_{TS} moves the broad peaks to lower *m/z*, but sharp peaks are observed at much lower *m/z* for the desolvated ions.

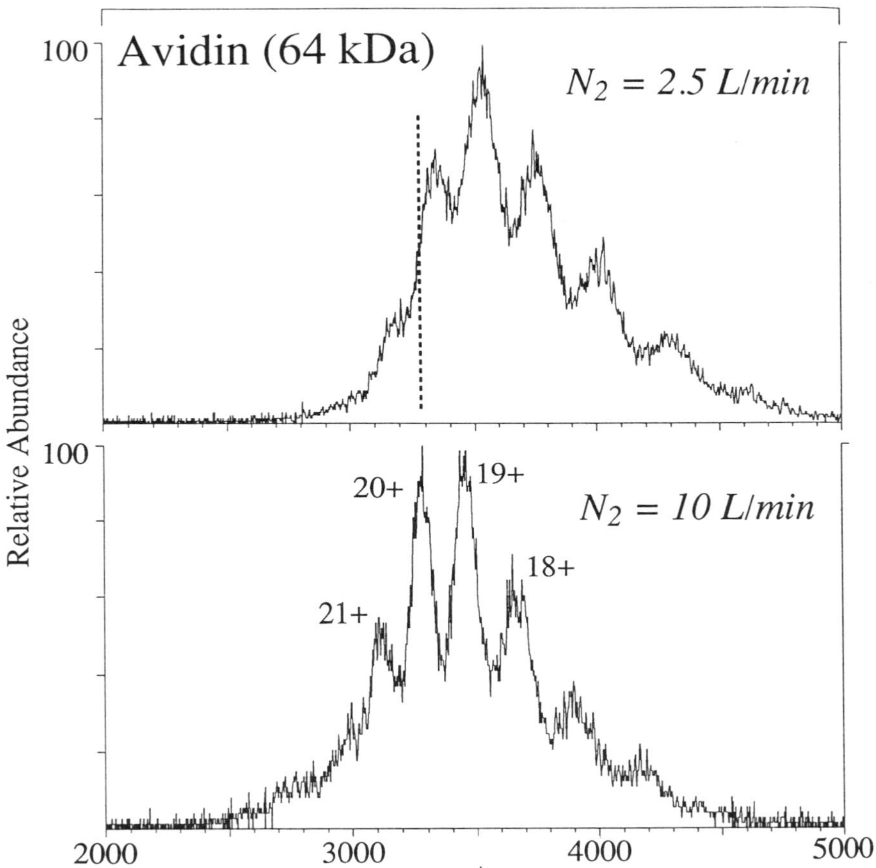

Figure 1. Electrospray ionization mass spectra of the tetrameric protein complex, avidin (64 kDa), in aqueous solution with different countercurrent nitrogen gas flowrates.

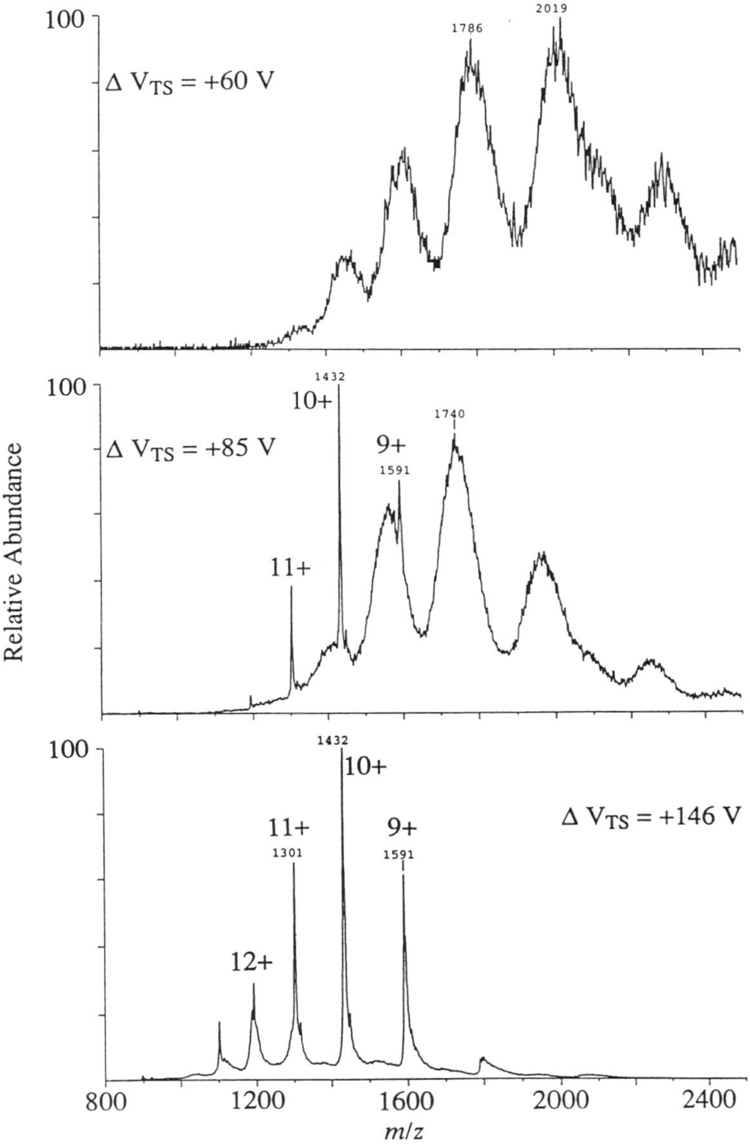

Figure 2. ESI mass spectra of hen egg white lysozyme (M_r 14 306) in water with varying voltage differences between the metalized glass capillary inlet and the first skimmer element of the ESI interface.

3.3. EFFECT OF ELECTROSPRAY SOURCE POSITION

The position of the electrospray source can also influence the amount of solvation observed for the protein ions. Optimum sensitivity with our system for the myoglobin complex (16.9 kDa polypeptide noncovalently-bound to heme) is achieved with the spray situated approximately 5.2 degrees "off-axis" (off the axis of the capillary inlet) (Figure 3). Directing the ESI source "on-axis" results in very broad peaks and an increase in low *m/z* background ions. The use of an on-axis electrospray results in "wet" ions because larger droplets are sampled by the MS inlet.

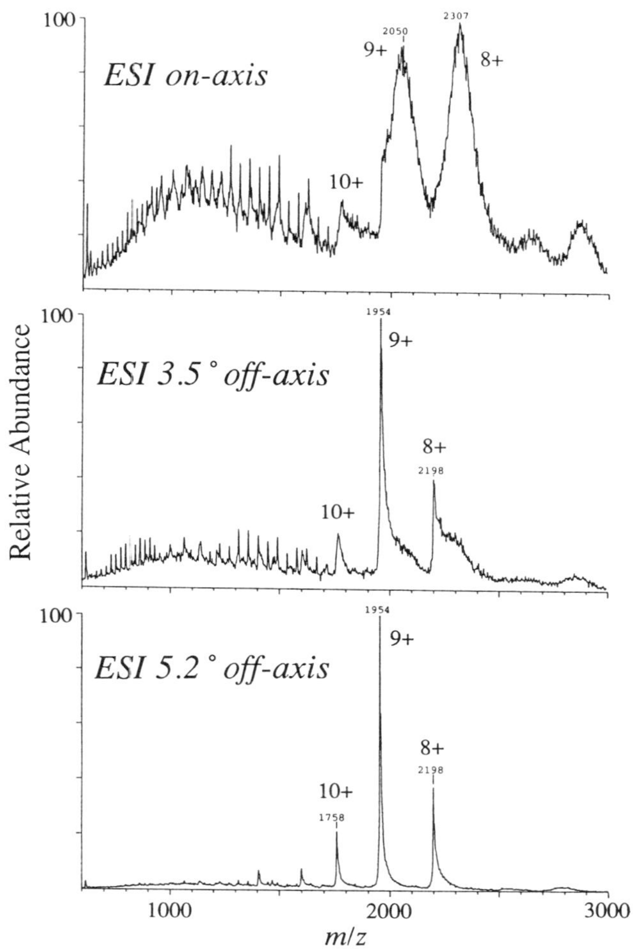

Figure 3. ESI mass spectra of equine myoglobin complex (holo form, M_r 17 568) in pH 6 solution acquired with different ESI spray positions. Using an off-axis spray yields more desolvated ions for the complex.

4. Studying Tat Peptide-TAR RNA and Drug Binding

To assess the usefulness of ESI-MS for studying noncovalent interactions, and in particular, protein-ribonucleic acid (RNA) recognition, we have examined the binding of various protein-RNA systems important in the replication cycle of human immunodeficiency virus type 1 (HIV-1) and the effect of targeted drug inhibitors on these complexes. HIV gene expression is controlled by the binding of viral regulatory proteins to specific RNA target sequences. However, the basis of protein-RNA recognition is poorly understood. Relatively few protein-RNA complex structures have been determined [26, 27]. Studying protein-RNA complexes could provide additional insight on their structures. Several results from our laboratory suggest that ESI mass spectra of protein-RNA complexes reflect the specific interactions found in solution.

Tat protein from HIV is a viral trans-activator that is essential for viral replication [28]. Tat is required to increase the rate of transcription from the HIV long terminal repeat (LTR) and its action is dependant on the region near the start of transcription in the viral LTR called the trans-activation responsive (TAR) element, which is located at the 5'-end of mRNA [29]. Tat protein contains an arginine-rich region that is essential for RNA binding (Figure 4). Fragments of Tat protein containing this arginine-rich region specifically bind to TAR RNA [30-32]. A C-terminal 40-residue peptide containing this basic region has very similar binding characteristics to TAR RNA compared to the protein. TAR RNA is a small RNA hairpin consisting of a stem-loop structure with a pyrimidine bulge. This three nucleotide bulge is essential for Tat binding and activity [33].

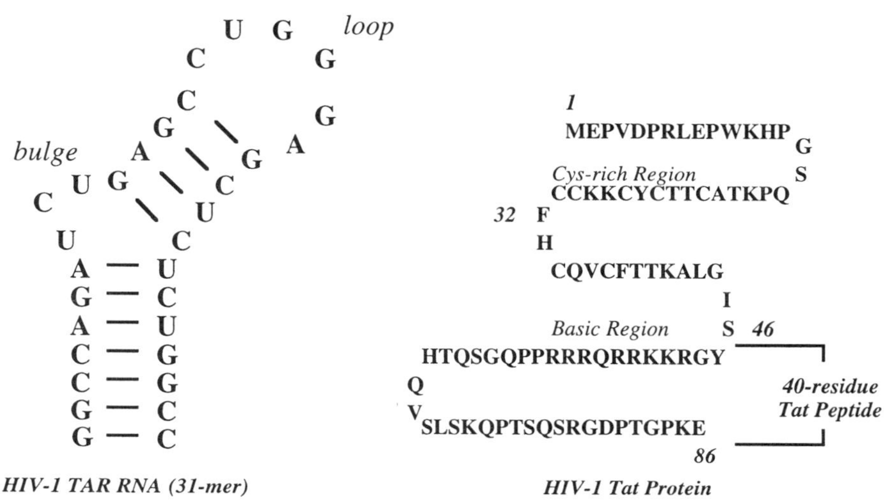

Figure 4. Sequence of HIV-1 TAR RNA (31-mer) and Tat peptide and protein used for this study.

Negative ion ESI-MS of a solution containing Tat peptide (40 amino acid residues, M_r 4644) and TAR RNA (31-mer, M_r 9941) yields a peptide-RNA complex consistent for a 1:1 stoichiometry (Figure 5). Increasing the Tat peptide relative concentration yields a 2:1 Tat peptide-RNA complex. However, the 2:1 complex easily dissociates to the 1:1 complex under relatively mild conditions in the ESI interface (i.e., relatively low ΔV_{TS} voltages), suggesting that the 1:1 complex is a much tighter interaction than the 2:1 complex. Similarly, a solution containing Tat protein (86 residues, 9.8 kDa) and TAR yields a 1:1 complex (Figure 6). Interestingly, the addition of a reducing agent (e.g., dithiothreitol, DTT) was necessary for observing the Tat protein-RNA complex. Apparently, Tat protein needs to be in the reduced state to form a complex with TAR RNA.

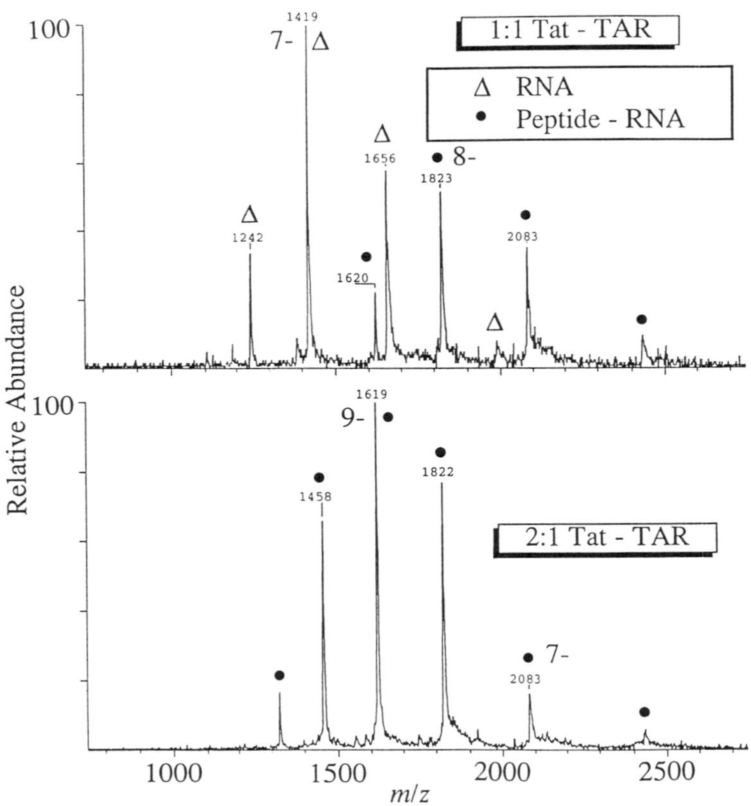

Figure 5. Negative ion ESI mass spectra of the 1:1 Tat peptide (40 residues, M_r 4644) - TAR RNA (31-mer, M_r 9941) complex. Tat peptide was added at different molar ratios to a solution containing previously annealed TAR (13 pmol μL^{-1}), 0.3 mM CDTA, 10% (*v/v*) methanol, and 10 mM ammonium acetate, pH 6.9.

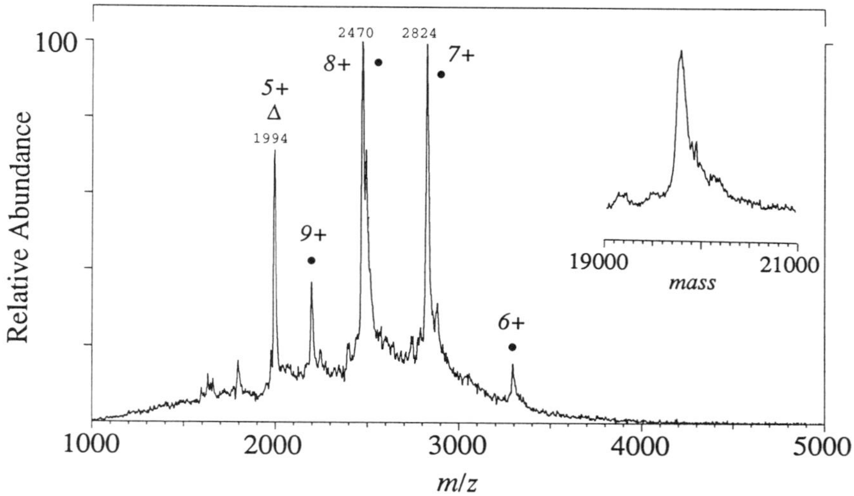

Figure 6. Positive ion ESI mass spectrum of the 1:1 Tat protein (86 residues, M_r 9778) - TAR RNA complex in 10 mM ammonium acetate, pH 6.9, 10% (*v/v*) methanol, and 1 mM DTT. The peaks labeled with "•" represent the Tat-TAR complex and peaks with the "Δ" symbol represent ions for free TAR RNA.

To test whether the specificity of the peptide-RNA interaction is mirrored by the ESI-MS data, competitive binding experiments were performed. Under competitive binding conditions where the total RNA concentration equals the protein concentration in solution, ESI-MS mass spectra show that Tat peptide affinity for TAR RNA is greatly reduced for a bulge-less 28-mer TAR RNA, consistent with solution phase measurements [33].

Because the interaction of Tat protein to TAR RNA is critical for HIV replication, drugs that interfere with this complexation might inhibit HIV replication. ESI-MS provides an ideal method to determine drug binding stoichiometry and potentially, binding constants, for biochemical systems. ESI-MS has been used to examine the action of various compounds on the Tat peptide-TAR RNA complex. Calnan *et al.* [34] showed that Arg_9 binds TAR with the same affinity and specificity as the Tat peptides. Thus, the peptide Arg_{11} might effectively compete with Tat peptide for binding TAR. Figure 7 shows the result upon addition of Arg_{11} to a solution containing Tat peptide (40 residues) and TAR RNA. Dissociation of the Tat-TAR complex is evident by the ions representing the TAR-Arg_{11} complex. In addition, the ternary Tat-TAR-Arg_{11} complex was also observed.

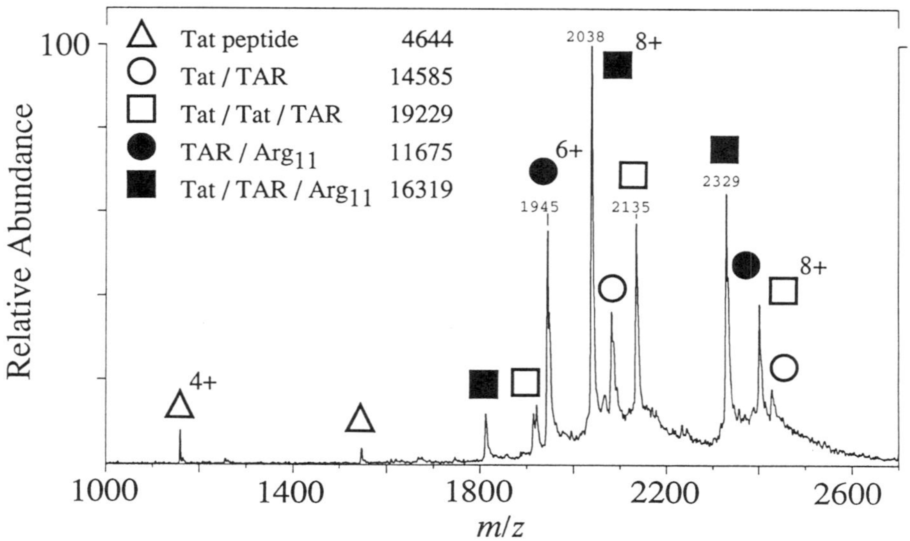

Figure 7. Positive ion ESI mass spectrum from a solution containing Tat peptide (40 residues, M_r 4644), TAR RNA (M_r 9941), and Arg_{11} (M_r 1736) at a molar ratio of 1.5:1:2, respectively, in 0.3 mM CDTA, 10% (*v/v*) methanol, and 10 mM ammonium acetate, pH 6.9. Addition of Arg_{11} results in dissociation of the Tat-TAR complex.

The aminoglycoside antibiotics, neomycin and streptomycin, are known to specifically bind to RNA molecules [35, 36] and might interfere with the complexation of Tat with TAR RNA. It has been proposed that a stem-loop structure like that found in TAR RNA is necessary for binding of the antibiotics [37]. Titrating the Tat peptide-TAR RNA complex with these compounds showed that both aminoglycosides bind to the complex as well as to TAR RNA itself. Neomycin was effective in inhibiting the complexation of Tat-TAR. Streptomycin did not appear to dissociate the Tat-TAR complex at the concentrations studied.

Positive and negative ion mode ESI-MS give different results for the titration of the Tat-TAR complex with the drugs. The negative ion mode results for the titration of Tat-TAR complex with neomycin and the titration of TAR with neomycin agree with gel shift assay studies [37]. Both studies indicate that neomycin binds to the Tat-TAR complex and then the neomycin-Tat-TAR complex dissociates to form the 1:1 neomycin-TAR complex. In addition, both the gel shift assays [37] and the ESI-MS results indicate that multiple neomycin molecules can bind to TAR RNA.

5. Monitoring Zinc Ejection from a Zinc-Finger Protein By ESI-MS

The zinc finger protein, nucleocapsid protein (NCp7), is believed to be involved in encapsulation and packaging of genomic RNA during human immunodeficiency virus (HIV) particle assembly [38, 39]. The interaction with metal ions is essential for function and structural stability for many proteins. The 55-residue NCp7 (6.4 kDa) contains two zinc finger motifs characterized by the sequence Cys-X_2-Cys-X_4-His-X_4-Cys (X, any amino acid residue) which functions as the zinc-binding domain (Figure 8). The mass spectrum of a zinc-containing acidic solution of NCp7 (pH 2.5) showed a measured mass consistent for the apoprotein. At pH 6.9, the expected protein-Zn_2 species was detected as the major species [18].

Certain aromatic 2,2'-dithiobisbenzamides and benzisothiazolone compounds induce extraction of zinc from the peptide and possess anti-HIV activity [40, 41]. A common structural feature of these agents is the presence of electron-poor sulfur atom(s), making it likely that the first step in zinc ejection is nucleophilic attack of the 2,2'-dithiobisbenzamide or benzisothiazolone sulfur by a Zn-coordinated Cys sulfur atom. The solution behavior of these compounds and the mechanistic events leading to removal of zinc from HIV nucleocapsid protein *in vitro* was studied by electrospray ionization mass spectrometry [18].

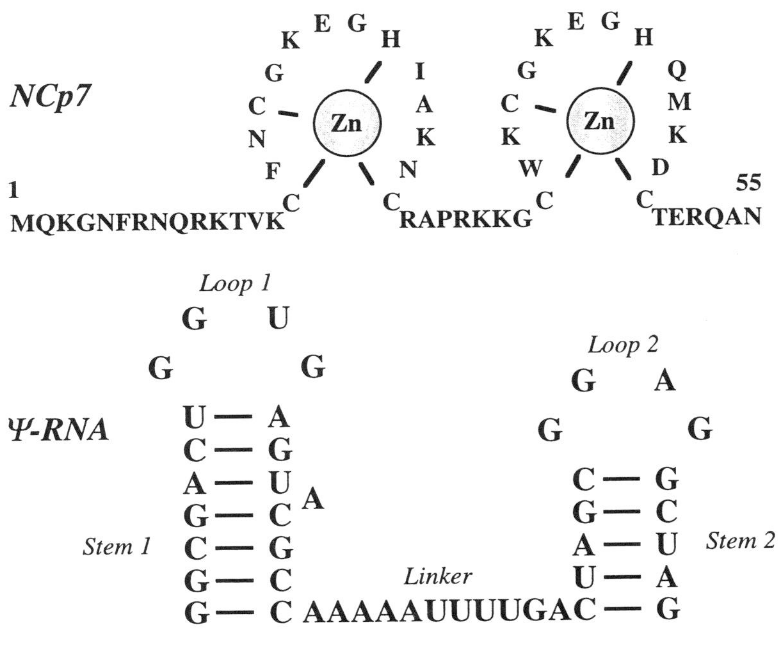

Figure 8. Sequence of HIV-1 nucleocapsid protein (NCp7) and the 44 nucleotide region from Ψ-RNA.

ESI-MS experiments were performed to monitor reaction products from the interaction of NCp7-Zn_2 with these disulfide benzamide compounds that show anti-HIV activity. Covalent complexation of these drugs results in zinc ejection. For example, two minutes after addition of 4 molar equivalents of disulfide A_2, the major products were the (NCp7+2Zn) and (NCp7+Zn+A) molecules, where A represents half of the disulfide drug molecule. At later time periods, the apo-protein and its various covalently-modified forms are the major products. The sites of these covalent modifications during the early reaction stages were identified by analyzing fragments resulting from trypsin digestion. From ESI-MS, LC/MS, and MS/MS results, the C-terminal zinc finger is more reactive to the disulfides than the N-terminal finger. The C-terminal finger Cys residues were found to be covalently modified with the drug. High resolution ESI mass measurements indicate that up to 3 intramolecular Cys-Cys bonds are formed upon complete zinc ejection. Figure 9 shows a simplified scheme for zinc ejection induced by disulfide compounds.

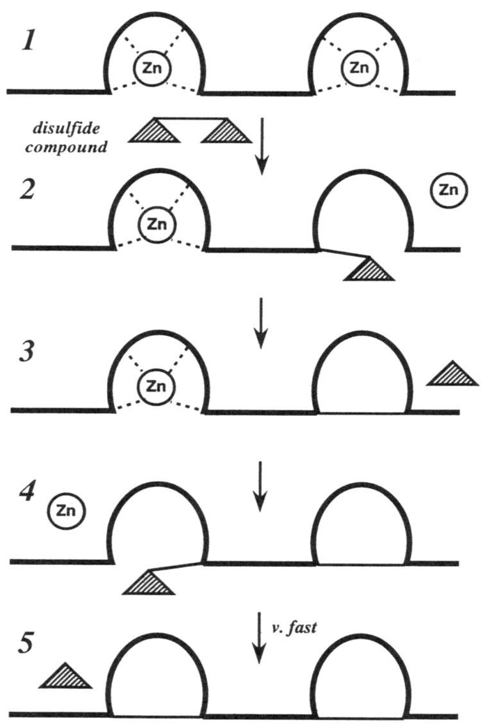

Figure 9. Schematic of the mechanism of zinc ejection from the zinc-fingers of nucleocapsid protein induced by disulfide and isothiazolone compounds. Addition of the drug compound to NCp7-Zn_2 (**1**) results in zinc ejection from the C-terminal zinc finger (**2**). Intramolecular Cys-Cys bond formation occurs (**3**) after zinc ejection. The process is repeated by drug action on the N-terminal zinc finger (**4** and **5**).

Complexation of RNA with NCp7 was also studied. NCp7 is thought to be responsible for packaging viral RNA by recognizing the Ψ-packaging site on the viral RNA genome. A 44-mer RNA containing the HIV-1 Ψ-region is folded into two stem-loop structures (Figure 8). Noncovalent complexes between zinc-bound NCp7 and RNAs representing each individual stem-loop structure were observed by ESI-MS (Figure 10).

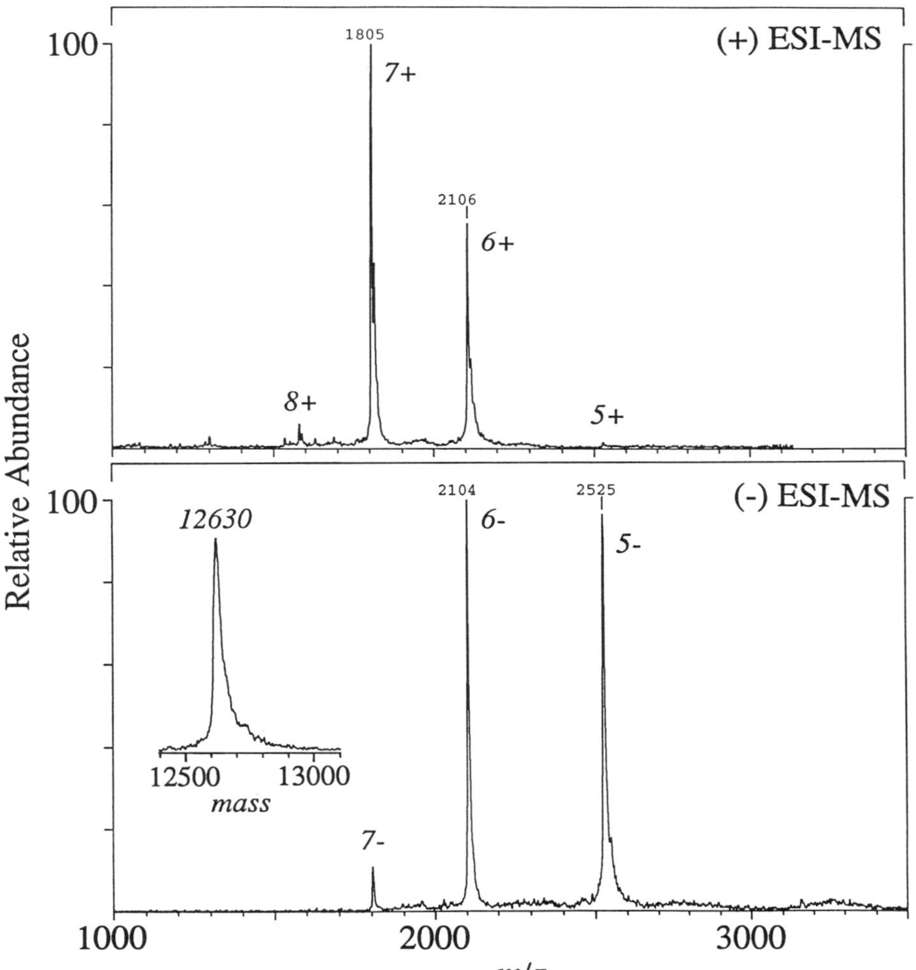

Figure 10. Positive and negative ion ESI mass spectra of the 1:1 complex formed between stem-loop 1 Ψ-RNA (M_r 6132, previously annealed) and NCp7-Zn_2 in 10 mM ammonium acetate, pH 6.9, and 10% (*v/v*) methanol.

Furthermore, ESI-MS competition experiments established the preference for a specific interaction of NCp7-Zn_2 with stem-loop 1 RNA (believed to be involved with specific interactions with the zinc-finger protein) compared to stem-loop 2, consistent with previously reported solution data [42]. Peptides comprising each zinc finger of NCp7 also were examined to assess their relative affinities to stem-loop 1 RNA. Again, under competitive binding conditions, the N-terminal zinc finger peptide forms a noncovalent complex with RNA. Addition of disulfide drugs to the NCp7-Zn_2/stem-loop 1 RNA complex yielded some dissociation of the complex to the apo-protein form in addition to covalent modification of the protein with the drug.

For both Tat-TAR and NCp7-RNA examples, protein-RNA complexes were detected in positive- and negative ion modes. In addition, unbound RNA was detected in *both* ionization modes. Multiply charged positive ions for DNA's and RNAs as large as transfer RNA (25 kDa) have been observed at relative high *m/z* (~ 2000) [17]. For oligonucleotides between 5-10 kDa, the maximum charge observed was between 4-6 positive charges. The maximum charge observed for the 25 kDa tRNA sample was 10+. For the studies to date, the sequence or nucleotide base composition does not appear to affect the amount of charging observed in the positive ion mode. Best results have been obtained with solution pH's between 6-10 and buffered with ammonium acetate or ammonium hydroxide. ESI-MS of acidic pH aqueous solutions (with acetic acid) or solutions of near neutral pH without ammonium acetate produced much poorer results.

6. Conclusions

For a more accurate molecular weight determination of protein complexes, efficient desolvation is required. Solvated ions can yield higher mass values than expected. However, there is a fine balance in experimental conditions for producing desolvated complex ions versus the dissociated complex. Careful control of all experimental factors affecting droplet/ion desolvation is necessary for successful detection of weakly-bound complexes by ESI-MS.

However, not enough is known about all these factors. As a case in point, during the early days of ESI-MS of noncovalent complexes (1990-1991), two scientists were in the mass spectrometry laboratory, trying to optimize experimental conditions to detect the complex between ribonuclease S-protein (11555 Da) and ribonuclease S-peptide (2166 Da) [8, 19]. Each scientist separately constructed their own ESI sprayer using the same design and parts. However, during an ESI "spray-off" that afternoon, using the same quadrupole mass spectrometer (and instrument tuning conditions) and the same ribonuclease S solution, the abundance of the ions representing the intact complex relative to the free protein and peptide and the amount of solvation were different for each scientist. This experience is still consistent with the current situation in our laboratory. Each researcher has his/her own "touch" with the instrument when establishing conditions to observe a noncovalent complex. Everybody optimizes these conditions their own way.

Regardless, many examples in the literature have demonstrated that ESI-MS can be used to measure *specific* noncovalent interactions. Henion's group [43] recently demonstrated that data from ESI-MS experiments can be used to construct Scatchard plots for measuring the binding constants of vancomycin antibiotics with tripeptide ligands. Their gas phase measurements were in reasonable agreement with previously reported solution phase values. Griffey and coworkers [44] determined solution phase dissociation constants for oligonucleotide-serum albumin complexes by using ESI-MS titration data. In a similar fashion, solution phase dissociation constants for the interactions between Src SH2 (Src homology 2) domain protein (12.9 kDa), critical in the signal transduction pathways of the tyrosine kinase growth factor receptors, and various phosphotyrosine-containing peptides were measured by ESI-MS [45, 46]. As the methodology matures, more research laboratories will be using electrospray ionization MS to study important weakly-bound complexes.

Acknowledgements

We gratefully acknowledge John Domagala, Donald Hupe, David Moreland, Peter Tummino, and Michael Reily for numerous insightful discussions, Rocco Gogliotti, Joseph Sanchez, and Michael Stier for supplying the NCp7 inhibitors, and the assistance of Scott Buckel and Dana DeJohn with the proteolysis experiments.

References

1. Fenn, J.B., Mann, M., Meng, C.K., Wong, S.F., and Whitehouse, C.M. (1989) Electrospray ionization for mass spectrometry of large biomolecules, *Science* **246**, 64-71.
2. Karas, M., Bahr, U., Ingendoh, A., and Hillenkamp, F. (1989) Laser desorption/ionization mass spectrometry of proteins of mass 100 000 to 250 000 Dalton, *Angew. Chem. Int. Ed. Engl.* **28**, 760-761.
3. Smith, R.D., Loo, J.A., Barinaga, C.J., Edmonds, C.G., and Udseth, H.R. (1990) Collisional activation and collision induced dissociation of multiply charged ions of large peptides and proteins produced by electrospray ionization, *J. Am. Soc. Mass Spectrom.* **1**, 53-65.
4. Loo, J.A., Edmonds, C.G., and Smith, R.D. (1991) Tandem mass spectrometry of very large molecules: Serum albumin sequence information from multiply charged ions formed by electrospray ionization, *Anal. Chem.* **63**, 2488-2499.
5. Loo, J.A., Edmonds, C.G., and Smith, R.D. (1993) Tandem mass spectrometry of very large molecules. 2. Dissociation of multiply charged proline-containing proteins from electrospray ionization, *Anal. Chem.* **65**, 425-438.
6. Smith, R.D., Light-Wahl, K.J., Winger, B.E., and Loo, J.A. (1992) Preservation of noncovalent associations in electrospray ionization-mass spectrometry: Multiply charged polypeptide and protein dimers, *Org. Mass Spectrom.* **27**, 811-821.

7. Smith, R.D. and Light-Wahl, K.J. (1993) Perspectives - The observation of non-covalent interactions in solution by electrospray ionization mass spectrometry - Promise, pitfalls and prognosis, *Biol. Mass Spectrom.* **22**, 493-501.
8. Ogorzalek Loo, R.R., Goodlett, D.R., Smith, R.D., and Loo, J.A. (1993) Observation of a noncovalent ribonuclease S-protein S-peptide complex by electrospray ionization mass spectrometry, *J. Am. Chem. Soc.* **115**, 4391-4392.
9. Katta, V. and Chait, B.T. (1991) Observation of the heme-globin complex in native myoglobin by electrospray-ionization mass spectrometry, *J. Am. Chem. Soc.* **113**, 8534-8535.
10. Drummond, J.T., Ogorzalek Loo, R.R., and Matthews, R.G. (1993) Electrospray mass spectrometric analysis of the domains of a large enzyme: Observation of the occupied cobalamin-binding domain and redefinition of the carboxyl terminus of methionine synthase, *Biochemistry* **32**, 9282-9289.
11. Ganem, B., Li, Y.-T., and Henion, J.D. (1991) Observation of noncovalent enzyme-substrate and enzyme-product complexes by ion-spray mass spectrometry, *J. Am. Chem. Soc.* **113**, 7818-7819.
12. Ganem, B., Li, Y.-T., and Henion, J.D. (1991) Detection of noncovalent receptor-ligand complexes by mass spectrometry, *J. Am. Chem. Soc.* **113**, 6294-6296.
13. Loo, J.A., Ogorzalek Loo, R.R., and Andrews, P.C. (1993) Primary to quaternary protein structure determination with electrospray ionization and magnetic sector mass spectrometry, *Org. Mass Spectrom.* **28**, 1640-1649.
14. Loo, J.A. and Pesch, R. (1994) Sensitive and selective determination of proteins with electrospray ionization magnetic sector mass spectrometry and array detection, *Anal. Chem.* **66**, 3659-3663.
15. Stults, J.T. and Marsters, J.C. (1991) Improved electrospray ionization of synthetic oligodeoxynucleotides, *Rapid Commun. Mass Spectrom.* **5**, 359-363.
16. Limbach, P.A., Crain, P.F., and McCloskey, J.A. (1995) Molecular mass measurement of intact ribonucleic acids via electrospray ionization quadrupole mass spectrometry, *J. Am. Soc. Mass Spectrom.* **6**, 27-39.
17. Sannes-Lowery, K.A., Mack, D.P., Hu, P., Mei, H.-Y., and Loo, J.A. (in press) Positive ion electrospray ionization mass spectrometry of oligonucleotides, *J. Am. Soc. Mass Spectrom.*.
18. Loo, J.A., Holler, T.P., Sanchez, J., Gogliotti, R., Maloney, L., and Reily, M.D. (1996) Biophysical characterization of zinc ejection from HIV nucleocapsid protein by anti-HIV 2,2'dithiobis[benzamides] and benzisothiazolones, *J. Med. Chem.* **39**, 4313-4320.
19. Goodlett, D.R., Ogorzalek Loo, R.R., Loo, J.A., Wahl, J.H., Udseth, H.R., and Smith, R.D. (1994) A study of the thermal denaturation of ribonuclease S by electrospray ionization mass spectrometry, *J. Am. Soc. Mass Spectrom.* **5**, 614-622.
20. Loo, J.A. (1995) Observation of large subunit protein complexes by electrospray ionization mass spectrometry, *J. Mass Spectrom.* **30**, 180-183.
21. Light-Wahl, K.J., Schwartz, B.L., and Smith, R.D. (1994) Observation of the noncovalent quaternary associations of proteins by electrospray ionization mass spectrometry, *J. Am. Chem. Soc.* **116**, 5271-5278.
22. Schwartz, B.L., Light-Wahl, K.J., and Smith, R.D. (1994) Observation of noncovalent complexes to the avidin tetramer by electrospray ionization mass spectrometry, *J. Am. Soc. Mass Spectrom.* **5**, 201-204.
23. Schwartz, B.L., Bruce, J.E., Anderson, G.A., Hofstadler, S.A., Rockwood, A.L., Smith, R.D., Chilkoti, A., and Stayton, P.S. (1995) Dissociation of tetrameric ions of noncovalent streptavidin complexes formed by electrospray ionization, *J. Am. Soc. Mass Spectrom.* **6**, 459-465.
24. Tang, X.-J., Brewer, C.F., Saha, S., Chernushevich, I., Ens, W., and Standing, K.G. (1994) Investigation of protein-protein noncovalent interactions in soybean agglutinin by electrospray ionization time-of-flight mass spectrometry, *Rapid Commun. Mass Spectrom.* **8**, 750-754.
25. Eckart, K. and Spiess, J. (1995) Electrospray ionization mass spectrometry of biotin binding to streptavidin, *J. Am. Soc. Mass Spectrom.* **6**, 912-919.
26. Draper, D.E. (1995) Protein-RNA recognition, *Annu. Rev. Biochem.* **64**, 593-620.
27. Aboul-ela, F., Karn, J., and Varani, G. (1995) The structure of the human immunodeficiency virus type-1 TAR RNA reveals principles of RNA recognition by Tat protein, *J. Mol. Biol.* **253**, 313-332.
28. Frankel, A.D. (1992) Activation of HIV transcription by Tat, *Curr. Opin. Gen. Develop.* **2**, 293-298.

29. Dingwall, C., Ernberg, I., Gait, M.J., Green, S.M., Heaphy, S., Karn, J., Lowe, A.D., Singh, M., and Skinner, M.A. (1990) HIV-1 tat protein stimulates transcription by binding to a U-rich bulge in the stem of the TAR RNA strcture, *EMBO J.* **9**, 4145-4153.
30. Weeks, K.M., Ampe, C., Schultz, S.C., Steitz, T.A., and Crothers, D.M. (1990) Fragments of the HIV-1 Tat protein specifically bind TAR RNA, *Science* **249**, 1281-1285.
31. Frankel, A.D. (1992) Peptide models of the Tat-TAR protein-RNA interaction, *Protein Sci.* **1**, 1539-1542.
32. Long, K.S. and Crothers, D.M. (1995) Interaction of human immunodeficiency virus type 1 Tat-derived peptides with TAR RNA, *Biochemistry* **34**, 8885-8895.
33. Roy, S., Delling, U., Chen, C.-H., Rosen, C.A., and Sonenberg, N. (1990) A bulge structure in HIV-1 TAR RNA is required for Tat binding and Tat-mediated *trans*-activation, *Genes Develop.* **4**, 1365-1373.
34. Calnan, B.J., Biancalana, S., Hudson, D., and Frankel, A.D. (1991) Analysis of arginine-rich peptides from the HIV Tat protein reveals unusual features of RNA-protein recognition, *Genes Develop.* **5**, 201-210.
35. Wallis, M.G., von Ahsen, U., Schroeder, R., and Famulok, M. (1995) A novel RNA motif for neomycin recognition, *Chem. Biol.* **2**, 543-552.
36. Wang, Y. and Rando, R.R. (1995) Specific binding of aminoglycoside antibiotics to RNA, *Chem. Biol.* **2**, 281-290.
37. Mei, H.-Y., Galan, A.A., Halim, N.S., Mack, D.P., Moreland, D.W., Sanders, K.B., Truong, H.N., and Czarnik, A.W. (1995) Inhibition of an HIV-1 Tat-derived peptide binding to TAR RNA by aminoglycoside antibiotics, *Bioorg. Med. Chem. Lett.* **5**, 2755-2760.
38. Gorelick, R.J., Chabot, D.J., Rein, A., Henderson, L.E., and Arthur, L.O. (1993) The two zinc fingers in the human immunodeficiency virus type 1 nucleocapsid protein are not functionally equivalent, *J. Virol.* **67**, 4027-4036.
39. Darlix, J.-L., Lapadat-Tapolsky, M., de Rocquigny, H., and Roques, B.P. (1995) First glimpses at structure-function relationships of the nucleocapsid protein of retroviruses, *J. Mol. Biol.* **254**, 523-537.
40. Rice, W.G., Supko, J.G., Malspeis, L., Buckheit, R.W., Clanton, D., Bu, M., Graham, L., Schaeffer, C.A., Turpin, J.A., Domagala, J., Gogliotti, R., Bader, J.P., Halliday, S.M., Coren, L., Sowder, R.C., Arthur, L.O., and Henderson, L.E. (1995) Inhibitors of HIV nucleocapsid protein zinc fingers as candidates for the treatment of AIDS, *Science* **270**, 1194-1197.
41. Tummino, P.J., Scholten, J.D., Harvey, P.J., Holler, T.P., Maloney, L., Gogliotti, R., Domagala, J., and Hupe, D. (1996) The in vitro ejection of zinc from human immunodeficiency virus (HIV) type 1 nucleocapsid protein by disulfide benzamides with cellular anti-HIV activity, *Proc. Natl. Acad. Sci. USA* **93**, 969-973.
42. Sakaguchi, K., Zambrano, N., Baldwin, E.T., Shapiro, B.A., Erickson, J.W., Omichinski, J.G., Clore, G.M., Gronenborn, A.M., and Appella, E. (1993) Identification of a binding site for the human immunodeficiency virus type 1 nucleocapsid protein, *Proc. Natl. Acad. Sci. USA* **90**, 5219-5223.
43. Lim, H.-K., Hsieh, Y.L., Ganem, B., and Henion, J. (1995) Recognition of cell-wall peptide ligands by vancomycin group antibiotics: Studies using ion spray mass spectrometry, *J. Mass Spectrom.* **30**, 708-714.
44. Greig, M.J., Gaus, H., Cummins, L.L., Sasmor, H., and Griffey, R.H. (1995) Measurement of macromolecular binding using electrospray mass spectrometry. Determination of dissociation constants for oligonucleotide-serum albumin complexes., *J. Am. Chem. Soc.* **117**, 10765-10766.
45. Loo, J.A., Hu, P., and Thanabal, V. (1995) Studying noncovalent protein-peptide interactions by ESI-MS. *Proceedings of the 43rd ASMS Conference on Mass Spectrometry and Allied Topics.* Atlanta, GA, p. 35, American Society for Mass Spectrometry: Santa Fe, NM.
46. Loo, J.A., Hu, P., McConnell, P., Mueller, W.T., Sawyer, T.K., and Thanabal, V. (in press) A study of Src SH2 domain protein-phosphopeptide binding interactions by electrospray ionization mass spectrometry, *J. Am. Soc. Mass Spectrom.*.

MEASUREMENT OF NONCOVALENT COMPLEXES WITH HIGH M/Z BY ELECTROSPRAY TIME-OF-FLIGHT MASS SPECTROMETRY

I.V. CHERNUSHEVICH, W. ENS and K.G. STANDING
University of Manitoba, Physics Department
Winnipeg, Manitoba R3T 2N2 Canada

1. Introduction

In the last few years the electrospray (ES) ionization technique has proved to be gentle enough to obtain information on the higher-order structure of biomolecules [1]. However, most ES-MS studies with proteins have been carried out in acidic solutions. Such acidic solutions, particularly in combination with organic solvents, denature biopolymers, which then acquire sufficient charge to be detected within the limited *m/z* range of most commercial quadrupole mass spectrometers. These conditions are usually sufficiently severe so as to destroy any higher order structure, which must be examined closer to physiological conditions. There the charge acquired may be much lower, resulting in *m/z* values beyond the range of normal quadrupole instruments (see for example [2]).

An ES-TOF mass spectrometer built at University of Manitoba has in principle an infinite *m/z* range, moderate resolution (up to 10,000) and a sensitivity in the femtomole range [3,4]. These features make it an instrument of choice for studying noncovalent associations. About thirty noncovalent complexes have been studied up to now in our instrument in collaboration with other groups, and the results on some of them have been or will be published [2,5-7]. Most of our collaborations have arisen from the inability of other instruments and techniques to solve specific biochemical problems. Below we give a brief overview of some of these studies. More information on the results obtained with this instrument can also be found in several other papers in this volume: by Amster *et al.*, by Ayed *et al.*, by Fitzgerald *et al.*, and by Love *et al.*

2. Experimental

The ES-TOF spectrometer has been described previously [3,4]. In brief, an electrospray ion source was coupled to a reflecting time-of-flight mass spectrometer using the orthogonal-injection method introduced by Dodonov *et al.* The ES-TOF instrument

W. Ens et al. (eds.), New Methods for the Study of Biomolecular Complexes, 101–116.

(Fig. 1) may be operated either in the linear mode or in the reflecting mode. Spectra are recorded with a time-to- digital converter. All spectra presented here were recorded in the reflecting mode. Most of the studies were done with a conventional ES source. However, the nanospray technique developed by Mann and Wilm [9, 10] was used in several experiments, as will be described later. Ions produced by ES are transported to the storage region via 3 stages of differential pumping. Collision-induced dissociation spectra were obtained in a standard manner by increasing the voltage between the capillary and the first skimmer. Recent incorporation of an RF-quadrupole ion guide has led to further improvements in sensitivity and resolution and minimized mass and *m/z* discrimination [11].

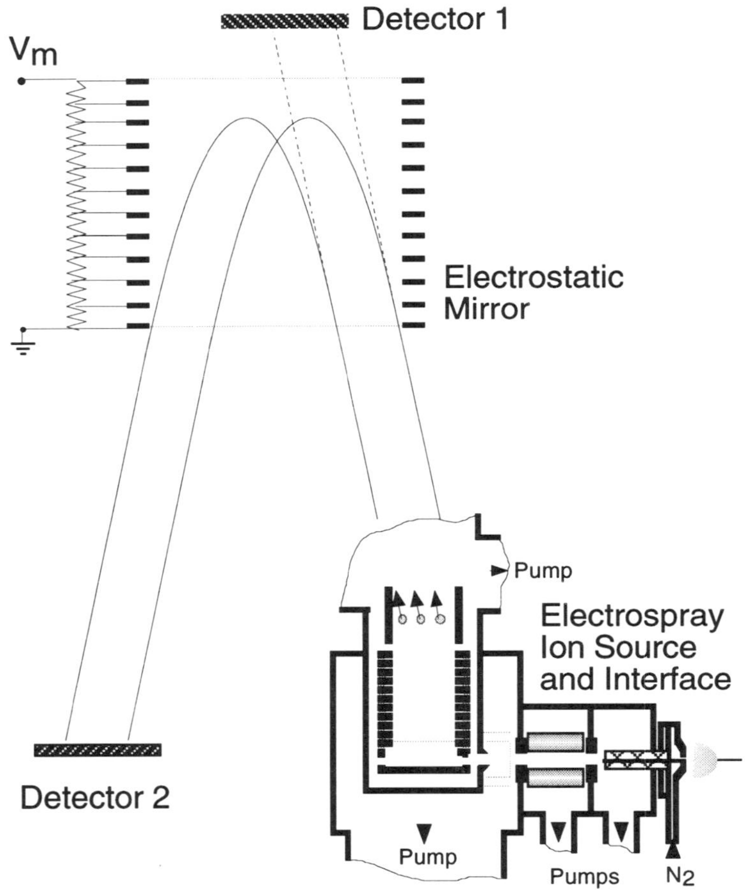

Figure 1 Schematic diagram of the electrospray time-of-flight mass spectrometer at Manitoba

3. Results and Discussion

3.1. OXALOCROTONATE TAUTOMERASE: ON THE ORIGIN OF IONS

Several studies were carried out with an enzyme 4-Oxalocrotonate Tautomerase ("4OT") and its analogs, prepared by total chemical synthesis at the Scripps Research Institute [5,6; also see this volume]. We have shown (Fig. 2a) that the ES spectrum of 4OT contains mostly monomer and hexamer ions when recorded from near neutral solution (pH ~7.5). The hexamer disappeared from the solution when its pH was reduced to ~4. Several mutants of 4OT have been studied in our instrument, and it was shown that very subtle differences in covalent structure can lead to large changes in higher order structure. We have also found that the oligomeric state of mutants is correlated with their enzymatic activity [6]. Here we confine our discussion to those studies which helped to understand the origin of the monomer and hexamer ions.

It was not clear initially whether the monomers were present in the sample solution or they resulted from decomposition of the hexamer during the ES process. Monomers were observed even at lowest declustering voltages. In an effort to reduce possible fragmentation of the hexamer during the electrospray process we also analyzed 4OT by nanospray. The native ES-TOF mass spectrum of 4OT acquired by nanospray is shown in Fig. 2b. Under these conditions the hexamer was the predominant species observed, indicating that nanospray *was* in fact a more gentle ionization process than electrospray under conventional conditions.

Nanospray is also able to answer another important question: are hexamers formed at the initial solution concentration, or they are produced after the concentration is increased during evaporation of droplets in the electrospray process? In other words, does the observed monomer-oligomer equilibrium reflect the situation in solution, or is it mainly an artifact of the electrospray process? The original idea of Wilm and Mann [5,6] was to use nanospray with a low analyte concentration corresponding to less than one analyte molecule per droplet. For a typical diameter of an initial nanospray droplet ~180 nm we calculate a concentration ~$5x10^{-7}$ M for this condition. The spectrum in Fig. 2b recorded from a solution with monomer concentration ~$6x10^{-7}$ M exhibits mainly hexamer ions. Thus, the hexamer ions were not formed during droplet evaporation; the hexamers were already present in solution.

3.2. EARLY PREGNANCY FACTOR PROTEIN: PROBLEM OF HETEROGENEITY

Two samples of early pregnancy factor (EPF) protein were synthesized at the University of Queensland, Australia. The biochemical side of this study is fully described in another paper by Love *et al.* in these proceedings. However, these samples are also interesting from an instrumental point of view, as they demonstrate some limitations of the method.

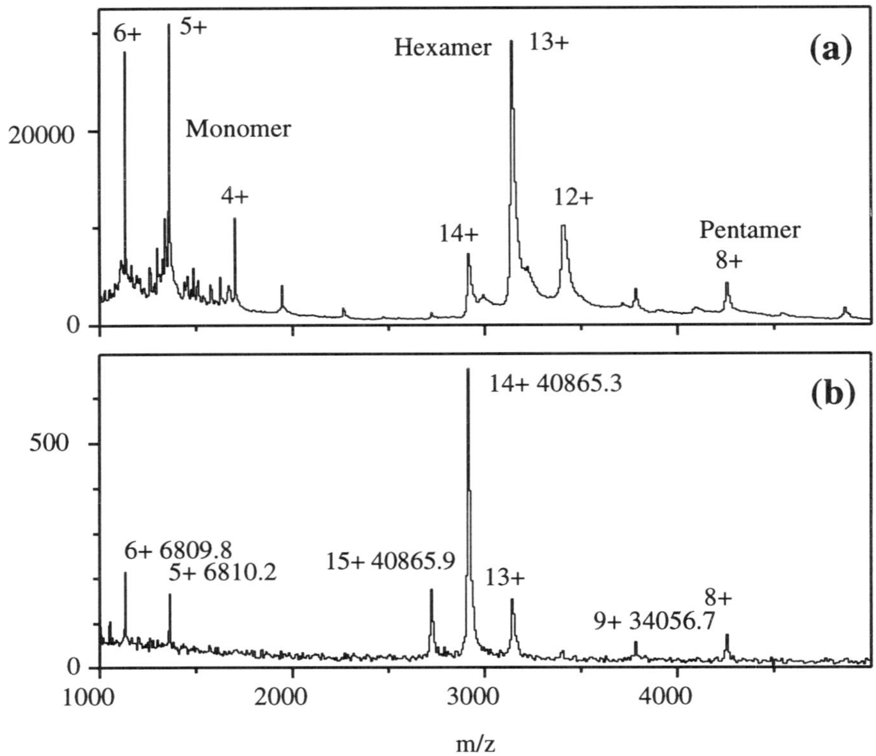

Figure 2. ES mass spectra of 4-oxalocrotonate tautomerase recorded from 5 mM ammonium bicarbonate solution (pH ~7.5). (a) - with normal ES; (b) - with nanospray.

The two samples of EPF protein differed in the way they were synthesized: one was produced by sequential fragment ligation ("ligated"), and another was made by stepwise solid phase synthesis ("nonligated"). The spectra of the denatured protein samples together with their deconvolutions are shown in Fig. 3. It is interesting to compare the degree of heterogeneity of the two samples: several peaks of truncated protein are present in the spectrum of the nonligated protein (Fig. 3d), while the ligated one is rela-tively homogeneous (Fig. 3b). In general, if a protein forms an oligomer under near physiological conditions, heterogeneity is amplified significantly in the oligomer. The number N of different n-mers formed from m different monomers can be calculated from the formula:

$$N = \frac{(m+n-1)!}{(m-1)!n!} \tag{1}$$

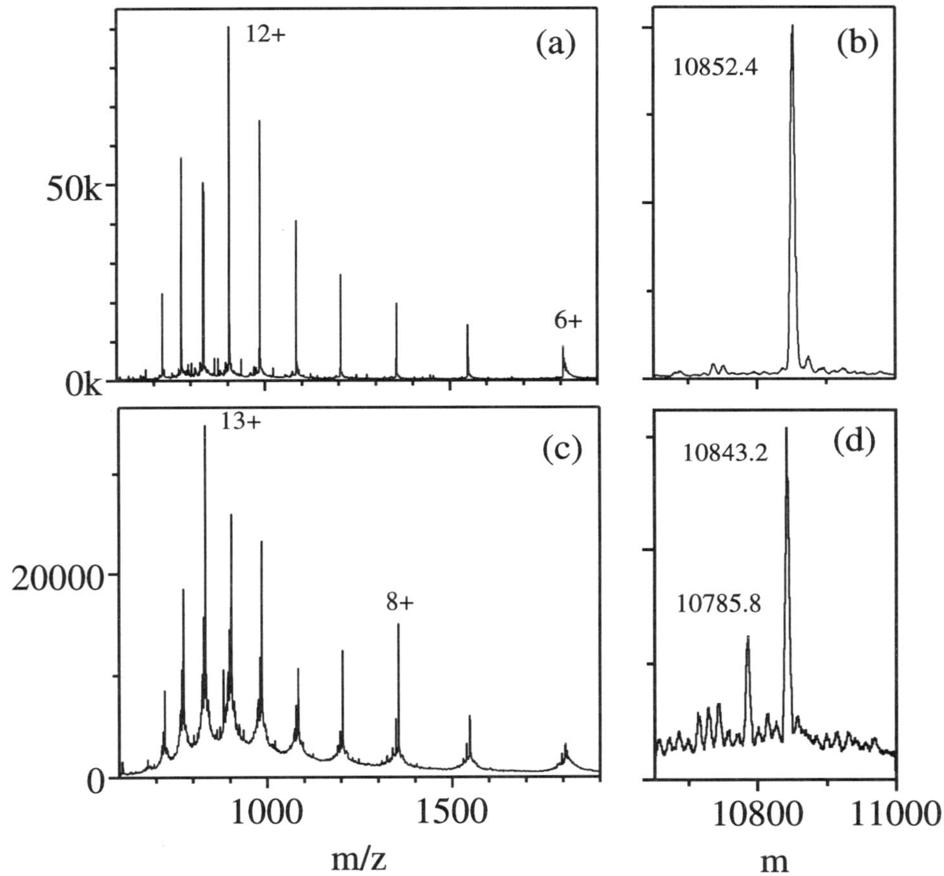

Figure 3. ES mass spectra (a, c) and deconvoluted spectra (b, d) of early pregnancy factor protein recorded ffom denaturing conditions (pH ~2.5). (a, b) - ligated protein; (c, d) - nonligated protein.

The oligomeric state of the protein has been studied by gel electrophoresis; it was shown to form a heptamer in solution. The above formula shows that 8 types of heptamer may in principle be constructed out of only 2 types of monomer. Therefore, the results shown in Fig. 4b are not unexpected since the spectrum quality of the nonligated protein is poor. In contrast, the narrow peaks in the spectrum of the ligated form (Fig. 4a) allowed an accurate determination of mass, and hence unambiguous determination of the oligomeric state of the protein.

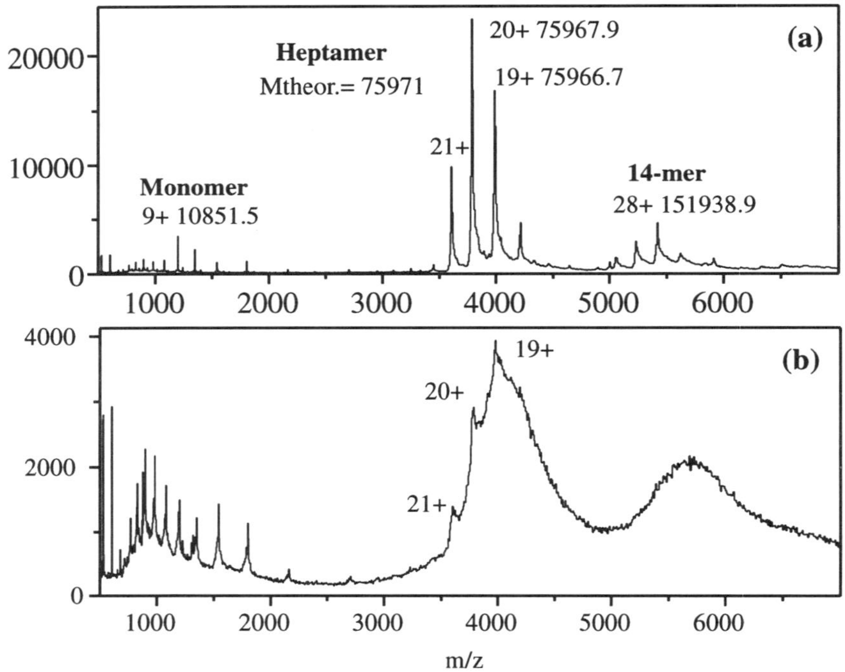

Figure 4. ES mass spectra of ligated (a) and nonligated (b) EPF protein recorded from 5 mM ammonium acetate solution (pH ~ 6.0).

3.3. STREPTAVIDIN-BIOTIN COMPLEX

Tetramers of avidin and streptavidin and their complexes with biotin are classical examples of noncovalent associations. Moreover, streptavidin has become a classical example in mass spectrometry as well: it has been studied with different degrees of success in an extended range quadrupole [12], in a Fourier transform ICR instrument [13], and in a magnetic-sector instrument with an array detector [14]. Measurements in a TOF mass spectrometer provide a useful comparison.

Streptavidin from streptomyces avidinii (Sigma) was used in our first and unsuccessful experiments. Only a broad hump centered around *m/z* ~ 3700 was observed in spectra of this compound recorded at low (non-fragmenting) declustering voltage ΔU = 100 V (not shown), indicating that the sample was either contaminated or heterogeneous. Dissociation of tetramers into monomers at higher voltage (ΔU = 200 V) allowed closer examination of the monomer composition. It turned out that the sample was very heterogeneous: it contains more than 15 different versions of truncated protein with comparable abundances. The largest of them were those corresponding to (14-136), (14-137), (14-138) and (13-138) amino acid residues of the full-length protein sequence. Taking into account the above-mentioned consequences of heterogeneity, it became clear that the sample was not suitable for mass spectrometric studies.

Core streptavidin from streptomyces avidinii (Boehringer Mannheim) was used for further investigations. This protein was known to be homogeneous [14], and corresponds to (14-136) amino acid residues of natural streptavidin. The protein used in [13] was slightly different and corresponded to residues (13-139).

Two spectra of core streptavidin recorded at low and high declustering voltages at pH ~ 8.0 are shown in Fig. 5. Only tetramer ions with mass 51,880 Da are observed at ΔU = 100 V; monomers and trimers start to appear at higher voltages. In general, the quality of the spectra is similar to that obtained in the FT ICR instrument [13]: resolution and mass accuracy are about the same, but the sensitivity seems to be higher in TOF.

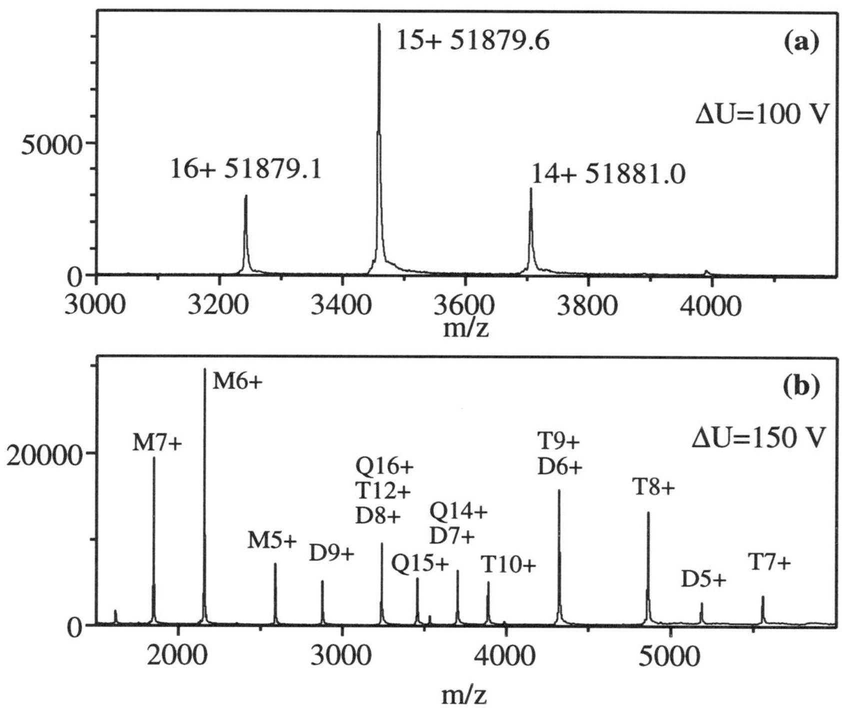

Figure 5. ES mass spectra of core streptavidin (Boehringer Mannheim) recorded at two different declustering voltages ΔU. Only tetramers are observed at ΔU=100 V. Labels in (b) are M, D, T and Q for monomer, dimer, trimer and tetramer correspondingly.

Experiments on stepwise addition of biotin (Sigma) to streptavidin were done in the same manner as in the studies performed in the magnet mass spectrometer [14]. Fig. 6 shows segments of spectra recorded with different concentrations of biotin corresponding to biotin/streptavidin ratios 0 (a), 2:1 (b) and 6:1 (c). The results are similar to those shown in Fig. 3(a), (c) and (e) of ref. [13]. However, the superior resolution of the TOF instrument allowed almost complete separation of ions containing different amounts of biotin. The spectra shown in Fig. 6 can be also found on the internet at: http://chait-sgi.rockefeller.edu/cgi-bin/massbank .

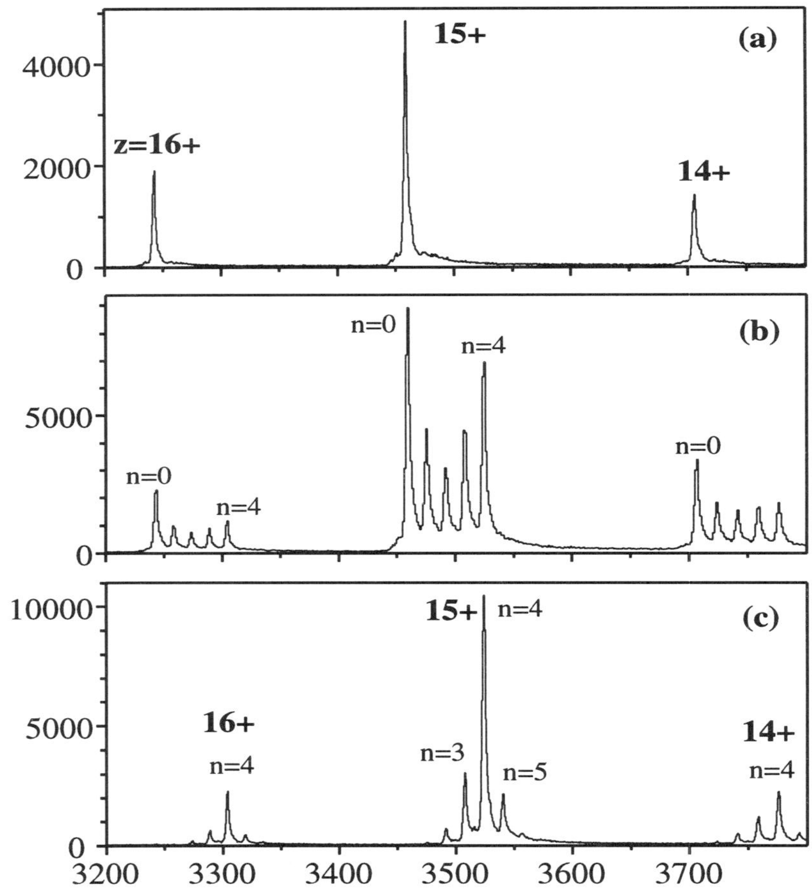

Figure 6. ES mass spectra of streptavidin (10 μM) at pH 8 with different concentrations of biotin: (a) - no biotin; (b) - 20 μM; (c) - 60 μM.

3.4. METALLOPROTEINS

Several non-heme iron-containing proteins have been studied in our instrument in cooperation with Jon Amster [7]. These proteins are of interest as models to examine the process of metal atom insertion into proteins, and to study the effect of metal atoms on protein folding. The metalloproteins examined are hemerythrin and rubrerythrin.

Hemerythrin proteins are composed of eight subunits, each of which contains an oxo-bridged diiron site that is coordinated by seven amino acid residues. Two spectra recorded for homo-HR at different declustering voltages are shown in the Fig. 7. Mostly octamer is observed at 100 V; it is fragmented into monomers, heptamers and hexamers at ΔU=150 V. The measured mass is in good agreement with that predicted from calculations. However, the most interesting results in this case were obtained when the monomer region of the spectra was examined.

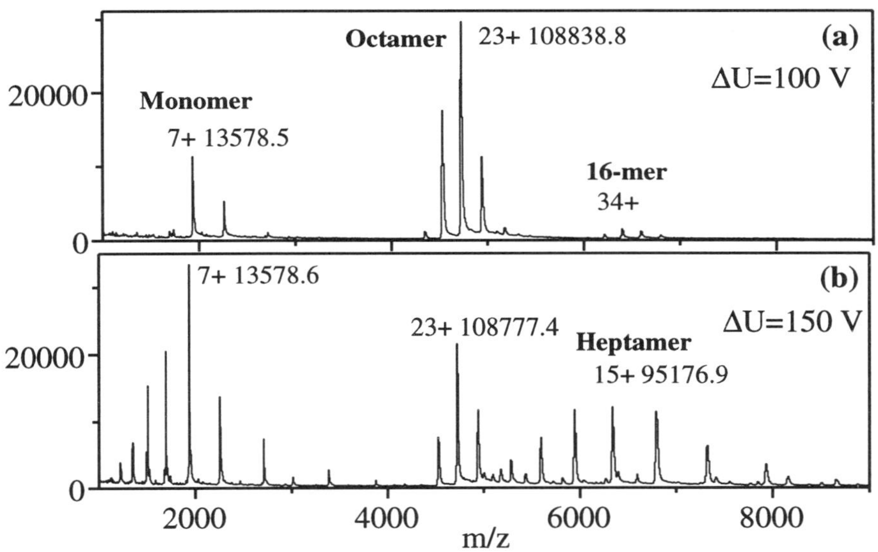

Figure 7. ES mass spectra of *Phascolopsis gouldii* hemerythrin recorded at different declustering voltage ΔU: (a) - 100 V; (b) - 150 V.

Three deconvoluted spectra of monomers are shown in Fig. 8. The first two were recorded from pH 6 at ΔU = 150 and 250 V, while the third one was recorded from denaturing conditions. The difference in mass between the main two peaks present in these spectra is ~106 Da, which implies that two irons in the upper spectrum are replaced by 6 protons in the lower spectrum. Interestingly, the low abundance of the peak in the middle suggests that the two irons are lost simultaneously.

Also examined in this study were hetero-hemerythrin composed of two types of subunits, and rubrerythrin. The latter contains 3 irons per monomer, and gel perfusion measurements suggest that it exists as a dimer in solution. However, we have determined its oligomeric state to be a tetramer (not shown). A possible explanation for this fact is that our mass spectrometric studies required the protein to be dissolved in a solution of much lower ionic strength than that used in gel perfusion studies.

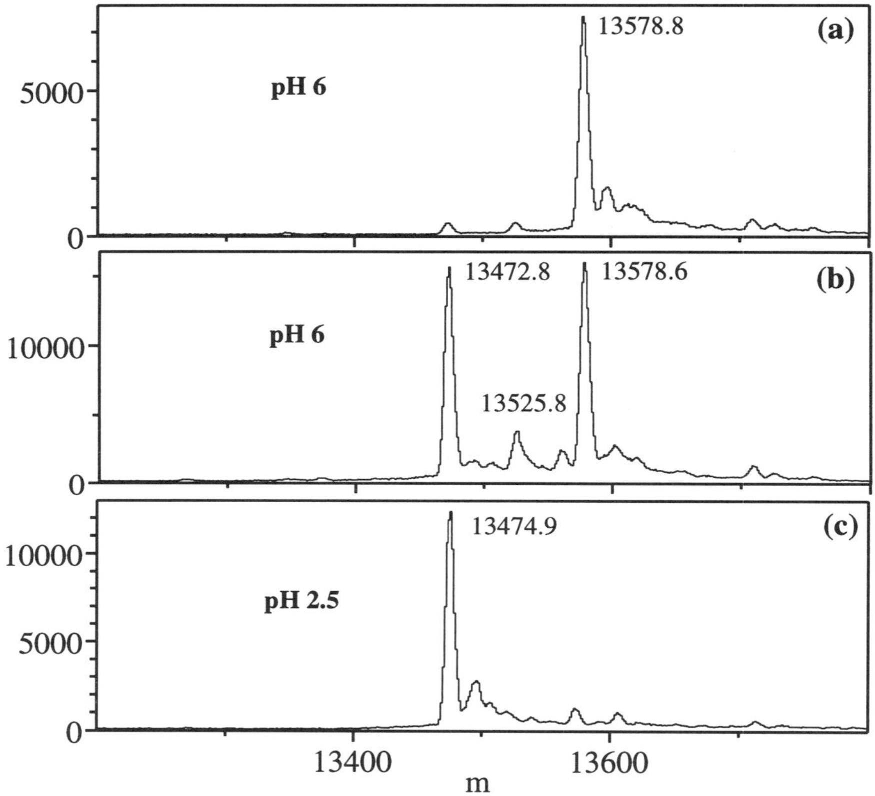

Figure 8. Deconvolution for monomers from the mass spectra of hemerythrin recorded at different declustering voltage and pH: (a) - ΔU = 150V, (b) - 250 V (both pH 6); (c) - 100 V, pH 2.5.

3.5. CATALASES HP I and HP II

Samples of catalases HP I and II were kindly provided by Peter Loewen from the Microbiology Department of the University of Manitoba. Our main interest in catalases was to determine differences in oligomeric state of their different types, and to study the high-mass limitations of the ES-TOF mass spectrometer.

Mass spectra of two types of catalase HP I recorded from 5 mM ammonium acetate aqueous solution are shown in Fig. 9a, b. HP I from *mycobacterium tuberculosis* was recorded as a dimer with mass ~161,573 Da, while HP I from *E. coli* was a tetramer with mass ~322,265 Da under the same conditions. As already seen in the studies of 4OT, the oligomeric state of a protein may affect its enzymatic activity [6]. These effects have still to be studied in case of catalases. Some heterogeneity of the main components in these spectra is explained by the different number of hemes attached to the oligomer.

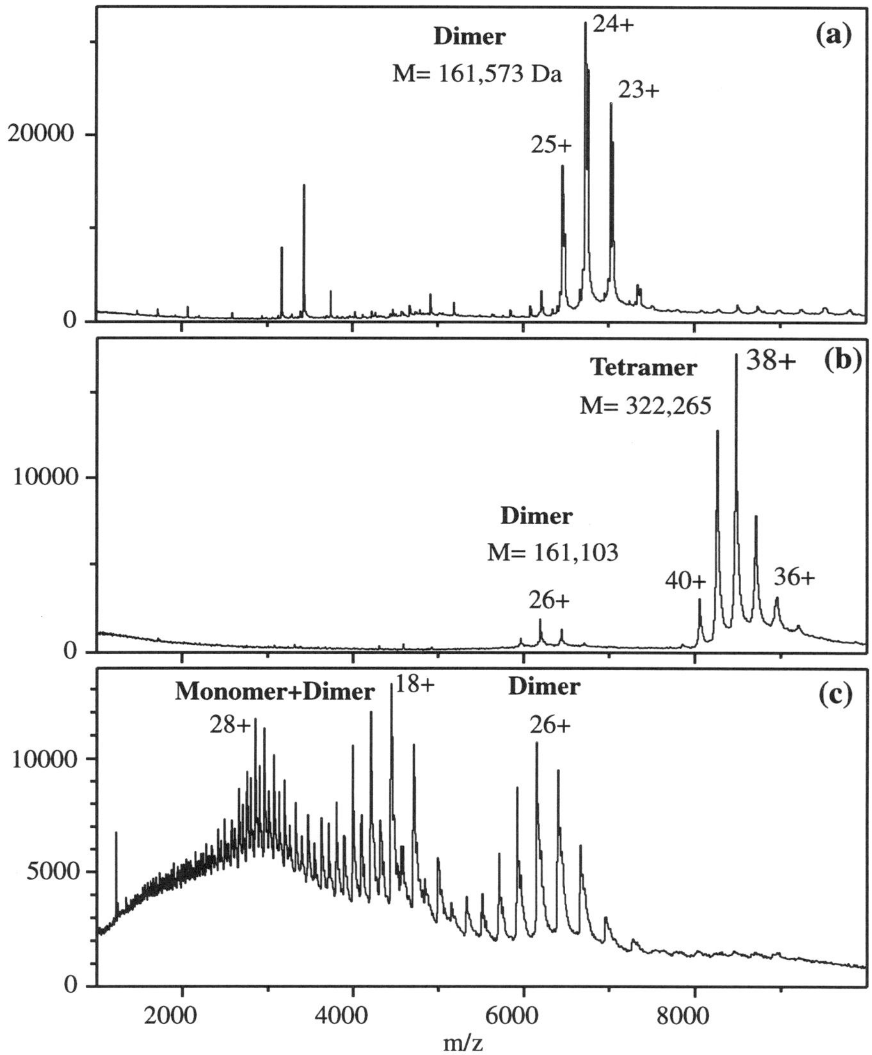

Figure 9. ES mass spectra of catalase HP I from (a) *mycobacterium tuberculosis* and (b, c) *E. coli*; (a) and (b) were recorded from pH 6, (c) - from pH 2.5.

We have become accustomed to the fact that it is easier to obtain a good spectrum of a denatured protein than that of the protein in its native state. However, there are several examples when this is not the case, and catalases are among them. Fig. 9 gives a comparison of spectra of HP I (*E. coli*) recorded at pH 6 (b) and pH 2.5 (c). Both dimer and monomer are present after addition of acid, and the peaks become wider due to formation of adducts. At even smaller pH (below 1.5) the dimer disappeared, but so did any resolved peaks of the monomers. We have no ready explanation for this effect, but we suspect that after denaturing, the hydrophobic parts of the proteins, originally inside, become exposed to solution and try to pick up any garbage to secure their hydrophobicity.

The heaviest complex we have successfully studied up to now was catalase HP II from *E. coli*, a tetrameric protein with mass 339,100 Da. The narrow peaks of HP II (Fig. 10) allowed determination of the mass with the accuracy of 0.03%. To our surprise, the spectrum of HP II did not change when the pressure in the vacuum chamber was increased: both resolution and sensitivity remained the same [4, 15]. This observation implies that the cross-section for collision-induced dissociation is not determined only by the dimensions of the ion but also by other factors, and it gives us some hope to record spectra of even higher masses. It is also worthwhile to emphasize that the charge state distribution of catalase HP II is centred around *m/z* 10,000, a range not normally available in other existing mass analyzers.

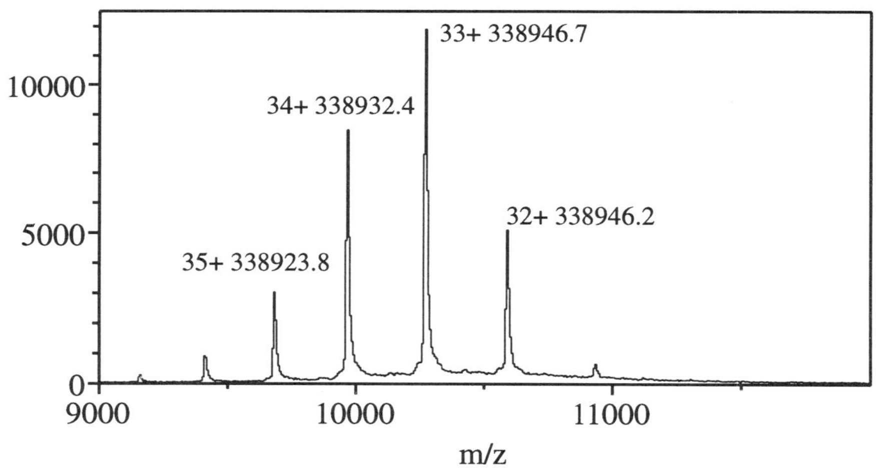

Figure 10. Mass spectrum of catalase HP II from *E. coli* recorded with the normal ES source; ΔU = 250 V.

A segment of a spectrum recorded with nanospray is presented in Fig. 11. The few additional peaks which appear to the left of the main peaks were those corresponding to the loss of one or two hemes at the highest declustering voltage. Note that the centre of the charge state distribution shifted from ~10,000 to ~9,000 in nanospray, presumably due to a smaller time available for the proton exchange that normally reduces the average charge state of complexes.

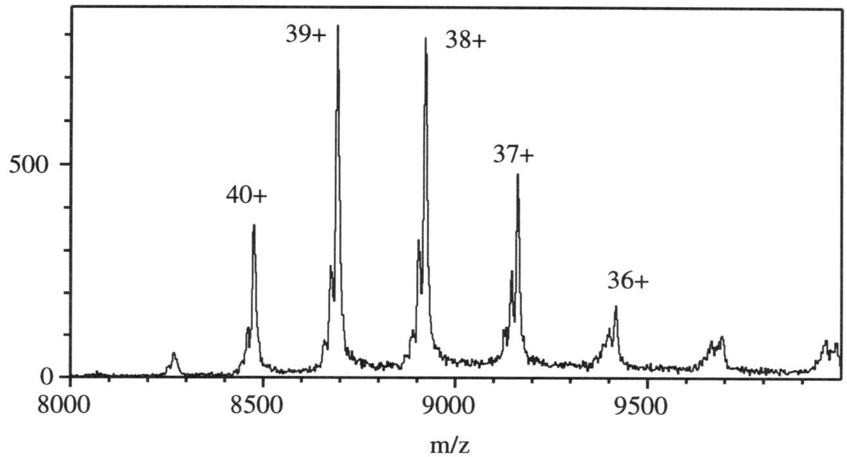

Figure 11. Mass spectrum of catalase HP II from *E. coli* recorded with nanospray; ΔU=350 V.

At low values of the declustering potential, clusters of the tetramer appear in the spectrum [4], presumably due to less specific association. The mass of some of these clusters is more than 1 Megadalton and the *m/z* reaches 17,000.

As already mentioned, about 30 noncovalent complexes have been studied in our instrument up to now, and their average *m/z* ratio was found to increase with mass. A plot of the most abundant charge state Z versus mass M for all protein complexes studied at pH from 6 to 7 in our instrument is shown in Fig. 12. A linear approximation on a log-log plot with slope equal to ~0.55 fits the data well. A similar relation but with the slope 0.52 was obtained for clusters of complexes of soybean agglutinin and catalase HP II having masses up to 1.37 MDa. Extrapolation of the Z-vs-M plot to larger masses allows an estimate of the charge carried by larger entities. When we recorded ions from a sample of brome mosaic virus (BMV, ~4.65 MDa), only a wide hump centered around *m/z* ~27,000 was observed as shown in Fig. 13. However, this value correlates well with the charge predicted for the known mass of the virus, as shown in Fig. 12, supporting the idea that the ions recorded indeed correspond to the intact virus.

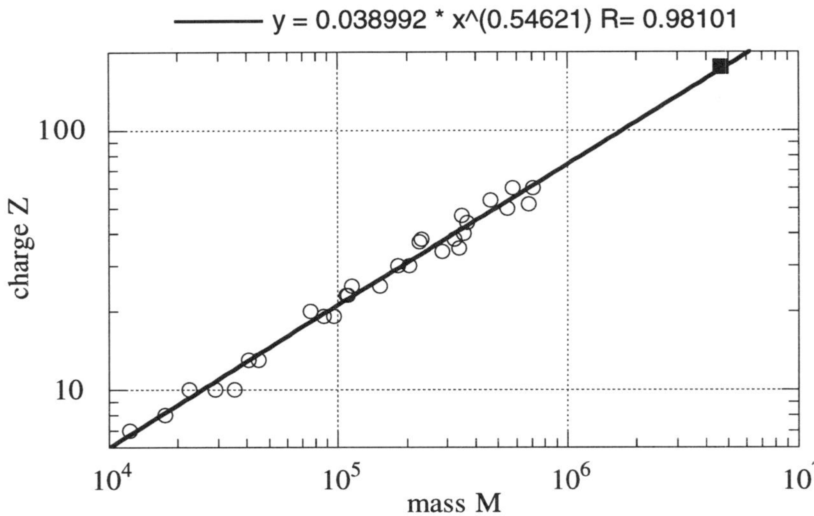

Figure 12. Charge of proteins and noncovalent complexes at pH ~6 corresponding to the most abundant peaks in the mass spectra, versus their mass; the full square represent brome mosaic virus.

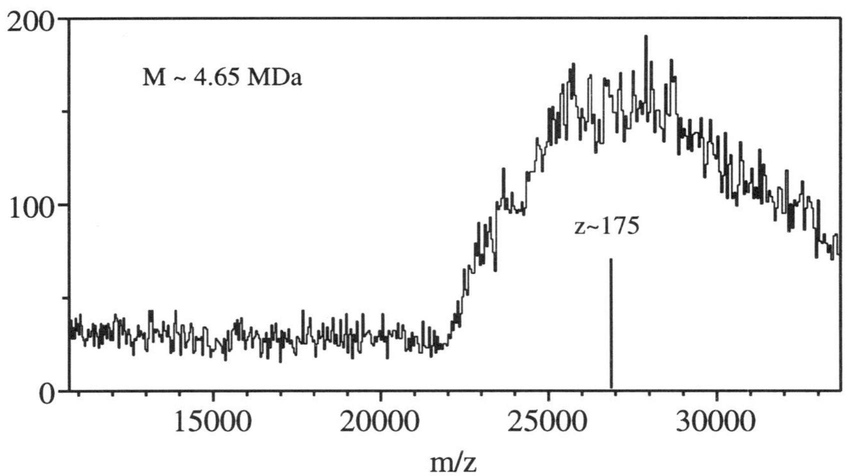

Figure 13. ES mass spectrum recorded from a solution of intact brome mosaic virus in 10 mM ammonium acetate.

4. Conclusions

The high *m/z* range and sensitivity of the TOF instrument together with the mild ionization conditions of ES provide an ideal technique for investigating noncovalent associations of biomolecules.

Most of the difficulties in these studies arise not from the instrumental performance, but from the preparation of samples: purification, heterogeneity and choice of solution conditions. The presence of impurities (salts, acids, etc.) was found to be especially critical when the solutions under investigation were at pH ~7: a large number of unresolved or partially resolved adducts is then observed in the spectra. Unfortunately, these adducts cannot be removed by increasing the declustering potential as the complex itself may be too unstable to survive declustering. On the other hand, some water and/or buffer molecules may become an integral part of a complex, and the complex cannot exist without them. Therefore, the accuracy of mass determination for many noncovalent associations is worse than that for proteins alone, and the experimentally determined mass may be larger by several hundred daltons than the calculated value.

5. Acknowledgments

This work was supported by grants from the US National Institutes of Health (GM-30605) and from the Natural Sciences and Engineering Research Council of Canada. We are grateful to all our collaborators who provided interesting and high quality samples for our studies: Paul Alewood, Jon Amster, Darrel Englebretsen, Michael Fitzgerald, Steve Haber, Steve Kent, Don Kurtz, Steve Love and Peter Loewen. We thank Andrew Krutchinsky for assistance in creation of the rf-quadrupole ion guide with which most of the data presented here were obtained. We are indebted to Matthias Mann and Matthias Wilm for sharing their nanospray experience and capillaries with us. We are grateful to Victor Spicer for his technical assistance. I.C also thanks his wife Elena for her patience.

6. References

1. Light-Wahl, K.J.; Schwartz, B.L.; Smith, R. D. (1994) Observation of the Noncovalent Quaternary Associations of Proteins by Electrospray Ionization Mass Spectrometry, *J. Am. Chem. Soc.* **116**, 5271-78, and references therein.
2. Tang, X.-J.; Brewer, C.F.; Saha, S.; Chernushevich, I.V.; Standing, K.G.; Ens, W. (1994) Investigation of Protein-Protein Noncovalent Interactions in Soybean Agglutinin by Electrospray Ionization Time-of-Flight Mass Spectrometry, *Rapid Commun. Mass Spectrom.* **8**, 750-754.
3. Verentchikov, A.N.; Ens, W.; Standing, K. G. (1994) Reflecting Time-of-Flight Mass Spectrometer with an Electrospray Source and Orthogonal Injection, *Anal. Chem.* **66**, 126-133.
4. Chernushevich, I.V.; Ens, W.; Standing, K.G. (1997) Electrospray Ionization Time-of-Flight Mass Spectrometry, in R. Cole (ed.), *Electrospray Ionization Mass Spectrometry: Fundamentals, Instrumentation & Applications,* John Wiley & Sons, pp 203 - 234.

5. Fitzgerald, M.C.; Chernushevich, I.V.; Standing, K.G.; Kent, S.B.H.; Whitman, C.P. (1995) Total Chemical Synthesis and Catalytic Properties of the Enzyme Enantiomers L- and D-4-Oxalocrotonate Tautomerase, *J. Am. Chem. Soc.* **117**, 11075-80.
6. Fitzgerald, M.C.; Chernushevich, I.V.; Standing, K.G.; Kent, S.B.H.; Whitman, C.P. (1996) Probing the Oligomeric Structure of an Enzyme by Electrospray Ionization Time-of-Flight Mass Spectrometry, *Proc. Natl. Acad. Sci USA*, **93**, 6851-6856.
7. Lei, Q.P.; Cui, X.; Kurtz, D.M., Jr.; Amster, I.J.; Chernushevich, I.V. and Standing, K.G., Electrospray Mass Spectrometry Studies of Non-Heme Iron-Containing Proteins, submitted to *J. Am. Chem. Soc*.
8. Dodonov, A. F.; Chernushevich, I. V.; Laiko, V. V.; in R. J. Cotter (Ed.) Time-of-Flight Mass Spectrometry, Am. Chem. Soc. Washington DC 1994, pp. 108 - 123.
9. Wilm, M.S.; Mann, M. (1994) Electrospray and Taylor-Cone theory, Dole's beam of macromolecules at last?, *Int. J. Mass Spectrom. Ion Processes* **136**, 167-180.
10. Wilm, M.S.; Mann, M. (1996) Analytical Properties of the Nanospray Ion Source, *Anal. Chem*., **68**, 1-8.
11. Krutchinsky, A.N.; Chernushevich, I.V.; Spicer, V.; Ens, W.; Standing, K. G. (1995) Collisional Focusing of Ions in the Electrospray TOF Mass Spectrometer, *Proceedings of the 43rd ASMS Conference on Mass Spectrometry and Allied Topics*, p.126.
12. Schwartz, B.L.; Gale, D.C.; Smith, R.D.; Chilkoti, A.; Stayton, P.S. (1995), Investigation of Non-covalent Ligand Binding to the Intact Streptavidin Tetramer by Electrospray Ionization Mass Spectrometry, *J. Mass Spectrom.*, **30**, 1095-1102.
13. Schwartz, B.L.; Bruce, J.E.; Anderson, G.A.; Hofstadler, S.A.; Rockwood, A.L.; Smith, R.D.; Chilkoti, A.; Stayton, P.S. (1995), Dissociation of Tetrameric Ions of Noncovalent Streptavidin Complexes Formed by Electrospray Ionization, *J. Am. Soc. Mass Spectrom*., **6**, 459-465.
14. Eckart, K.; Spiess, J. (1995), Electrospray Ionization Mass Spectrometry of Biotin Binding to Streptavidin, *J. Am. Soc. Mass Spectrom*., **6**, 912-919.
15. Chernushevich, I.V.; Verentchikov, A.N.; Ens, W.; Standing, K. G. (1996), Effect of Ion-Molecule Collisions in the Vacuum Chamber of an Electrospray Time-of-Flight Mass Spectrometer on Mass Spectra of Proteins, *J. Am. Soc. Mass Spectrom.* 7, 342-349.

NON-COVALENT COMPLEXES OF OLIGONUCLEOTIDES OBSERVED USING ELECTROSPRAY IONIZATION MASS SPECTROMETRY

Richard H. Griffey, Michael J. Greig, Henri Sasmor,
Lendall L. Cummins, Sheri L. Manalili, and Hans J. Gaus
Isis Pharmaceuticals
2280 Faraday Ave, Carlsbad, CA 92008

1. Introduction

The development of electrospray (ESI) and matrix-assisted laser desorption (MALDI) ionization techniques for the transformation of proteins and nucleic acids into gas phase ions has reinvigorated mass spectrometry in the biological sciences. Gas-phase ions generated using electrospray ionization are believed to retain their solution structures during the desolvation process.(1) As a result, it is now possible to detect and characterize the structure and properties of molecules too large for study via solution techniques such as NMR spectroscopy. ESI and MALDI ionization provide high sensitivity and transmission efficiency for detection of fmole quantities using FT-ICR, quadrupole ion trap, or conventional quadrupole mass spectrometers.

A natural evolution of the methodology has lead to experiments designed to detect non-covalent complexes between peptides and proteins, proteins and their substrates, and nucleic acids. Typically, ions of these complexes can be generated from buffered solutions of macromolecules using a variety of ion source geometries including pneumatically assisted electrospray, micro and nanospray, heated capillary sources, and glass capillaries.(2-22) The observation of complexes *always* requires some optimization of instrument parameters, including sprayer position, countercurrent gas flow, capillary temperature, and skimmer/lens potentials. However, once optimized, the instrument conditions are stable and allow reproducible detection of non-covalent complexes. Solution conditions are equally important, and require prior optimization of macromolecule and ligand concentrations, buffer type and concentration, and content of organic solvent used to stabilize the electrospray. As with solution experiments, proper control experiments must be designed to insure the accuracy of ESI-MS measurements.

Mass spectrometeric analysis of macromolecular complexes provides several unique advantages over solution methods. Each of the constituents in a mixture can be identified (given unique masses for the solution constituents) and quantitated via

W. Ens et al. (eds.), New Methods for the Study of Biomolecular Complexes, 117–133.

integration over the charge envelope observed with electrospray ionization. Hence, binding events that are hard to measure using solution methods such as allosteric effects or ordered binding should be observable using ESI-MS.(22-24) In addition, the order of a reaction can be followed as sequential complex formation among macromolecules and various ligands can be observed. ESI-MS is suited for observation of weak interactions between complexes which have high nM-μM dissociation constants (K_D). Tighter interactions (low nM) may be observed with more sensitive FT-ICR detection.

The weaker types of interactions are common and important for many classes of small and large biomolecules, including nucleic acids and proteins. Formation of weak complexes have particular relevance to the pharmaceutical industry, where the pharmacokinetics and toxicology of drug substances may be governed by weak interaction with abundant serum or cellular proteins. Examples include the binding of aspirin, barbiturates, and benzodiazepines to serum albumin.(25)

Phosphorothioate (PS) oligonucleotides represent a new class of therapeutic agents targeted against mRNA targets. Antisense oligonucleotides are structurally derived from natural nucleic acids, and are expected to bind to a variety of cellular and serum proteins. Of particular importance are the interaction with serum proteins. Thrombin and Factor IX have high affinities for PS oligomers, and these interactions produce dose-limiting toxicities.(26) Binding of oligonucleotides to serum albumin may govern their pharmacokinetic properties. The affinity of the complex between human or bovine serum albumin and a 20mer phosphorothioate oligonucleotide has proven difficult to measure. Gel shift assays and equilibrium dialysis experiments have yielded ambiguous results, suggesting the binding stoichiometry was not a simple bimolecular interaction.

In general, radiolabeling, optical, and biosensor (SPR) approaches are not well-suited for study of weak complexes. Experiments must be performed with one member present at a concentration ~10-fold below the anticipated K_D. The study of weak complexes can be difficult, since non-specific binding to solid surfaces, or exogenous "anchors" such as streptavidin can alter the true K_D values. It is difficult to determine binding stoichiometries for complexes with solution techniques when two K_D values are similar. Recent developments in electrospray mass spectrometery have allowed these types of interactions to be monitored or read out in the gas phase. While this may seem like a curioius choice of methodology for characterizing solution macromolecular interactions, the gentle desolvation conditions of the electrospray process preserves non-covalent complexes. Hence, ESI-MS can provide a "snapshot" of solution conditions, including levels of multiple intermediates along an enzymatic pathway or complexes which can not be observed directly in solution. The accurate measurement of molecular mass from multiply charged ions provides unambiguous information on stoichiometries and identities of substrates. Improvements in ESI-MS sensitivity from microspray ion sources and ion trap mass spectrometers has made ESI-MS the most powerful tool available to biochemists.

Another important aspect of ESI-MS measurement of K_D is the model-free nature of the calculations. Other solution techniques suffer from model-dependent parameters which may be multivalued, for example, bimolecular off-rates or NMR estimates of dynamics. Estimates of K_D from solution techniques such as SPR or CE are based on the assumption that binding is bimolecular and diffusion-controlled, and fail when rebinding occurs. The direct observation of the intermediates in a reaction pathway or sequential binding events using ESI-MS allows determination of K_D with an exact model or calculation. The order of a reaction can be established from direct ESI-MS observation of the intermediates. This capability provides an accurate fit, and makes a larger number of biologically interesting kinetic problems amenable to study.

We describe the determination of dissociation constants between a family of modified oligonucleotides and BSA. These dissociation constants are related to the pharmacokinetic properties of the oligonucleotides observed *in vivo*. The results demonstrate that ESI-MS and ESI-MS/MS provide unique and powerful tools for characterization of macromolecular interactions in solution. Application of these methods to problems of biological significance will open a new era in structural and molecular biology and biochemistry.

2. Materials and Methods

2.1 GENERAL

Bovine serum albumin was purchased from ICN (Costa Mesa, CA) and dialysed against 0.1 mM DTT in 10 mM ammonium acetate prior to use. Amino acids and buffers were purchased from Sigma (St. Louis, MO).

2.2 OLIGONUCLEOTIDE AND PNA SYNTHESIS

PS and phosphodiester oligonucleotides were prepared using conventional phosphoramidite chemistry and Applied Biosystems (Foster City, CA) 380B synthesizers. The structures of the modified oligonucleotides are shown in Figure 1. Novel phosphoramidites were prepared in house. Oligomers were purified by polyacrylamide gel chromatography or trityl-on using HPLC with C18 media. PNA molecules were prepared using standard conditions.(27) BSA was purchased from ICN (Costa Mesa, CA) and purified before use. The BSA was dissolved in a 30 mM imidazole buffer containing 10 mM ammonium acetate, pH 7.5.

2.3 ESI-MS

The N_2 nebulizing gas was maintained at 70 psi, with a 12 L/min flow of N_2 and a 2 L/min flow of O_2 mixed as drying gas at a temperature of 180^oC. The BSA solution was infused at a rate of 33.5 ng/μL at 7 μL/min. Spectra were obtained as the averaged sum of 15 scans from *m/z* 1500 to 2600 at a scan rate of 32 sec/scan. The raw data was processed with a 0.5 Da gaussian filter prior to display. Lens and skimmer voltages were optimized for the BSA:oligonucleotide complex and did not differ from those used for BSA alone or the BSA:oligo^2 complex. The concentration of BSA in solution was held fixed at 0.2 or 0.5 μM. The oligonucleotide concentration was taken from the known concentration added to the solution.

2.4 CALCULATION OF DISSOCIATION CONSTANTS

Ion abundances have been integrated directly as a function of oligonucleotide concentration using Hewlett-Packard software or Caeser (Analytical Devices; Alameda, CA). Where signals from the charge states of protein and complex overlap, the intensity has been partitioned into the respective bins in relation to the relative intensities of the closest neighboring higher and lower charge states. The integrated areas for each species were summed for each concentration of added oligonucleotide.

2.5 PHARMACOKINETICS

The characterization of pharmacokinetic parameters for modified oligonucleotides has been described.(28)

Compound	Composition
3082	5'-TsGsCsAsTsCsCsCsCsCsAsGsGsCsCsAsCsCsAsT
9044	5'-UoGoCoAoCoCoCoCoCoAoGoGoCoCoAoCoCoAoT (underline = 2'-O-propylriboside)
9045	5'-UsGsCsAsUsCsCsCsCsCsAsGsGsCsCsAsCsCsAsT (underline = 2'-O-propylriboside)
9047	5'-T*sGsCsAsTsCsCsCsCsCsAsGsGsCsCsAsCsCsAsT (T*=5'-octadecylaminothymidine)
8005	5'-U*sGsCsAsTsCsCsCsCsCsAsGsGsCsCsAsCsCsAsT (U*=5'-(2'-O-hexylamino-carbonyl-oxo-cholesterol)uridine)

Figure 1. List of oligonucleotide sequences and their chemical modifications

3. Results

The pI of BSA is approximately 4.5, and the protein is negatively charged in solution at pH 7.4. As shown in Figure 2, negative ionization ESI-MS of BSA generates a series of ions from the $(M\text{-}26H)^{26-}$ through the $(M\text{-}40H)^{40-}$ charge states between *m/z* 1600 and 2700. Deconvolution of these signals yields a mass of 66,497 Da for the BSA.

Mixing BSA with increasing concentrations (0.3 to 33 μM) of a 20mer phosphorothioate oligonucleotide (GCCCAAGCTGGCATCCGTCA; MW 6367 Da) generates a new series of ions (Figures 3-6) corresponding to free oligomer (highlighted with an O; *m/z*=1590.9; $(M - 4\ H)^{4-}$ charge state of oligomer) and the BSA:oligo complex (highlighted with a C). Deconvolution of the new set of signals from the complex yields a mass of 72,870 Da, corresponding to a 1:1 ratio of BSA:oligo. A second complex (MW 79,273 Da; expected MW 79,231 Da) corresponding to a disodium adduct of BSA:oligo^2 (labeled C^2), appears at higher concentrations of oligonucleotide.

In the complex between BSA and oligonucleotide, the most abundant charge state is $(M\text{-}32H)^{32-}$ while the most abundant charge state for the BSA is $(M\text{-}31H)^{31-}$. Since the oligonucleotide has a net charge of (4-), the charge of the BSA is reduced by

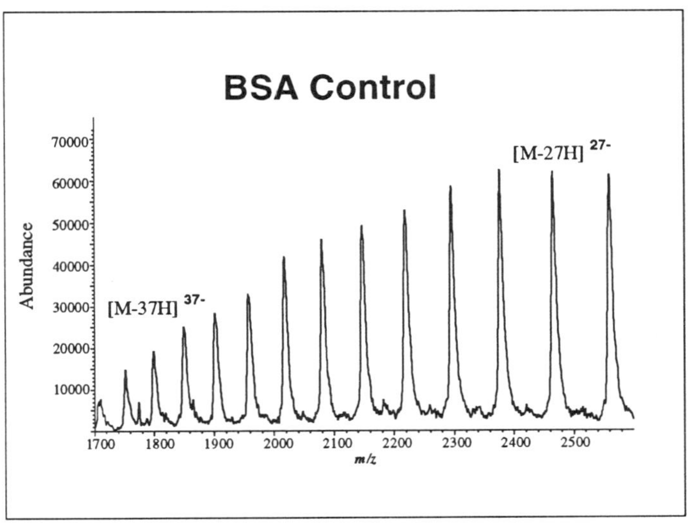

Figure 2. ESI-MS spectrum of BSA obtained using negative ionization.

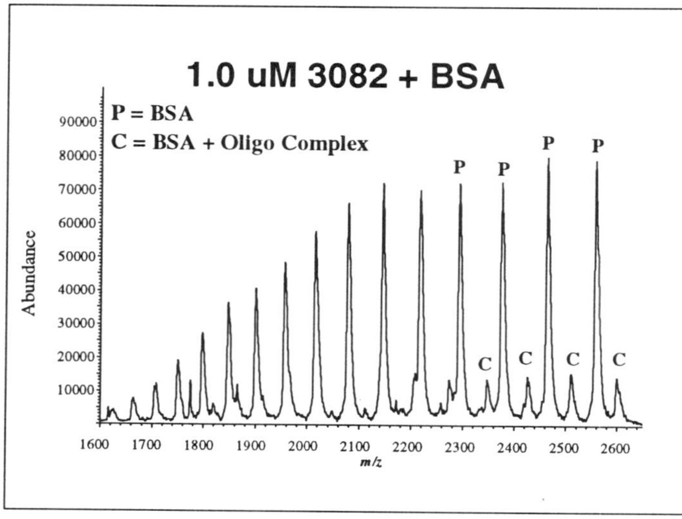

Figure 3. ESI-MS spectrum of BSA and the 3082:BSA complex in the presence of 1.0 μM 3082. The signals from BSA ions are marked P, while the signals from complex ions are marked C.

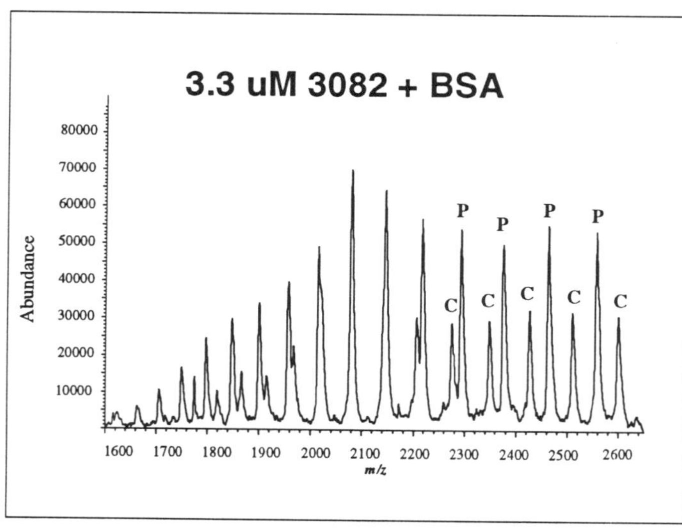

Figure 4. ESI-MS spectrum of BSA and the 3082:BSA complex in the presence of 3.3 μM 3082.

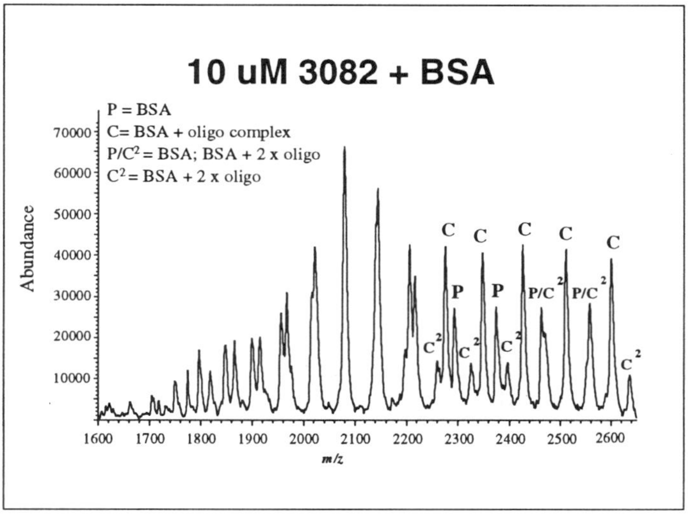

*Figure 5.*ESI-MS spectrum of BSA and the 3082:BSA complex in the presence of 10.0 μM 3082. Signals from the bis-oligo:BSA complex are marked C^2

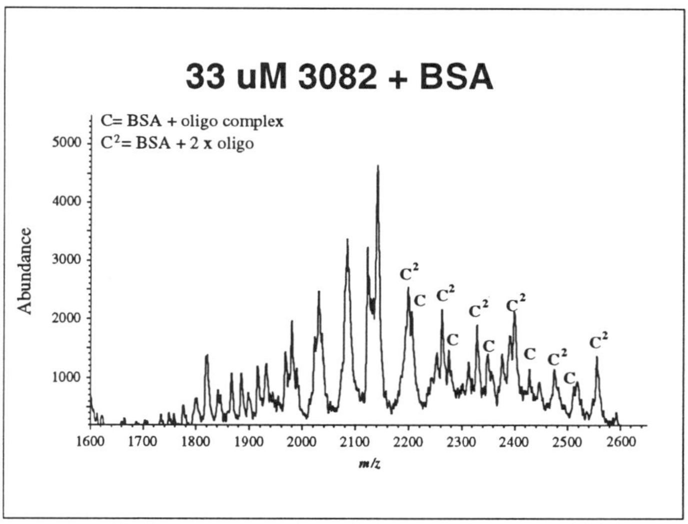

Figure 6. ESI-MS spectrum of BSA and the 3082:BSA complex in the presence of 33 μM 3082.

3 in the complex. In the bis-oligo complex, the most abundant charge state is $(M-34H)^{34-}$. This result suggests that the first oligo binding may increase the charge on amino acid side chains proximate to the oligo, or may reduce the pKa of carboxyl groups in the gas phase. A more detailed analysis of the changes in charge state as a function of buffer concentration might provide a window on the electrostatics of the interaction which is not available from solution studies.

3.1 CONTROL EXPERIMENTS

An analogous peptide nucleic acid 20mer (net charge +1) demonstrated no binding to BSA. This result confirms that the interaction of the oligonucleotides with BSA occurs predominantly through electrostatic interactions. However, the structure of BSA also is important, since no detectable binding was observed between heat-denatured BSA and the PS oligonucleotide.

Decomposition of complexes during transfer from the initial charged droplet to the gas phase could alter the measurement of K_D. However, two-fold variation in the voltages of the ion optics did not alter relative ion intensities for any species, suggesting that little induced dissociation of the complexes occurs under the conditions described above. In addition, the effective pH of the ESI droplet could change during the desolvation process. The binding of 3082 to BSA is pH-sensitive, since increasing the solution pH to 10.0 reduced the K_D to ~15 μM.

3.2 DETERMINATION OF DISSOCIATION CONSTANTS

For an interaction between a macromolecule and a substrate which binds at a single site, the kinetic parameters can be determined via a conventional Scatchard analysis. The fraction of bound complex divided by the concentration of substrate is plotted against 1/[S], and the dissociation constant can be determined from the slope of the line. Implicit in this analysis is the assumption that the concentration of macromolecule is well below the K_D.

If two molecules of substrate bind to the protein or macromolecule and their abundances can be determined directly, the kinetic parameters can be determined from a polynomial fit. In this case, the polynomial binding constant Q is defined as $Q = ([P]+[C]+[C^2])/[P]$, where [P] is the concentration of protein, [C] is the concentration of complex, and $[C^2]$ is the concentration of the bis complex. Q is plotted as a function of substrate concentration. K_D1 and K_D2 are determined from the fit to a 2nd order polynomial equation. This type of analysis is facilitated by ESI-MS, where the abundances of the individual species can be determined directly.

Additional information can be gleaned from the abundances of each species in the ESI-MS spectrum. For example, the ratio of K_D1 to K_D2 can be calculated from measurement of the abundance of the bis-complex to free protein in the ESI-MS data:

$$[C] = K_D1*[O]/[P]$$
$$[C^2] = K_D2*[O]/[C]$$
$$K_D1/K_D2 = [C^2]*[P]/ [C]^2$$

3.3 DETERMINATION OF K_D FOR OLIGONUCLEOTIDE:BSA COMPLEXES

Integrated ion abundances from the various charge states have been summed for BSA and each complex formed with 3082 as a function of the concentration of added oligomer. A plot of measured ([P]+[C]+[C^2]/[P]) versus added [O] has been fit to a second-order polynomial function (Figure 7). Dissociation constants of 3.1$\pm$0.3 (K_D1) and 11.9$\pm$0.6 μM (K_D2) have been calculated (R^2>.998) at pH 7.5. Independently, the ratio of K_D1/ K_D2 was determined to be 0.26$\pm$0.01 from direct measurement of [P][C^2]/[C]^2 for five values of [O].(22)

The value for K_D1 measured using ESI-MS is sensitive to the ionic strength of the solution buffer, varying from 0.7 mM in the absence of added buffer to $\geq$100 μM in the presence of 100 mM ammonium acetate:30 mM imidazole. For an oligo concentration of 10 μM, a plot of log(K_D1) versus log[NH_4OAc] yields a straight line with a slope of 1.3 and R^2=0.975 (Figure 8). The K_D1 is sensitive to pH, decreasing to 15 μM at pH 10.0. The dependence of K_D on the ionic strength and pH of the buffer suggests that electrostatic forces contribute significantly to the binding energy of the complex. The observed value of K_D also is a function of nucleotide sequence and backbone chemistry.

A second ESI-MS binding experiment was performed using a phosphodiester 20mer oligonucleotide. The K_D measured under standard conditions was 11.6 μM. None of the bis-oligo:BSA complex was observed. This study was extended to a larger series of chemically modified oligos. As shown in Table 1, K_D values were measured at 10 and 320 mOsm.

TABLE 1. K_D values for oligonucleotide:BSA complexes measured using ESI-MS as a function of ionic strength.

Compound	10 mOsm		320 mOsm	
	K_D1 (μM)	K_D2 (μM)	K_D1 (μM)	K_D2 (μM)
9044	3.5	15	250	
9045	2.9	13	3.5	20
9047	1.3	8.5	3.2	14
8005	1.5	5.5	2.4	9
3082	2.4	10.8	140	

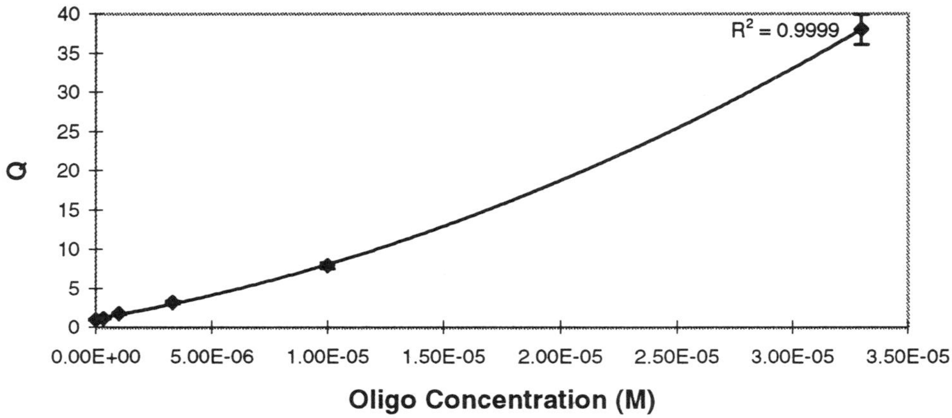

Figure 7. Fit of a second-order polynomial to the plot of Q versus 3082 concentration. The concentration of BSA was held at 0.5 μM.

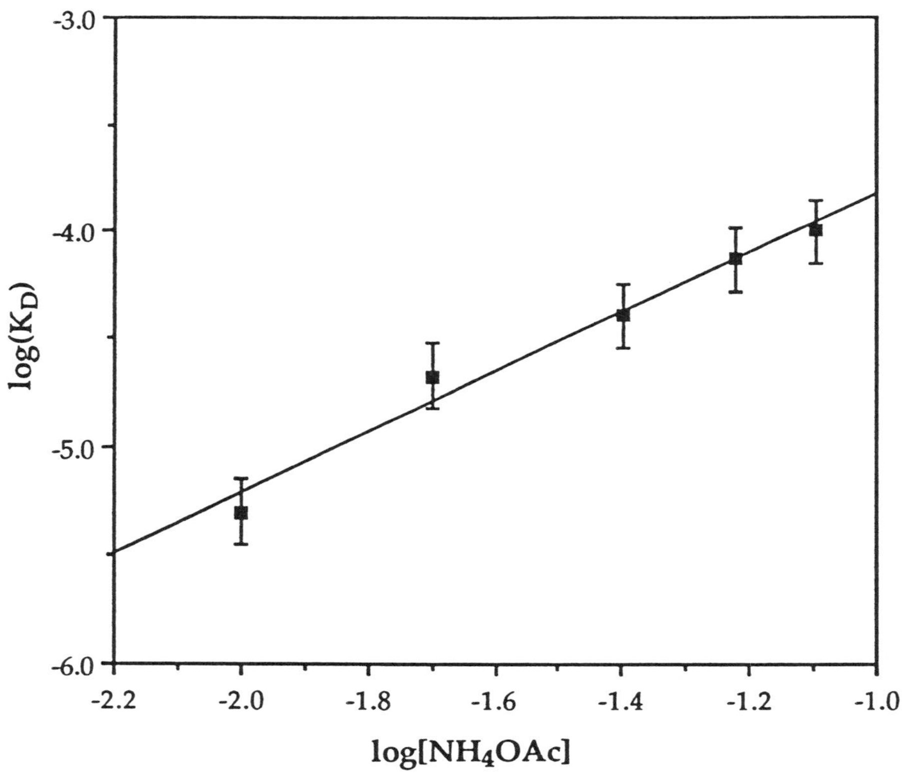

Figure 8. A plot of $\log(K_D)$ versus the $\log[NH_4OAc]$ for the 3082:BSA complex. Data were acquired for increasing concentrations of ammonium acetate in the buffer while holding the imidazole concentration at 33 mM.

The measured K_D values for the cholesterol adduct (8005), the 5'-C18 oligo (9047) and the 2'-O-propyl PS oligo (9045) are not salt-dependent. While this result is easily rationalized for the two molecules with lipophilic appendanges, the result for 9045 was unexpected. It should be noted that the binding of the 2'-O-propyl PO compound (9044) was sensitive to the ionic strength of the buffer, although the binding constant was tighter that the corresponding deoxy PO compound.

3.4 COMPETITION EXPERIMENTS

A competition experiment between 8005 and 3082 was performed to determine whether the two oligos bind to the same site on BSA or to different sites. The concentrations of BSA and 8005 were fixed at 0.5 and 3 μM, respectively, while the concentration of 3082 was varied from 0.5 to 33 μM. As shown in Figure 9, when the concentrations of the two oligos are both 3 μM, the integrated abundance of the 8005:BSA species is greater than the 3082:BSA species in proportion to their relative K_D values. Increasing concentrations of 3082 could not displace the 8005 from the BSA.

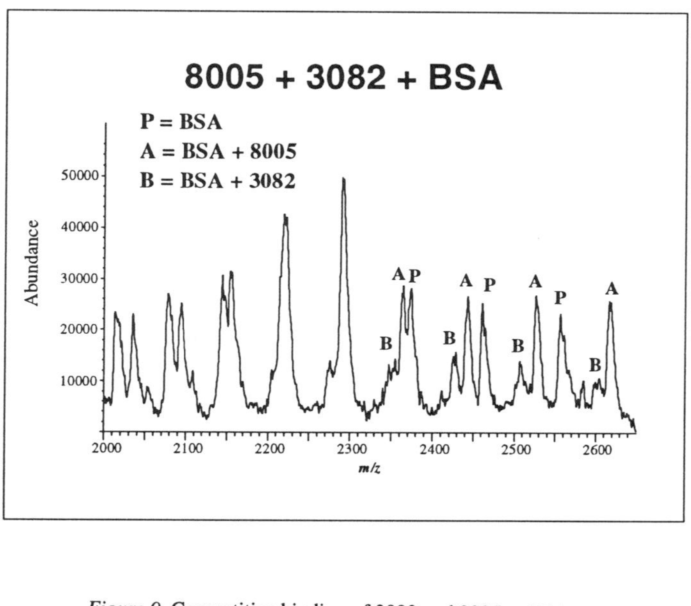

Figure 9. Competitive binding of 3082 and 8005 to BSA.

The cholesterol adduct may bind at a site on the BSA through hydrophobic rather than electrostatic interactions. If so, the electrospray and desolvation processes must be mild to preserve the hydrophobic interactions. This augers well for observation of other hydrophobic interactions among peptides and proteins, and small molecules with receptors.

3.5 COMPARISON OF ESI-MS WITH OTHER TECHNIQUES

The K_D for the oligonucleotide and BSA was determined independantly using capillary electrophoresis.(22) The ionic strengths of the buffers were matched to 45 mOsm in both cases. A fixed 100 nM concentration of BSA was employed, and the change in the migration time for the protein was measured as a function of oligo concentration (0.5-40 μM). A Scatchard analysis of the data showed that two oligo molecules bound to the protein. However, only the K_D of 10.4 μM for the weaker interaction could be calculated directly from the slope of the asymptotic plot. The value of K_D1 could only be estimated from the limited number of data points. K_D values are listed in Table 2.

The K_D for the complex also has been determined using surface plasmon resonance (SPR) spectoscopy with a Biacore 2000 instrument. Although the SPR data was acquired in the presence of higher salt than the ESI-MS results, the two K_D values agree well, as shown in Table 2.

TABLE 2. Measured values of the dissociation constants for 3082 with serum albumin as a function of analytical technique

Method	K_D1 (μM)	K_D2 (μM)
CE	<2.8	10.4
ESI-MS	2.4	10.8
SPR[a]	8	40

a. From solution studies performed at 40 mM $NaPO_4$.

3.6 CORRELATION WITH PHARMACOKINETIC PARAMETERS

The K_D values measured for the modified oligonucleotides using ESI-MS were predictive of their pharmacokinetic properties *in vivo*. The grouping of 9045 with the more lipophilic 8005 and 9047 from the ESI-MS binding data was borne out in the plasma pharmacokinetic parameters.(28) The maximum plasma concentration for 9045 was 6-fold greater that 3082, and similar to 8005 and 9047. The $T_{(1/2)}$ in plasma was increased two-fold for all three lipophilic compounds. The volume of distribution and clearance rates for these three oligos were smaller than for the PO oligo and 3082.

3.7 BINDING OF OLIGONUCLEOTIDES WITH ARGININE

We have studied the interaction of PS and PO oligonucleotides with the ethyl ester of arginine (RRE) as an example of a weak interaction between a small, charged molecule and a macromolecule. Addition of 10 μM RRE to a 0.5 μM solution of 3082 results in binding of three arginines to the oligonucleotide, as shown in Figure 10. The relative dissociation constants are not equivalent, as evidenced from the differences in the relative abundances of the poly RRE:oligomer species. The experiment was repeated as a function of RRE concentration, and a K_D1 of 5.5 μM was determined from a fit to a third-order binding polynomial.

4. Discussion

For all ESI-MS experiments which evaluate non-covalent interactions of macromolecules, the ionization mode shoudl be selected based on the pI (proteins) or pK (nucleic acid). Generally, these choices match the ionization mode with the highest sensitivity (positive for poteins; negative for nucleic acids). BSA has a pI of 4.5, and we have selected negative ionization for the study of the interaction with 3082 and derivatives. If the pI of the protein of interest is basic relative to physiological conditions, then positive ESI would be the best choice. For oligonucleotides, negative ESI best mimics physiological conditions. In most experiments, the concentration of one macromolecule will be fixed at a low

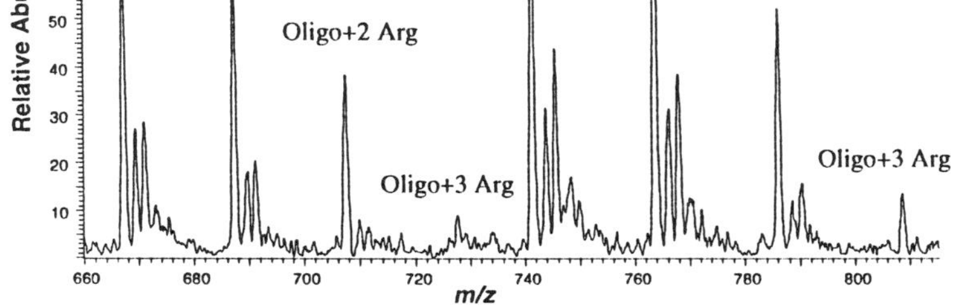

Figure 10. ESI-MS spectrum demonstrating the binding of arginine methyl ester (RRE) to 3082. Three adducted specise are observed for the $(M\text{-}9H)^{9-}$ and $(M\text{-}8H)^{8-}$ charge states. Lower abundance ions from sodium-adducted species also are seen for the oligo and oligo:RRE complex.

concentration. Instrument sensitivity and performance should be optimized for the charge states of this species.

4.1 BUFFER

The selection of an appropriate buffer is critical. The buffer of choice should be suficiently volatile to allow removal during the desolvation of the macromolecule. A monoionic buffer with opposite charge from the ionization method or a zwitterionic buffer will minimize loss of sensitivity during the ionization process. Loading of the droplet with buffer may produce dramatic shifts in effective droplet pH during the electrospray process. For negative ESI studies of oligonucleotides, imidazole and glycine represent good choices.(29) Ammonium acetate is popular, although this buffer can supress ionization on some instruments. Ammonium bicarbonate can be used as an alternative.

4.2 SAMPLE PREPARATION AND INSTRUMENTAL PARAMETERS

Other intrumental parameters or artefacts of experimental design may alter the apparent equilibrium constants. One must insure as complete as possible an integration of the entire charge state distribution for the macromolecule. It is impossible to optimize the linearity of instrument response as a function of *m/z* for the entire mass range, and tuning of the lens and skimmers should be selected for the molecule whose concentration is fixed. The temperature of the capillary interface should be reduced to 120-150°C. The flow of countercurrent drying gas can be increased to minimize adduction of volatile anions and cations to the macromolecule. The instrument should be operated with nozzle-skimmer or declustering voltages adjusted to minimize "non-specific" complexes often detected at high macromolecular concentrations. In these experiments, the ion intensity from free oligonucleotide was not integrated, since the intstrument parameters were not optimized for the smaller oligo ions. The macromolecules should be purified and non-volatile cations must be removed by HPLC, dialysis, or filtration before MS experiments. Residual sodium, potassium, and chloride ions adducted to macromolecules reduce the intensity of the main charge states and disperse the signal across additional *m/z* values.(30,31) The proportion of organic solvent which is added to facilitate the electrospray process should be varied to minimize effects on the proportion of detected complex.

4.3 BIOCHEMICAL CONSIDERATIONS AND CONTROLS

The biochemical factors which may alter the equilibrium binding between the macromolecule and substrate or cofactor must be considered in the experimental design. First, the solution concentrations of macromolecule must be matched to the sensitivity of the MS instrument. Typically, modern ESI-MS spectometers can generate an adequate signal from ~0.1-1 pmol/μL of analyte. Greater sensitivity can be realized with Fourier Transform mass spectrometers. The concentration of the mcromolecule must be matched to the approximate K_D which is to be determined. In general, the concentration of the fixed species should be ~10-fold below the K_D. It is impossible to measure a nM K_D using μM concentrations of macromolecule. At these higher concentrations, the apparent K_D is a complex combination of the true K_D and K_{off} for the system. In addition, the order of the reaction or binding is an important part of the experimental design.

Consideration should be given to the type(s) of forces which govern the interaction. For example, is the interaction electrostatic or hydrophobic? Our results suggest that both types of interaction can survive the electrospray process. However, experimental variables such as buffer concentration and proportion of added organic solvent may require adjustment to optimize detection of non-covalent complexes.

Execution of proper control experiments is critical for proper interpretatioin of results on any type of instrument. The presence of non-specific associations must be reduced through instrumental tuning or changes in buffer and organic. An appropriate molecule should be selected which is known not to bind, or which has the ability to ablate the affinity between protein and substrate via chemical or biochemical modification. One should insure that the complex demonstrates the proper sensitivity to salt, organic solvents, or denaturation. With a suitable experimental protocol in hand, systematic errors in the intensities of ions introduced by instrument must be ruled out or corrected. In many cases, one probably should compare K_D values determined using positive and negative ionization to bracket the K_D, although the K_D from the "best" ionization mode should be the most accurate. When all these conditions have been fulfulled, you can have confidence that the ESI-MS determination of K_D accurately reflects solution conditions.

4.4 DETERMINATION OF K_D VALUES AND UTILITY

K_D values can be determined accurately from integrated intensities of ions from macromolecule and complex(es). The data can be converted into K_D values using conventional Scatchard plots or using a polynomial binding function in cases where higher order complexes are detected. In both cases, direct observation of the proportion of each species based on unique molecular mass facilitates the analysis. The speed and ease of analysis allows other factors to be studied, such as the salt dependance of the binding process and competition between substrates. For oligonucleotides binding to albumin, the ESI-MS K_D values successfully predicted the unusual tight binding properties of a 2'-O-propyl PS oligonucleotide, which were confirmed *in vivo*.

5. Conclusions

ESI-MS can be used to determine the solution binding properties among biomolecules, their substrates, and their targets. The method is particularly useful for measurement of dissociation constants for bimolecular and higher order complexes. The measurement of molecular mass for individual species provides an unambiguous determination of binding stoichiometry. Changes in binding affinity produced by altered solution conditions, such as salt concentration, are reflected as changes in the abundances of macromolecular complexes in the gas phase using ESI-MS. Care must be taken in designing appropriate experiments and their controls, and potential errors generated by instrumental performance and the transfer of complexes from solution to the gas phase must be minimzed. The described methods should be applicable broadly to analyses of a variety of important biological systems.

6. Acknowledgements

We wish to thank Stan Crooke for helpful discussions and support.

7. References

1. Suckau, D.; Shi, Y.; Beu, S.C.; Senko, M.W.; Quinn, J.P.; Wampler, F.M.; McLafferty, F.W. *Proc. Natl. Acad. Sci USA* **1993**, *90*, 790.
2. Ganem, B.; Li, Y.T.; Henion, J.D. *J. Am. Chem. Soc.* **1991**, *113*, 6294.
3. Ganem, B.; Li, Y.T.; Henion, J.D. *J. Am. Chem. Soc.* **1991**, *113*, 7818.
4. Smith, R.D.; Light-Wahl, K.J.; Winger, B.E.; Loo, J.A. *Org. Mass Spectrom.* **1992**, *27*, 811.
5. Baca, M.; Kent, S.B.H. *J. Am. Chem. Soc.* **1992**, *114*, 2992.
6. Ganguly, A.K.; Pramanik, B.N.; Tsarbopoulos, A.; Covey, T.R.; Huang, E.; Fuhrman, S.A. *J. Am. Chem. Soc.* **1992**, *114*, 6559.
7. Smith, R.D.; Light-Wahl, K.J. *Biol. Mass. Spectrom.* **1993**, *22*, 493.
8. Loo, J.A.; Ogorzalek Loo, R.R.; Andrews, P.C. *Org. Mass Spectrom.* **1993**, *28*, 1640.
9. Camilleri, P.; Haskins, N.J. *Rapid Commun. Mass Spectrom.* **1993**, *7*, 603.
10. Jaquinod, M.; Potier, N.; Klarskov, K.; Reymann, J.M.; Sorokine, O.; Kieffer, S.; Barth, P.; Andriantomanga, V.; Biellmann, J.F.; Van Dorsselaer, A. *Eur. J. Biochem.* **1993**, *218*, 893.
11. Jaquinod, M.; Leize, E.; Potier, N.; Albrecht, A.M.; Shanzer, A.; Van Dorsselaer, A. *Tet. Letters* **1993**, *34*, 2771.
12. Li, Y.T.; Hsieh, Y.L. Henion, J.D.; Senko, M.W.; McLafferty, F.W.; Ganem, B. *J. Am. Chem. Soc.* **1993**, *115*, 8409.

13. Ganem, B.; Li, Y.T.; Henion, J.D. *Tet. Letters* **1993**, *34*, 1445.
14. Light-Wahl, K.J.; Springer, D.L.; Winger, B.E.; Edmonds, C.G.; Camp, II, D.G.; Thrall, B.D.; Smith, R.D. *J. Am. Chem. Soc.* **1993,** *115*, 803.
15. Goodlett, D.R.; Camp, D.G. Hardin, C.C.; Corregan, M.; Smith, R.D. *Biol. Mass Spectrom.* **1993**, *22*, 181.
16. Huang, E.C.; Pramanik, B.N.; Tsarbopoulos, A.; Reichert, P.; Ganguly, A.K.; Trotta, P.P.; Nagabhushan, T.L.; Covey, T.R. *J. Am. Soc. Mass Spectrom.* **1993**, *4*, 624.
17. Ogorzalek Loo, R.R.; Goodlett, D.R.; Smith, R.D.; Loo, J.A. *J. Am. Chem. Soc.* **1993**, *115*, 4391.
18. Hsieh, Y.L.; Li, Y.T.; Henion, J.D.; Ganem, B. *Biol. Mass Spectrom.* **1994**, *23*, 272.
19. Bacznskyj, L.; Bronson, G.E.; Kubiak, T.M. *Rapid Commun. Mass Spectrom.* **1994**, *8*, 280.
20. Bayer, E.; Bauer, T.; Schmeer, K.; Bleicher, K.; Maier, M.; Gaus, H.J. *Anal. Chem.* **1994**, *66*, 3858.
21. Anderegg, R.J.; Wagner, D.S. *J. Am. Chem. Soc.* **1995**, *117*, 1374.
22. Greig, M.J.; Gaus, H.J.; Cummins, L.L.; Sasmor, H.; Griffey, R.H. *J. Am. Chem. Soc.* **1995,** *117,* 10765.
23. Schwartz, B.L.; Gale, D.C.; Smith, R.D.; Chilkoti, A.; Stayton, P.S. *J. Mass Spectrom.* **1995**, *30*, 1095.
24. Lim, H.K.; Hsieh, Y.L.; Ganem, B.; Henion, J. *J. Mass Spectrom.* **1995,** *30*, 708.
25. Fehske, K.J.; Schlafer, U.; Wollert, U.; Muller, W.E. *Mol. Pharmacol.* **1982**, *21*, 387-393.
26. Galbraith, W.M.; Hobson, W.C.; Giclas, P.C.; Schechter, P.J.; Agrawal, S. *Antisense Res. Dev.* **1994**, *4*, 201-207.
27. Christensen, L.; Fitzpatrick, R.; Gildea, B.; Petersen, K.H.; Hansen, H.F.; Koch, T.; Egholm, M.; Buchardt, O.; Nielsen, P.E.; Coull, J.; Berg, R.H.; *J. Peptide Sci.* **1995**, *3*, 175-183.
28. Crooke, S.T.; Graham, M.J.; Zuckerman, J.E.; Brooks, D.; Conklin, B.S.; Cummins, L.L.; Greig, M.J.; Guinosso, C.J.; Kornbrust, D.; Manoharan, M.; Sasmor, H.M.; Schleich, T.; Tivel, K.L.; Griffey, R.H. *J. Pharm. Exptl. Ther.* **1996**, *277*, 923-937.
29. Greig, M.J.; Griffey, R.H. *Rapid Comm. Mass Spectrom.* **1995**, *9*, 97.
30. Potier, N.; Van Dorsselaer, A.; Cordier, Y.; Roch, O.; Bischoff, R. *Nucl. Acids Res.* **1994**, *22*, 3895.
31. Bleicher, K.; Bayer, E. *Biol. Mass Spectrom.* **1994**, *23*, 320.

OBSERVATION OF NON-COVALENT COMPLEXES OF CITRATE SYNTHASE AND NADH BY ESI/TOF MASS SPECTROMETRY

A. AYED[1], A. KRUTCHINSKY[2], I. V. CHERNUSHEVICH[2], W. ENS[2], H. W. DUCKWORTH[1], AND K. G. STANDING[2]
Department of Chemistry[1] and Department of Physics[2], University of Manitoba. Winnipeg, Manitoba, Canada. R3T 2N2

1. Introduction

Electrospray ionization/time-of-flight (ESI/TOF) mass spectrometry is a powerful tool for the study of large noncovalent complexes. The gentle nature of the electrospray technique preserves higher order structures, allowing their accurate mass measurement by time-of-flight mass spectrometry over a wide m/z range. Here we demonstrate the application of ESI/TOF MS in the detection of noncovalent complexes and quaternary structural changes as a result of ligand binding to an allosteric enzyme.
Citrate synthase (CS) is the enzyme that initiates the citric acid cycle, catalyzing the reaction of oxaloacetate and acetyl coenzyme A to form citrate and coenzyme A. Citrate synthases from gram-negative bacteria (such as *Escherichia coli*) can exist as dimers and hexamers, and are inhibited specifically by NADH (nicotinamide adenine dinucleotide, reduced form) through an allosteric mechanism, unlike citrate synthases from eukaryotes and gram-positive bacteria, which exist as dimers only and are not subject to inhibition by NADH. *E. coli* CS exhibits a dimer-hexamer equilibrium at low ionic strength that is pH dependent, the oligomeric state increasing as pH is decreased [1]. The number of NADH binding sites in *E. coli* CS is also pH dependent, increasing with decreasing pH to a maximum of six sites per hexamer as determined previously by fluorescence methods [2]. Although the effect of NADH on the activity of CS is known, the mechanism of inactivation of CS by NADH remains elusive.

2. Experimental Procedures

Our standard method of preparation [3] yields *E. coli* CS in 20 mM TRIS 1 mM EDTA, pH 7.8. For mass spectrometry, the sample is washed 6 to 8 times with two ml aliquots of 20 mM ammonium bicarbonate (pH 8.0) or 20 mM ammonium acetate (pH 6.5) in a

W. Ens et al. (eds.), New Methods for the Study of Biomolecular Complexes, 135–139.

Centricon 30 concentrator (Amicon). NADH (disodium salt) was dialyzed against 20 mM ammonium bicarbonate or 20 mM ammonium acetate in a 500 MW cut-off membrane for 24 hours at 4°C. Samples for electrospray were prepared such that the final concentration of the buffer component was 5 mM, and the protein component 10 μM. The concentrations of CS and NADH were determined spectrophotometrically using their known extinction coefficients.

The time-of-flight mass spectrometer (TOF III) with electrospray ion source and orthogonal extraction used for these experiments has been described elsewhere [4, 5].

3. Results and Discussion

The mass spectra of CS in 5 mM ammonium bicarbonate or ammonium acetate (Figs. 1 and 3) exhibit two sets of charge distributions, centred at m/z values of 5000 and 8500 with average charges of +19 and +35, respectively. These correspond to dimeric and hexameric forms of CS as shown in the deconvoluted spectra in Figs. 2 and 4, with molecular weights of about 95 774 ± 10 and 287 352 ± 30, respectively, in reasonable agreement with the theoretical molecular weights (95 768 for dimer and 287 304 for hexamer). The hexameric species is favored at pH 6 (in ammonium acetate), while approximately equal amounts of hexamer and dimer are observed at pH 7.7 (ammonium bicarbonate), in agreement with previous findings by analytical ultracentrifugation [1].

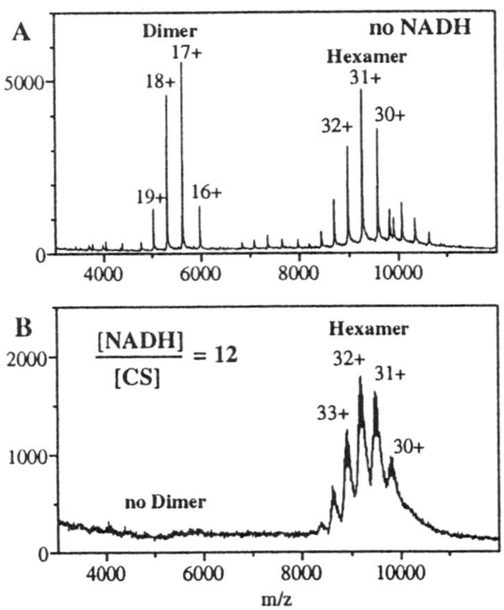

Figure 1. ESI/TOF spectra of citrate synthase in 5 mM ammonium bicarbonate in the absence (A) and presence (B) of NADH. The molar ratio of NADH to CS is shown.

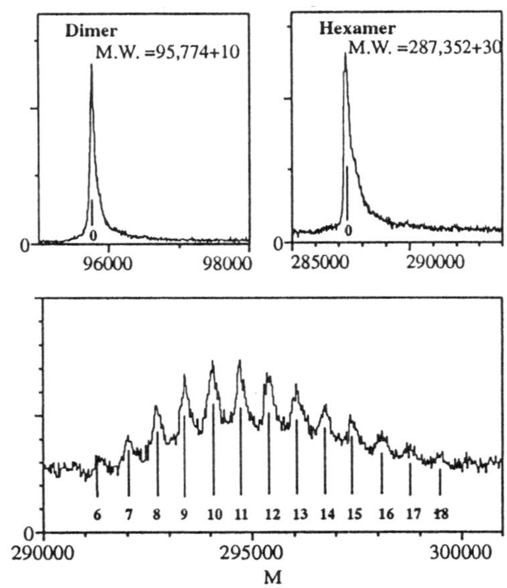

Figure 2. Deconvolutions of the spectra in Fig.1. The labels in the lower panel (B) indicate the number of NADH molecules bound.

NADH, a known allosteric inhibitor of CS, is shown here to induce a transition from a state where a population of dimers and hexamers is present to one that is almost entirely hexameric. At pH 6, the shift to hexamer is complete when the concentration of NADH is equivalent to the concentration of CS subunits, while a higher NADH concentration is required for the shift at pH 7.5 (see Figs. 1 - 4). In terms of an allosteric mechanism, these results suggest that NADH regulates the activity of citrate synthase by converting it to a hexameric, NADH-bound inactive conformation.

Clearly resolved peaks corresponding to incremental binding of NADH to the hexamer at pH 7.5 are observed in the deconvoluted spectra (Fig. 2B). The peaks are also evident at pH 6, but are not as well resolved (Fig. 4B). Although it was previously found by solution techniques that a maximum of 6 NADH molecules bind per CS hexamer, up to 18 NADH molecules are observed to bind the hexamer at pH 7.5 by ESI/TOF (Fig. 2B). A plot of the molecular weights from the various hexamer-NADH complexes versus the number of NADH molecules bound at pH 7.5 (Fig. 5) yields an average molecular weight for NADH of 668, in good agreement with the theoretical mass of 665. The extra mass in the calculated average is likely caused by additional protons associated with the NADH.

At intermediate concentrations of NADH where the dimer species is still present, the dimer exhibits NADH binding (spectra not shown). However, the concentration of NADH required for binding to the dimer is higher than for the hexamer, suggesting that the dimer binds NADH weakly or perhaps less specifically than the hexamer does.

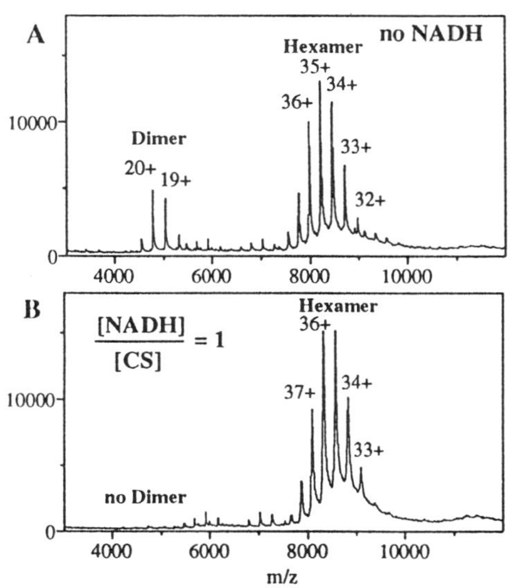

Figure 3. ESI/TOF spectra of citrate synthase in 5 mM ammonium acetate in the absence (A) and presence (B) of NADH. The molar ratio of NADH to CS is shown.

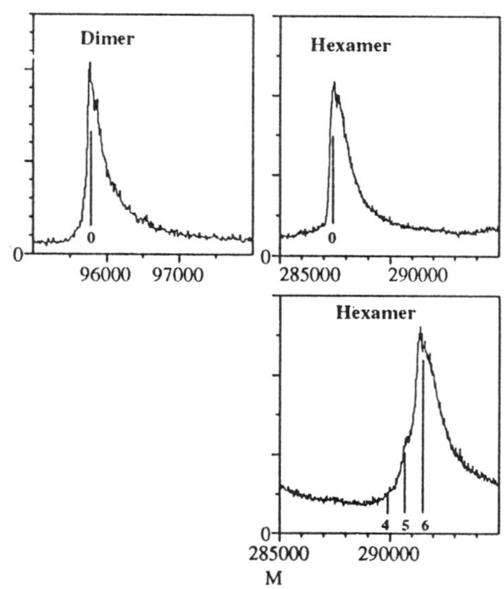

Figure 4. Deconvolutions of the spectra in Fig.3. The labels in the lower panel (B) indicate the number of NADH molecules bound.

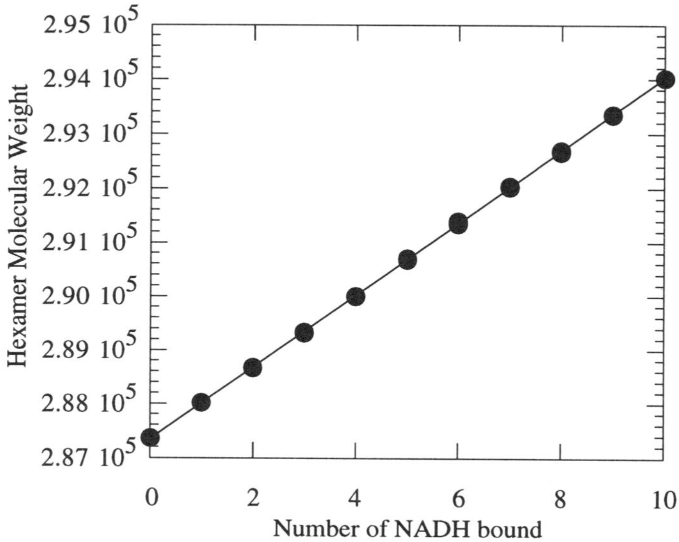

Figure 5. The change in the molecular weight of the hexamer of citrate synthase as a function of NADH binding. Linear least squares fitting yielded the equation MW=287 342 (±7) + 668 (±1)[number of NADH].

4. Conclusions

ESI/TOF mass spectrometry has been applied in the observation of hexameric and dimeric assemblies of *E. coli* citrate synthase subunits that resemble those found in solution. Non-covalent complexes of NADH and citrate synthase are clearly resolved with the hexamer and the dimer forms of CS. Binding of NADH to the hexamer occurs at a lower concentration range than binding to the dimer form, indicating that binding to the dimer is weaker or less specific. NADH shifts the dimer/hexamer equilibrium from a state where dimers and hexamers are present to one in which only hexamers exist, suggesting a mechanism of inhibition of CS involving quaternary structural changes such that NADH-bound hexamer is favored. This state is assumed to be the inactive conformation of the enzyme.

5. Work In Progress

Work is now underway in assessing the specificity of NADH binding to the dimeric and hexameric forms of CS by using other adenylates as competitors for binding. ESI/TOF is also being used to study substrate binding to CS (oxaloacetate and acetyl coenzyme A) in the presence and in the absence of NADH.

6. Acknowledgements

This work was supported by grants from the Natural Sciences and Engineering Research Council of Canada and from the US National Institutes of Health (GM30605).

7. References

1. Tong, E. K., and Duckworth, H. W. (1975) The Quaternary Structure of Citrate Synthase from *Escherichia coli* K12. *Biochemistry* **14**, 235 - 241.
2. Duckworth, H. W. and Tong, E. K. (1976) The Binding of Reduced Nicotinamide Adenine Dinucleotide to Citrate Synthase of *Escherichia coli* K12. *Biochemistry* **15**, 108 - 114.
3. Anderson, D. H, and Duckworth, H. W. (1988) *In Vitro* Mutagenesis of *Escherichia coli* Citrate Synthase to Clarify the Locations of Ligand Binding Sites. *The Journal of Biological Chemistry* **263**, 2163 - 2169.
4. Verentchikov, A., Ens, W., and Standing, K. (1994) Reflecting Time-of-Flight Mass Spectrometer with an Electrospray Ion Source and Orthogonal Extraction. *Analytical Chemistry* **66**, 126 - 133.
5. Krutchinsky, A. N., Chernushevich, I. V., Spicer, V., Ens, W., and Standing, K. G. (1995) Collisional Focusing of Ions in the Electrospray TOF Mass Spectrometer. *43rd ASMS Conference on Mass Spectrometry and Allied Topics.*

GAS-PHASE DISSOCIATION ENERGIES OF HYDROGEN-BONDED COMPLEXES IN A QUADRUPOLE ION TRAP MASS SPECTROMETER

E. C. KEMPEN, A. COLORADO, J. S. BRODBELT
Department of Chemistry and Biochemistry
University of Texas at Austin, Austin TX, 78712

1. Introduction

Intermolecular and intramolecular hydrogen-bonds are perhaps the most important interactions in molecular self assembly and molecular recognition in solution chemistry. These interactions are also significant factors in host-guest chemistry and site-selective complexation. Recent studies have shown that host-guest complexes can be constructed in the gas phase with the absence of solvent, and thus some of the intrinsic details of complexation can be probed. One of the most informative parameters of host-guest complexes is the strength of their binding interactions. There are numerous ways to evaluate binding energies in solution, and comparison to gas-phase binding would provide new insight into evaluating structural and solvation effects on molecular recognition. Collisionally activated dissociation (CAD) and photodissociation (PD) are two methods currently employed for obtaining the binding energy of gas-phase ionized complexes. In the CAD technique, the disappearance of the molecular ion of interest and appearance of a specific fragment ion is monitored as a function of the energy used to excite the precursor ion [1]. The energy required to induce dissociation is then utilized in the evaluation of the binding energy, along with an estimation of the internal energy distribution of the precursor ion and RRKM correction for kinetic shifts based on the dissociation time scale of the instrument. For the method of estimating binding energies by photodissociation, dissociation rate constants are measured as a function of wavelength of internal energy in the ion of interest [2]. These rate constants along with RRKM modeling allow dissociation energies to be calculated.

Presented here are binding energy measurements obtained from threshold activation voltage measurements in a quadrupole ion trap [3]. The threshold point is defined as the point where only 10 % of the precursor ions have undergone dissociation. This method does not provide the same degree of accuracy as the binding energy measurements accessible in conventional tandem mass spectrometers, but it offers greater versatility in the types of complexes which can be evaluated in an ion trap. This

W. Ens et al. (eds.), New Methods for the Study of Biomolecular Complexes, 141–147.

method will be used to evaluate model host-guest complexes formed from a variety of crown ethers and ammonium ions. The system of greatest interest is a series of lariat ethers [4]. These host were chosen because of their enhanced binding selectivities for small cations [4]. Lariat ethers have pendant arms which assist in organization of complexes [4].

2. Experimental

All experiments were performed in a Finnigan quadrupole ion trap mass spectrometer operated in the mass selective instability mode [5]. The trap temperature was held at 120 °C. Typical tickle voltages for collisionally activated dissociation were between 10-400 mV_{p-p}. The exact resonant frequency of the applied tickle voltage was optimized at each new activation voltage used to establish threshold behavior. The threshold activation voltage was defined as that voltage where the fragment ion intensity is 10% of the total ion intensity. The sample pressures of the hosts (crown ethers) and guests (amines) were 1-2 x 10^{-6} torr and 3-4 x 10^{-6} torr (uncorrected) respectively. Helium was introduced into the chamber at a pressure of 1.2×10^{-4} torr (uncorrected). Ion-molecule reaction times were in the range of 10 - 650 ms. Also, an activation time of 10 ms and a q_z-value of 0.18 or 0.20 were used. All samples were purchased from Aldrich Chemical Co. (Milwaukee, WI) with the exception of anhydrous ammonia (MG Industries). Lariat ethers were provided by Prof. Richard Bartsch, Texas Tech. University.

2.1 CALIBRATION

The dissociation behavior of a series of proton-bound complexes with well-characterized binding energies was examined in order to establish a calibration of energies in the ion trap. A plot of the Threshold activation vs. literature binding energies shows a near-linear correlation (Figure 1) [6]. A consistent standard deviation of ± 2 kcal/mole was obtained.

3. Results and Discussion

3.1 . CROWN ETHER / AMINE SYSTEMS

A series of [polyether-H^+-amine] complexes were evaluated to correlate the strength of hydrogen-bonding interactions with structural features. Ion-molecule reactions were used to generate proton-bound complexes, between 12-crown-4, 15-crown-5, 18-crown-6 and various amines. Activation experiments showed that the number of available hydrogen-bonding sites, the gas-phase basicities of the polyether and amine

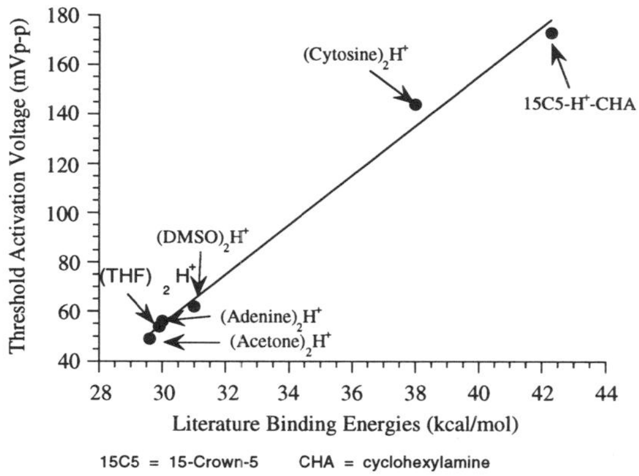

Figure 1. Binding Energy Calibration Curve for Proton Bound Dimers

components, and the geometry of the interactions correlated with the threshold values. Upon examination of the binding energies obtained by the threshold method, a noticeable trend exists between the number of possible binding interactions and binding energy. In all cases observed, the binding energies of (crown ether + H^+ + amine) increased when the size of the crown ether was increased while the amine guest was held constant. Not only do the number of possible binding sites increase from 12-crown-4 to 15-crown-5 to 18-crown-6, but the size of the ring is also substantially increased. Thus not only are a greater number of bonds possible for the larger structure but, the hydrogen-bonds involved in the (crown + H^+ + amine) complex are optimized because the greater flexibility of the larger polyether ring allows formation of near-linear hydrogen bonds. Since the 12-crown-4 molecule is smaller, more rigid and has fewer possible hydrogen-bond acceptors (i.e.. oxygen atoms), it may form non-linear, less optimal bonds with the amine guest and the 12-crown-4 may adopt a less favorable conformation. The steric bulk of the guest amine also plays a role in the stability of the (crown ether + H^+ + amine) complex. This is easily observed when pyridine, 2-picoline and 3-picoline are complexed with the same ether host.

The addition of a methyl at the 2-position of the pyridine in the cases of (crown ether + H^+ + 3-picoline) complexes shows no significant increase in binding energy compared to the (crown ether + H^+ + pyridine) complexes. The addition of a methyl at the 3-position of the pyridine ring in the cases of (crown ether + H^+ + 2-picoline) shows more significant increases in binding energies for the 12-crown-4 and 15-crown-5 systems. This is probably due to steric crowding of the single hydrogen which may participate in hydrogen-bonding. The increased flexibility of the 18-crown-6 system may be responsible for the opposite results which are observed for (18-crown-6 + H^+ + amine) systems. Table 1 shows the trends described above.

TABLE 1. Binding Energies of Crown Ether-Amine Complexes
(All values in kcal/mol)[b]

	NH_3	N,N-DEMA	N-MBA	Pyridine	2-picoline	3-picoline
12-crown-4 	32	n/a[a]	34	33	31	33
15-crown-5 	35	30	38	36	33	34
18-crown-6 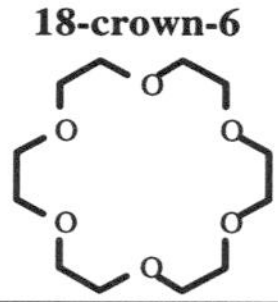	41	34	41	35	37	39

[a] n/a=dimer does not form
[b] the standard deviation is estimated as ± 2 kcal/mol

N,N-DEMA = N,N-diethyl methylamine
N-MBA = N-methyl butylamine

3.2. LARIAT ETHER / AMMONIA SYSTEMS

In an attempt to gain insight into some more complicated systems, lariat ethers **1-6** were complexed with ammonia to form (lariat ether + H^++ NH3) complexes. The binding energies for the loss of NH3 are reported in Table 2. The addition of the electron-donating methoxy group in **2** does not significantly increase the binding energy of its complex with ammonia vs. that of the complex formed from **1**. The methoxy group is not close enough to any of the other basic sites on the host molecule to have much inductive electron-donating character and is not connected through a long enough tether to allow it to come in contact with the guest were ammonia to bind in the cavity of the crown ether. Compounds **4-6**, however, have sufficiently long tethers that their capabilities should not be so hindered. No threshold value could be obtained for compound **3**, since more fragmentation occurred than the simple loss of neutral ammonia. This might indicate that the hydrogen bonds of the complex are as strong or stronger than some of the covalent bonds within the lariat ether. The complex of compound **4** with ammonia showed a binding energy 2 kcal/mol higher than that of the unmodified lariat ether (**1**) with ammonia. Both these results would seem to indicate that the basic group on the tether is indeed undergoing some interaction with the guest.

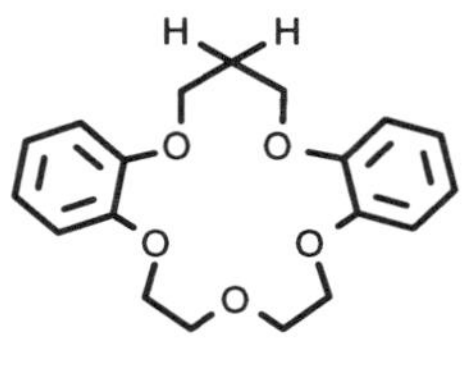

1

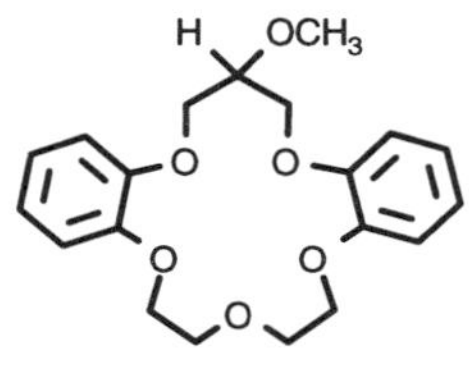

2

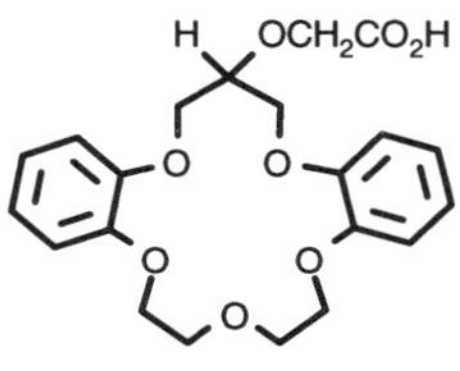

3

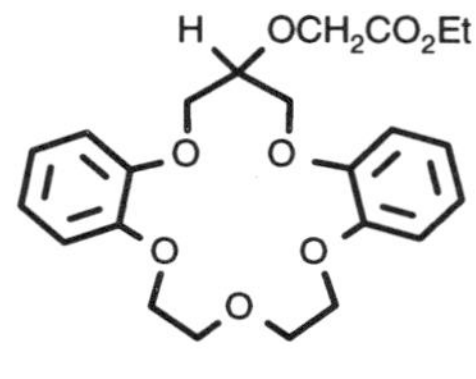

4

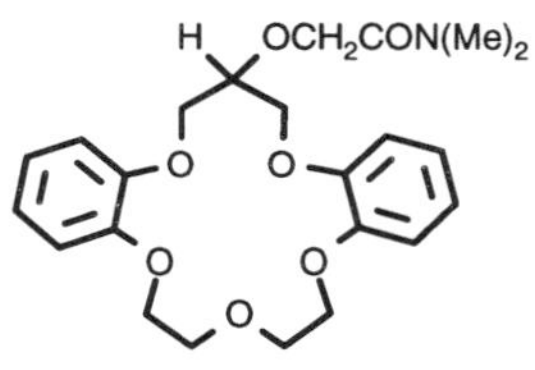

5

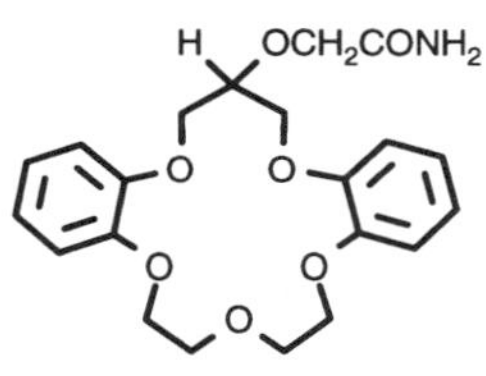

6

Figure 2. Structures of Lariat Ethers

In the case of the complexes formed from the amide compounds **5** and **6**, the neutral lariat ethers have such high gas-phase basicities that they spontaneously deprotonate the (lariat ether + H^+ + NH_3) complexes, and no direct measurements could be obtained. However, estimates from the complex of (**5** + H^+ + NH_3) indicate a binding energy of approximately 32 kcal/mol, significantly lower than the binding energies of the other complexes. This result could indicate that the tethered amide functionality is itself partially bound into the cleft of the lariat ether and the ammonia is simply bound to the amide functionality, not in the cavity of the ether.

TABLE 2. Binding Energies of Lariat Ether-Ammonia Complexes

Lariat Ether	Binding Energy (kcal/mol)	Threshold Tickle Voltage (mV)
1	37	122
2	36	114
3	n/a	n/a
4	39	141
5	P	P
6	P~32	P~80

n/a= Other fragmentation occured at low voltages and threshold values could not be obtained.
P= Protonation of neutral lariat ether occured more rapidly than activation, only very rough estimates were obtained, or no threshold estimates could be made.

4. Conclusions

Estimation of binding energies in a quadrupole ion trap by a threshold CAD method appears feasible. Binding energies can be estimated with accuracies ± 6 kcal/mol for most ions, with a standard deviation of ± 2 kcal/mol. This technique is relatively fast to perform, reproducible, and not affected by many changes in experimental parameters. The main limitation of this method is that the construction and extension of the calibration curve requires accurate binding energy values obtained from other methods, for systems that have similar sizes and bonding interactions to the ones of special interest.

5. Acknowlegements

This work was funded in part by the National Science Foundation, the Welch Foundation, the Dreyfus Foundation, the Sloan Foundation, and a grant from the Texas Advanced Research Program. ECK and AC acknowledge fellowships from the Department of Education.

6. References

1. (a) Su, C-X; Hales, D. A.; Armentrout, P. B. (1993) The Bond Energies of Cr_2 and Cr_2^+ *Chem. Phys. Lett.*, **201,** 199-204. (b) Dalleska, N. F.; Honna, K.; Armentrout, P. B. (1993) Stepwise Solvation Enthalpies of Protonated Water Clusters: Collision-Induced Dissociation as an Alternative to Equilibrium Studies, *J. Am. Chem. Soc.* , **115**, 12125-12131. (c) Armentrout, P. B. "*Thermochemical Measurements by Guided Ion Beam Mass Spectrometry*", in N. Adams, L. Babcock, eds., "*Advances in Gas Phase Ion Chemistry*", JAI Press Inc., Greenwich, (1992), **1,** 83. (d) Loh, S. K., Hlaes, D. A., Lian, L., Armentrout, P. B. (1989) Collision Induced Dissociation of Fe^+n (n=2-10) with Xe: Ionic and Neutral Iron Binding Energies, *J. Chem. Phys.* , **90,** 5466-5485.
2. (a) Weddie, G. H.; Dunbar, R. C.; Song, K.; Morton, T. H. (1995) Time-Resolved Photodissociation of the Molecular Ions of Propyl Phenyl Ethers, *J. Am. Chem. Soc.* ,***117,*** 2573-2580. (b) Dunbar, R. C. (1994) Kinetics of Thermal Unimolecular Dissociation by Ambient Infrared Radiation, *J. Phys. Chem.* , **98,** 8705-8712. (c) Lin, C-Y; Dunbar, R. C. (1995) Time-Resolved Photodissociation of Gas-Phase Nicklocene Cation: Dtermination of Bond Strength and Radiative Relaxation Rate, *J. Phys. Chem.* , **99** , 1754-1759.
3. Colorado, A.; Brodbelt, J. S., (1996) An Empirical Method for the Estimation of Critical Energies in a Quadrupole Ion Trap, *J. Am. Soc. Mass spec.* in press.
4. a)Strzelbicki, J.; Bartsch, R. A. (1981) Extraction of Alkali Metal Cations from Aqueous Solutions by a Crown Ether Carboxylic Acid, *Anal .Chem.* , **53,** 1894-1899. b) Bartch, R. A.; Heo, G. S.; Kang, S. I.; Kang, Y.; Strzelbicki, J. (1983) Synthesis of Highly Lipophilic Crown Ether Carboxilic Acid, *J. Org. Chem.* **48,** 4864-4869.
5. March, R. E.; Hughes, R. J.; Todd, J. F. J. (1989) *Quadrupole Ion Storage Mass Spectrometry*, Wiley: NY.
6. (a) Hiraoka, K.; Takimoto, H.; Yamabe, S. (1986) Gas-Phase Stabilities of Symetric Proton-Held Dimer Cations, *J. Phys. Chem.* , **90,** 5910-5914. (b) Dunbar, R. C. (1990) Kinetic Parameters for the Unimolecular Dissociation of Styrene Ion, *J. Phys. Chem* ., **94,** 3283-3286. (c) Bomse, D. S.; Beauchamp, J. L. (1981) Thermochemistry and Proton Bond Energies of Gas-Phase Proton-Bounnd Dimers of Aliphatic Alcohols, *J. Phys. Chem.*, **85**, 488-492. (d) Moet-Ner, M. (1983) The Ionic Hydrogen Bond. 3. Multiple NH---O and CH---O bonds. Complexes of Ammonium Ions with Polyethers and Crown Ethers, *J. Am. Chem. Soc.* , **105**, 4912-4915. (e) Moet-Ner, M. (1979) Ion Thermochemistry of Low-Volatility Compounds in the Gas Phase. 2. In trinsic Basicities and Hydrogen-Bonded Dimers of Nitrogen Heterocycles and Nucleic Bases, *J. Am. Chem. Soc.*, **101**, 2396-2403.

ELECTROSPRAY IONIZATION FTICR SPECTROMETRY OF METALLOPROTEINS

SHUBHADA S. KULKARNI , P. KRISTINA TAYLOR.,
DONALD M. KURTZ JR., AND I. JONATHAN AMSTER*
Department of Chemistry and Center of Metalloenzyme Studies,
University of Georgia, Athens, Georgia 30602-2556

Introduction

Metalloproteins play an important role in many biochemical processes such as respiration, metabolism, photosynthesis, nitrogen fixation, nerve transmission, muscle contraction, and signal transduction. Determining the details of the structure of the metal-center is central to understanding the structure and function of a metalloprotein, as the metal-center is usually the site of catalytic activity. For structure-function studies, metalloproteins are often produced by overexpression of recombinant genes. Frequently, a recombinant metalloprotein is produced in a denatured form, with its metal center absent. In such cases the protein must be activated by inserting the metal atoms and allowing the protein to fold into its native structure. There is a need for a rapid, accurate assay of these recombinant metalloproteins to determine the subunit amino acid composition, the metal stoichiometry, and the subunit stoichiometry. We have been investigating the capability of mass spectrometry to make such measurements for multimeric metalloproteins.

The model metalloproteins that we have been studying are the multimeric non-heme iron-containing proteins, hemerythrin and rubrerythrin. Hemerythrin is an oxygen-carrying protein found in a few marine invertebrates [1]. Despite its name, hemerythrin does not contain a heme group, but has a non-heme diiron site that reversibly binds one oxygen molecule. The two iron atoms are connected by an oxo-bridge in the active site. The holoprotein consists of eight subunits bound by noncovalent interactions. Each subunit is folded into a four helix bundle containing one diiron site. The other protein of interest, rubrerythrin, is also a non-heme iron-containing protein and is found in the anaerobic sulfate-reducing bacteria, *Desulfovibrio vulgaris* [2]. This protein contains three iron atoms per subunit. Two iron atoms are bound in an oxo-bridged, hemerythrin-like diiron site, surrounded by a four helix bundle and the third iron is bound in a rubredoxin-like FeS_4 site [3]. Rubrerythrin has ferroxidase activity, i.e. it catalyzes oxidation of Fe^{2+} to Fe^{3+} by dioxygen [4]. Gel filtration measurements of rubrerythrin suggest that the

W. Ens et al. (eds.), New Methods for the Study of Biomolecular Complexes, 149–156.

holoprotein is a homodimer in solution. However, recent studies of rubrerythrin with mass spectrometry have found evidence of a tetramer [5] which is also seen in the x-ray crystal structure of rubrerythrin [3].

Electrospray ionization mass spectrometry has become a powerful tool for the detection and characterization of large biomolecules including metal-protein complexes and noncovalent complexes. This gentle method of ionization produces intact, multiply protonated, gas-phase ions directly from protein molecules in solution [6-12]. More recently, it has been demonstrated that multimeric proteins composed of noncovalently-bound subunits can be examined by electrospray ionization [13-16]. ESI time-of-flight mass spectrometry measurements of rubrerythrin and hemerythrin have been recently made by Standing and coworkers which show that these proteins can be electrosprayed under nondenaturing conditions which allow multimeric holoproteins to be observed [17]. The multiply charged peaks for both proteins appear between m/z 4000 and 8000, and although the peaks are not isotopically resolved, the experimental masses correspond well with the theoretical masses of the metal-containing oligomers, allowing verification of the stoichiometry of the metals.

We are particularly interested in applying FTICR mass spectrometry to measure the multimeric nature of holoproteins and to examine the metal constitution of metalloenzymes. For characterization of multiply charged ions, Fourier transform ion cyclotron resonance (FTICR) has proven to be an efficient method, capable of ultrahigh resolution and high mass accuracy [18-21].

Determinations of the monoisotopic mass of large biomolecules has been reported [22]. We seek to apply these high resolution methods to metal-containing proteins. There are several challenges to carry out such measurements. Foremost is that the metals are usually bound within the protein by weak forces. The denaturing conditions typically used for ESI disrupt these interactions, leading to the loss of the metal center. We report here progress toward preserving these weak interactions.

Experimental

The instrument used for these experiments is a Bruker Bioapex 70 FTICR (Bruker Analytical Instruments, Billerica, MA) with a 7.0 tesla magnet. This instrument is equipped with an electrospray ionization source (Analytica, Branford, CT) with a glass capillary vacuum interface. Ions are transferred from the source region to the analyzer by using electrostatic ion optics to focus ions through the fringe field of the superconducting magnet. The analyte solution is injected through a syringe into a stainless steel needle which is held at ground potential. The front of the glass capillary tube is maintained at -3-4 kV relative to the needle. The solvent used to observe the metalloenzyme under denaturing conditions is a 49:49:2 mixture of methanol/water/acetic acid. Metal binding by the proteins was observed in less denaturing solutions, i.e. from 0 to 25% methanol concentration. An external calibration using ubiquitin as a standard was applied to all mass spectra.

Recombinant rubrerythrin and hemerythrin were produced by overexpression of the cloned genes in E. coli, insertion of iron and removal of denaturant and purified by ion exchange and gel filtration as described previously [23]. Before dissolving the proteins in methanol-water solution, the buffer (50 mM HEPES, 200 mM Na_2SO_4 at pH = 7.4) was removed by filtration through a 10 kDa molecular weight cut-off membrane (Millipore Co. Bedford, MA) and washed several times with 5 mM ammonium acetate (pH = 5.4). All other chemicals used were obtained from Sigma Chemical Co., (St. Louis, MO).

Results and Discussion

Our ultimate goal is to make high resolution measurements of multimeric metalloprotein ions by ESI-FTICR. As a first step toward this goal, we have investigated the conditions necessary to maintain the folded structure of the subunits of the metalloenzyme hemerythrin and rubrerythrin. When methanol/water/acetic acid solution in the proportion of 49:49:2 is used as a solvent, only the aposubunits of the metalloproteins are observed, as shown in Figure 1a for hemerythrin. The binding of iron atoms to the protein occurs through relatively weak coordinate covalent bonds, and when the protein is denatured, these attachments are easily broken. Protein concentration, pH, ionic strength, and the concentration of the organic solvent need to be carefully chosen in order to maintain metal attachment.

Metal attachment to the subunits of hemerythrin can be preserved by using a nondenaturing solution (10% methanol) of the protein for electrospray, as can be seen in Figure 1b. The charge state distribution in the mass spectrum has shifted to lower m/z values compared to the mass spectrum obtained with 49% methanol, with the 8+ charge state being the most abundant versus 14+ in the mass spectrum of the denatured protein. This shift is consistent with a more tightly folded structure, as has been noted by others [8]. The more tightly folded ion had retained its diiron center. The monoisotopic masses that were measured for denatured and nondenatured subunits of hemerythrin are 13466.3 Da and 13572.1 Da. The difference between the two masses is 105.8 Da which corresponds exactly to the mass of two iron atoms minus six protons (2 x 55.9 amu - 6 x 1.0 amu = 105.8 amu) indicating that hemerythrin binds two iron atoms in the 3+ charge state per subunit. Since each iron atom is bound to the protein in the 3+ charge state, three protons are displaced per iron atom, giving a mass difference of 105.8 amu rather than 111.8 amu. Although the protein was reconstituted with Fe^{2+}, the metal atoms have been converted to the more stable Fe^{3+} through exposure to air. There is no evidence of peaks corresponding to the mass of the subunit plus one iron atom, consistent with the known cooperative binding behavior of hemerythrin for the two iron atoms [24]. X-ray crystallographic data indicates that the oxygen bridge between the two iron atoms is provided by a water molecule in the nondenatured protein in solution [1]. However, there are no peaks in the mass spectrum corresponding to the mass of the diiron-containing subunit plus the mass of a water molecule as hydroxide ion or oxo ion. It is possible that the water molecule that provides the oxygen bridge is removed during the desolvation process. The difference between the measured and calculated values of hemerythrin is 42 ppm and is 37 ppm for the diiron-containing subunit of hemerythrin. If a water molecule were present, it would be easily discerned.

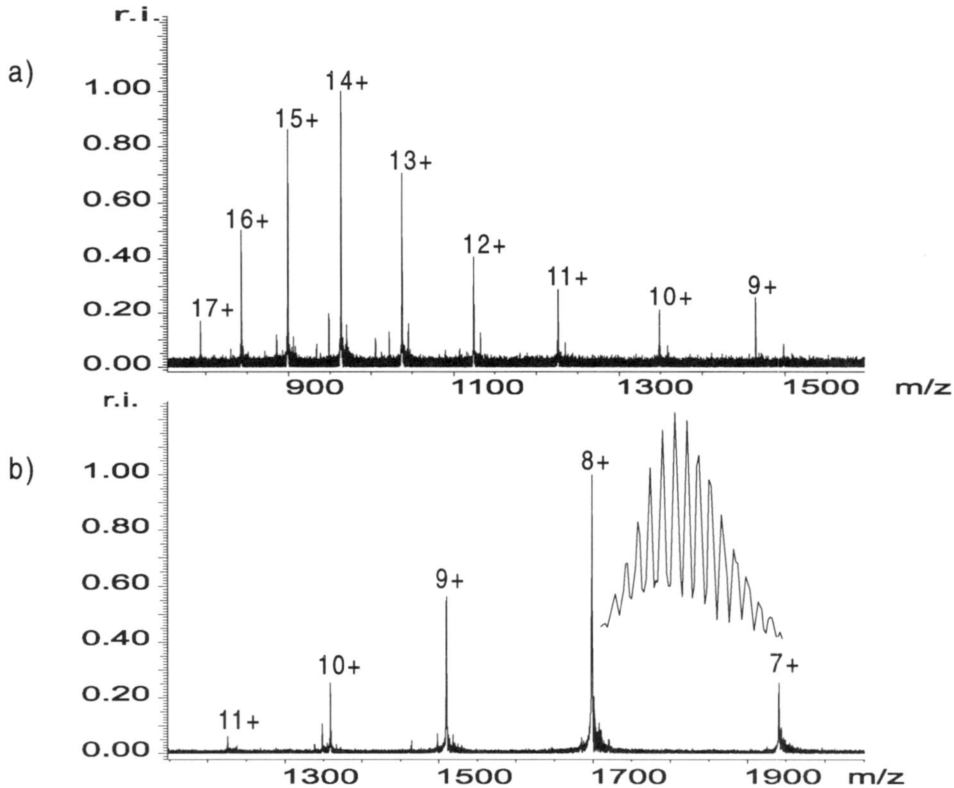

Figure 1: a) ESI-FTMS of hemerythrin under denaturing conditions (49:49:2 methanol/water/acetic acid), exhibiting 14+ charge state as most abundant state. b) ESI-FTMS of hemerythrin under non-denaturing conditions (10 % methanol), exhibiting the 8+ charge state as most abundant state. The inset is a mass scale expansion of the 8+ charge state, showing that isotopic resolution is achieved.

The behavior of rubrerythrin upon ESI-FTMS is similar to that of hemerythrin. Pneumatically assisted electrospray of pure aqueous solutions of rubrerythrin yields the mass spectrum shown in Figure 2a. Iron attachment to the aposubunit is observed in the mass spectrum.

Each peak in the mass spectrum exhibits four groups of isotopically resolved peaks. Figure 2b shows an expansion of the mass scale around 15+ charge state. The two most intense peaks correspond to the aposubunit and aposubunit plus two irons. All other charge states exhibit a similar pattern of isotopically resolved peaks. Table 1 lists the peaks, range of charge states, average molecular weight, and molecular weight difference for the four principal ions observed for the 15+ charge state. Up to three iron atoms are observed to be attached to the subunit of nondenatured rubrerythrin.

Table 1: Interpretation of the four peaks observed in each charge state of rubrerythrin FTMS under nondenaturing conditions.

peaks	Charge states	Average Molecular weight (z * m/z-z)	Mass difference from aposubunit
aposubunit	20-11	21545.4	0
apo + 1 Fe (III)	16-12	21598.7	53.3
apo + 2 Fe (III)	18-11	21651.3	105.9
apo + 3 Fe (III) water	16-12	21721.4	176.0

The lowest mass peaks in the grouping correspond to the aposubunit, and has an average mass of ions of 21545.7 Da and a monoisotopic mass of 21530.3 Da, with a difference of 22 ppm from the value predicted based on the known elemental composition. The second peak is assigned as the aposubunit plus one iron (III) with an average mass of 21598.7 Da. The mass difference of 53.3 Da from the aposubunit is equal to the mass difference between iron and three protons. As observed for hemerythrin, each iron atom replaces three protons, suggesting that the oxidation state of iron is 3+ in rubrerythrin. The third peak is assigned as the subunit plus two iron (III) with an average mass of 21651.3 Da and a mass difference of 105.9 Da from the aposubunit. The fourth peak can be interpreted as the aposubunit plus three iron (III) plus water with an average mass of 21721.4 Da and a mass difference of 176.0 Da from the aposubunit. A small peak is observed at m/z 1447.8 appears to be 53 mass unit higher than the aposubunit plus two iron (III). This peak indicate that the third iron atom is attached to the aposubunit. As with hemerythrin, rubrerythrin was reconstituted with iron (II) ions but the iron (II) is converted to iron (III) when the protein solution is exposed to air.

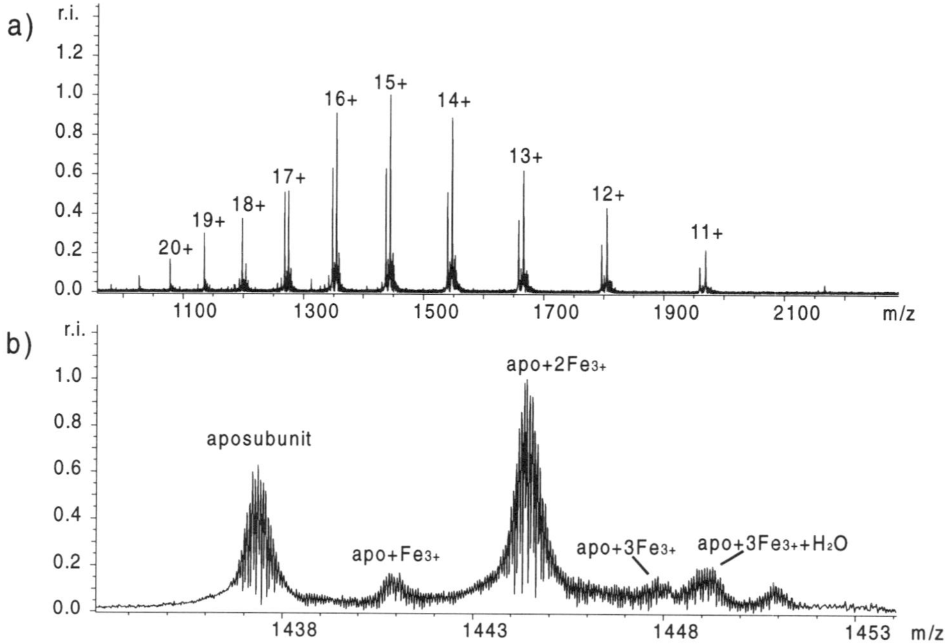

Figure 2: a) ESI-FTMS of rubrerythrin in pure water, obtained at a flow rate of 516 nL/min. b) Expansion of 15+ charge state exhibiting peaks corresponding to the aposubunit, +1 Fe (III), +2 Fe (III), and +3 Fe (III) + water.

Rubrerythrin solutions containing various concentrations of methanol were examined, to determine the effect of organic solvent on the denaturation of this metalloprotein. Even with methanol concentrations as high as 24%, metal binding is observed, as in Figure 3. The percentage of methanol was reduced from 49 % to 24% and the charge distribution was shifted to higher m/z. The charge state distribution ranged from 26+ to 15+ with 22+ as the most abundant with 49% methanol and for 24% of methanol the charge state distribution ranged from 25+ to 12+ with 19+ charge state as the most abundant, indicative of a more folded structure. Rubrerythrin solutions containing 10%, 6% and 4% methanol concentration have also been examined, and the results obtained are the same as those for 24% methanol. Our future experiments will utilize nanospray of hemerythrin and rubrerythrin with the goal of observing peaks at higher m/z, in order to examine the multimeric holoproteins by FTICR.

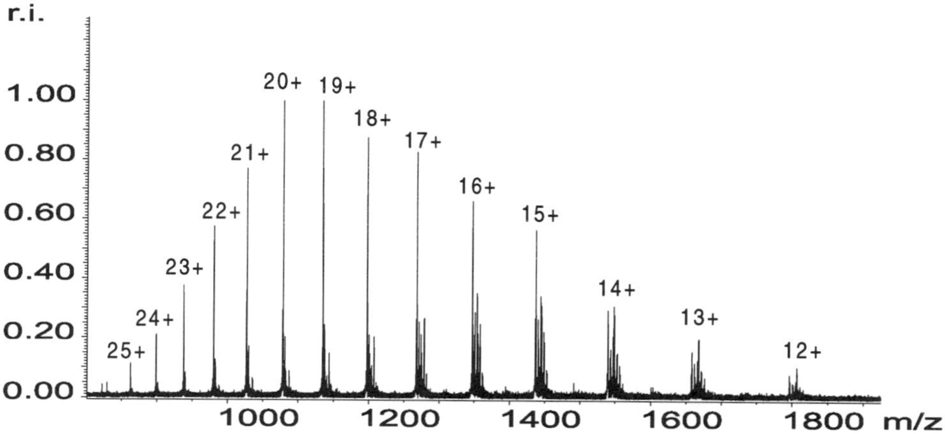

Figure 3: ESI-FTMS of rubrerythrin under denaturing conditions (in 24:74:2 methanol/water/acetic acid mixture) with flow rate = 250 nL/min, showing 19+ charge state as the most abundant state.

Acknowledgments

We gratefully acknowledge financial support from the National Science Foundation (BIR-9413918 and CHE-9412334) and from Bruker Analytical Instruments (IJA), and from the National Institute of Health, GM40388 (DMK).

References

1. Stenkamp, R. E. (1994) *Chem. Rev.* **94**, 715.
2. Moura, I., Tavares, P., and Ravi, N. (1994) *Meth. Enzymol.* **243**, 216.
3. deMarc', F., Kurtz, D. M., Jr., and Nordlund, P. (1996) *Nature Struct. Biol.* **3**, 539.
4. Bonomi, F., Kurtz, D. M., Jr., and Cui, X. (1996) *J. Biol. Inorg. Chem.* **1**, 69.
5. Lei, Q. P., Amster, I. J., Chernushevich, I., Standing, K., (1996) *The 44th ASMS Conference*, MOA, 9:50.
6. Fenn, J. B., Mann, M., Meng, C. K., Wong, S. F., and Whitehouse, C. M. (1989) *Science*, **64**, 246.
7. Covey, T., Bonner, R. F., Shushan, B. I., and Henion, J. (1988) *Rapid Commun. Mass Spectrum.* **2**, 289.
8. Smith, R. D., Loo, J. A., Edmonds, C. C., Barinaga, C. J., and Udseth, H. R. (1990) *Anal. Chem.* **62**, 882.
9. Henry, K. D., Williams, E. R., Wang, B. H., McLafferty, F. W., Shabanowitz, J., and Hunt, D. F. (1989) *Proc. Natl. Acad. Sci.* **86**, 9075.
10. Winger, B. E., Hein, R. E., Becker, B. L., and Campana, J. E. (1994) *Rapid Commun. Mass Spectrum.* **8**, 495.
11. Chowdhury, S. K., Katta, V., and Chait, B. T. (1990) *Rapid Commun. Mass Spectrum.* 3, 81.
12. Chowdhury, S. K., Katta, V., and Chait, B. T. (1990) *J. Am. Chem. Soc.* **112**, 9012.

13. Fitzgerald, M. C., Chernushevich, I., Standing, K. G., Whitman, C. P., and Kent, S. B. H., (1996) *Proc. Natl. Acad Sci.* **93**, 6851.
14. Tang, X. J., Brewer, C. F., Saha, S., Chernushevich, I., Ens, W., and Standing, K. G., (1994) *Rapid Commun. Mass Spectrum.* **8**, 750.
15. Verentchikov, A. N., Ens, W., and Standing, K. G., (1994) *Anal.Chem.* **66**, 126.
16. Fitzgerald, M. C., Chernushevich, I., Standing, K. G., Kent, S. B. H., and Whitman, C. P., (1995) *J. Am. Chem. Soc.* **117**, 11075.
17. Lei, Q. P., Cui, X., Kurtz, D. M., Amster, I. J., Chernushevich, I., and Standing, K., *J. Am. Chem. Soc.* (submitted).
18. Senko, M. W., Beu, S. C., and McLafferty, F. W. (1995) *J. Am. Soc. Mass Spectrum.* **6**, 229.
19. Hofstadler, S. A., Beu, S. C., and Laude, D. A. (1993) *Anal.Chem.* **65**, 312.
20. Chen, R., Cheng, X., Mitchell, D. W., Hofstadler, S. A., Wu, Q., Rockwood, A. L., Sherman, M. G., and Smith, R. D., (1995) *Anal. Chem.* **67**, 1159.
21. Yao, J., Dey, M., Pastoe, S. J., and Wilkins, C. L. (1995) *Anal. Chem.* **67**, 3638.
22. Valaskovic, G. A., Kelleher, N. L., Little, D. P., Aaserud, D. J., and McLafferty, F. W. (1995) *Anal. Chem.* **67**, 3802.
23. Gupta, N., Bonomi, F., Kurtz, D. M., Jr., Ravi, N., Wang, D. L., and Huynh, B. H. (1995) *Biochemistry* **34**, 3310.
24. Kurtz, D. M., Jr., (1996) "Mechanisms of Metallocenter Assembly" Hausinger, R. P., Eichhorn, G. L., and Marzilli, L. P., Eds., VCH, New York, pp 19-40.

ON THE DETERMINATION OF THE ION INTERNAL ENERGY IN AN ELECTROSPRAY SOURCE

MOONEN F. , COLLETTE C. , DE PAUW E.
Mass spectrometry laboratory (B6)
University of Liège, Sart-Tilman,
B-4000 Liège, Belgium

Abstract

Electrospray ionisation (ESI) is a "soft" ionisation method which produces ions from solution by nebulisation in an electric field . It allows the direct analysis of samples in solution with a mixture of usual solvents (H_2O, CH_3OH, CH_3CN), and has rapidly become an essential method for the analysis of biological compounds (peptides, proteins, organometalics, organic salts).

The widespread use of ESI would benefit from a database of ESI-spectra. This requires the harmonisation of methods for recording spectra and presupposes an understanding of phenomena underlying the method. Because the internal energy of the ions governs their fragmentation and therefore the mass spectral pattern, we developed a method for its determination. In this study, the fragmentation of a set of probe molecules (benzylpyridinium salts) was studied to determine the internal energy distribution of ions emitted by the electrospray source under various conditions (the voltage and the collision gas). The correlation of the internal energy with the experimental conditions will allow calibration of individual working conditions and therefore comparison of results obtained on different spectrometers.

1. Introduction

In the last few years, mass spectrometry has assumed an increasing role in analytical chemistry and biochemistry. It allows the determination of the molecular mass of analytes and provides structural information from measurements of their fragmentation. The choice of an appropriate ionisation method is based on the criterion of the sample volatility and on the nature of the information looked for. The development of soft-ionisation methods allowed the extension of mass spectrometry to

W. Ens et al. (eds.), New Methods for the Study of Biomolecular Complexes, 157–169.

thermally labile, non-volatile compounds. Electrospray ionisation (ESI) is a soft-ionisation method which produces ions from solution by nebulisation in an electric field [1] . It allows the direct analysis of samples in solution with a mixture of usual solvents (H_2O, CH_3OH, CH_3CN). Moreover, the ionisation conditions are soft enough to allow weakly-bonded supramolecular species to be observed [2 - 4]. This brings to mass spectrometry the ability to monitor noncovalent interactions in solution.

A central question however remains to be answered: does the nature of the ions in the gas phase reflect the solution composition [5]? This will be the case if the magnitude of the internal energy of the emitted ions is the same as that of ions in solution. It is essential to establish the internal energy content of ESI ions under different experimental conditions to allow the production of ions with a known internal energy content. This will not only allow optimisation of the conditions to observe labile species, but it will also allow a comparison of spectra obtained on different spectrometers. This could lead to the standardisation of experimental conditions and the development of a mass spectral data base.

According to the theories proposed to explain the ion emission in electrospray [6 - 7], it is generally agreed that the ionisation process does not require any transfer of vibrational energy to the sample, but resembles a particular electrolysis [8]. The ions are however emitted in a rather high pressure zone (1 mbar) where they can undergo collisions before they enter the mass analyser. They may acquire enough internal energy to fragment.

In this study, the fragmentation of a set of probe molecules (benzylpyridinium salts) was studied to determine the internal energy distribution of ions emitted by the electrospray source under various conditions involving different skimmer-cone voltages and different gases in the source.

2. Experimental

SOLUTION PREPARATION.

The organic salts were prepared by condensation of the substitued benzyl halide on pyridine. For the substituents on the pyridine ring, dried ether was used as solvent. The salts precipitated after a few hours of reflux heating and were recrystallized in ether. Their purity was checked by the fusion point test.

ELECTROSPRAY MASS SPECTROMETRY.

A quadrupole electrospray mass spectrometer (VG platform, FISONS) was used for all experiments. The electrospray ion source (Figure 1) has been described previously [9]. Samples were dissolved in a mixture of water/acetonitrile (50/50) and were introduced into the ionisation chamber (at atmospheric pressure) through a steel capillary (needle) with a flux of 0.04 ml/min. A gas flux (N_2) acts as a nebulising gas (15 l/h). A voltage difference (3 kV) is applied between the needle and the counter electrode . A drying gas (N_2) with a flow rate of 250 l/h, also enters in the source. The voltages applied to the sampling cone vary between 15 and 35V.

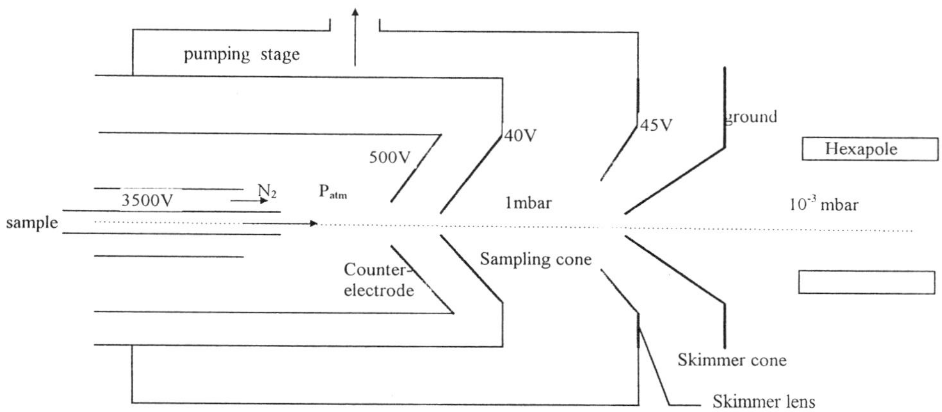

Figure 1. Schematic diagram of the electrospray source with the different pressure zone and the different voltages.

3. Results

3.1 REFERENCE MOLECULES.

To determine the internal energy distribution of ions emitted by the electrospray source, the ESI spectra of a set of probe molecules (benzylpyridinium salts) were recorded under various conditions . A similar method was developed to calculate the internal energy distribution of ions emitted by "soft" ionisation techniques such the LSIMS and Laser Desorption (10-11). The family of molecules was chosen as being organic salts ionised in solution. The fragmentation scheme is simple (Fig. 2), the C-N bond being the first reaction coordinate leading to fragmentation.

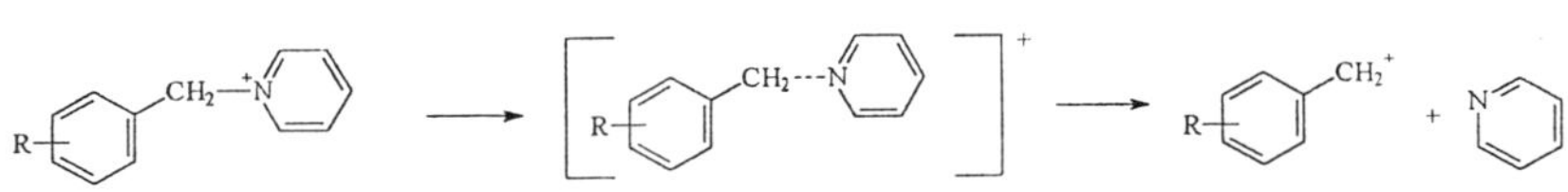

Figure 2. Reaction of benzyl-pyridine dissociation.

An increase of the voltage on the sampling cone leads to fragmentation corresponding to the breaking of the bond between the benzyl radical and the pyridine radical. At higher energies, fragmentation takes place at the level of the substituents and the cycle. In this paper we will limit our investigation to the first dissociation along the C-N bond.

3.2 MODEL

Fragmentation will occur when the internal energy acquired by the molecular ions M^+ is sufficient so that they cross that barrier (*Eb*) corresponding to the reaction ($M^+ \rightarrow F^+ + N$). A survival fraction of ions T is introduced, which gives the ratio of the number of molecular ions observed in the spectrum to the number of intact molecular ions leaving the source.

$$T = \frac{I(M^+)}{I(F^+) + I(M^+)}$$

In the set of benzylpyridinium salts, each different substituent corresponds to a known energy barrier *Eb* (12).

TABLE 1. Summary of the benzylpiridinium salts and the corresponding energy barrier (*Eb*) .

R	Eb (eV)
p-OCH_3	1.3
p-CH_3	1.6
o-CH_3	1.64
m-CH_3	1.65
m-OCH_3	1.68
p-Cl	1.73
p-F	1.8
p-NO_2	2.1

The correlation between the survival yield and the energy barrier will be used to get access to the internal energy distribution on the vibration mode of R(C-N) according to the RRKM theory . To achieve that purpose, we assume (i) that all ions are produced in the ground state, being already in ionic form in solution; (ii) that all types of ions undergo the same energy transfer process, being of quasi-equivalent structure; and (iii) that their masses are similar so that their center-of-mass references will be similar.

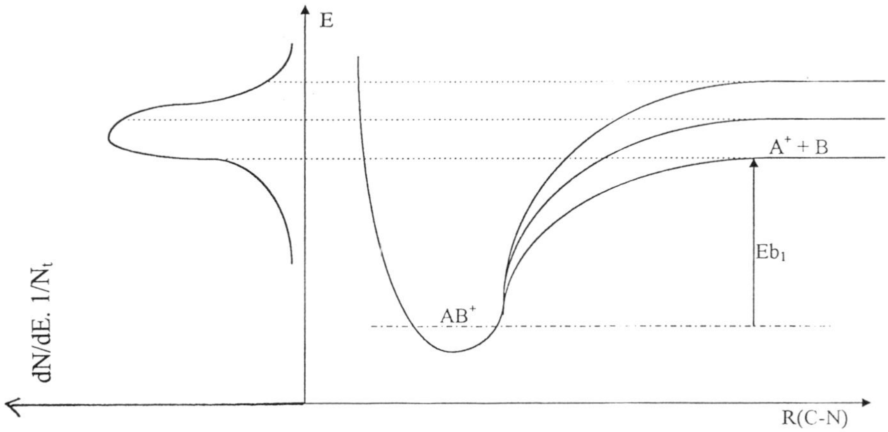

Figure 3. Qualitative representation of the model used to evaluate the internal energy in ESI.

3.3 ION INTERNAL ENERGY DISTRIBUTION AT FIXED VOLTAGE.

The full set of probe molecules was analysed at a fixed voltage, using nitrogen gas (N_2). The survival rate *T* for different probe molecules was calculated from the spectra taken at 15V. An example spectrum is shown in Fig. 4, and the survival rates are shown in Table 2.

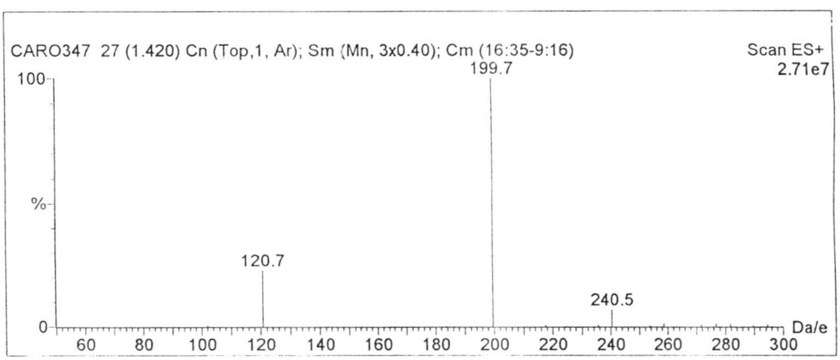

Figure 4. Spectrum of p-OCH_3 Bz-Pyr compound at 15V.

TABLE 2. The survival rate for the different benzylpyridinium salts at 15 V, and the corresponding energy barrier.

Substances	Eb (eV)	N_2 15V
p-OCH_3	1.3	0.81
p-CH_3	1.6	0.98
o-CH_3	1.64	0.99
m-CH_3	1.65	0.995
m-OCH_3	1.68	1
p-Cl	1.73	1
p-F	1.8	1
p-NO_2	2.1	1

The survival rates are plotted against the energy barrier values in Fig. 5. The survival rates for a fixed voltage show a dependence on the internal energy of the ions; a sigmoid is obtained. The larger the barrier, the greater the survival rate. The typical sigmoid equation is:

$$f(X) = \frac{A \exp(-B/X)}{1+A \exp(-B/X)}$$

In the present application, $X = Eb$ and $f(X) = N/N_T$.

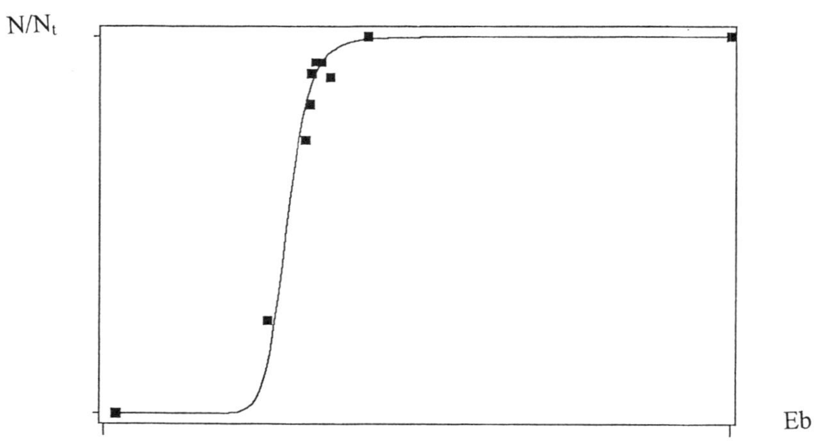

Figure 5. Sigmoid determined from the survival rate as a function of *Eb* at fixed voltage

The y-axis of the sigmoid corresponds to the fraction of the ions (N/N_t) having their internal energy lower than *Eb* and consequently remaining intact. The sigmoid is therefore an integral curve. Its derivative becomes a distribution function of the ions crossing the ESI source.

$$f'(X) = \frac{AB\exp(-B/X)}{X^2(1+A\exp(-B/X))^2} \quad ; \quad f'(X) = \frac{dN/dE}{N_T}$$

A non-symmetrical curve is obtained. Each point of the sigmoid corresponds to the integral under the distribution function from zero to the energy barrier.

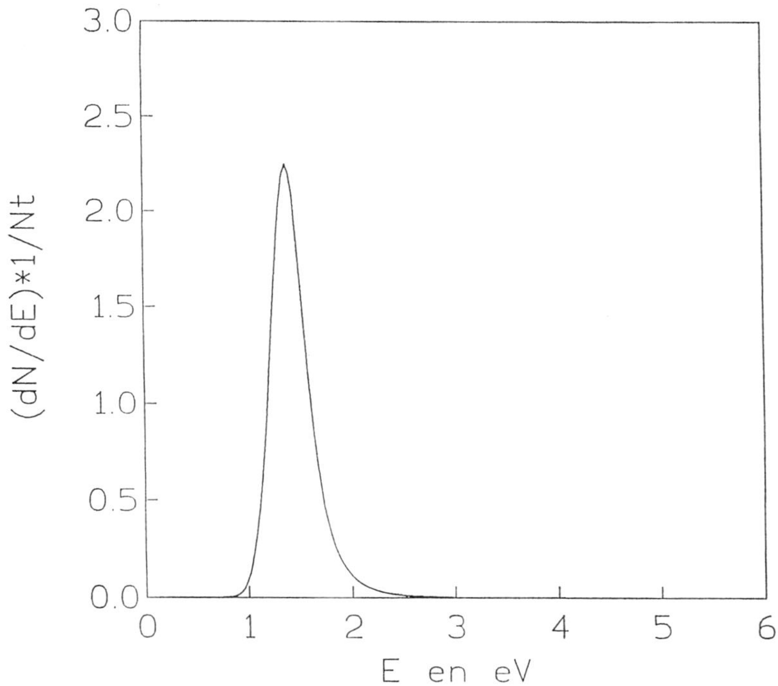

Figure 6. Energy distribution at 15 V.

3.4 EVOLUTION OF THE ION INTERNAL ENERGY DISTRIBUTION AS A FUNCTION OF DIFFERENT VOLTAGES ON THE SAMPLING CONE IN THE ELECTROSPRAY SOURCE.

The ESI spectra were recorded at different voltages (15, 25, 35 V) for the whole set of probe molecules. A typical spectrum of p-OCH_3 Bz-Pyr at 15 and 25 Volts is represented at Fig. 7. The survival rates *T* for the various benzylpyridinium salts at different voltages are shown in Table 3.

15V

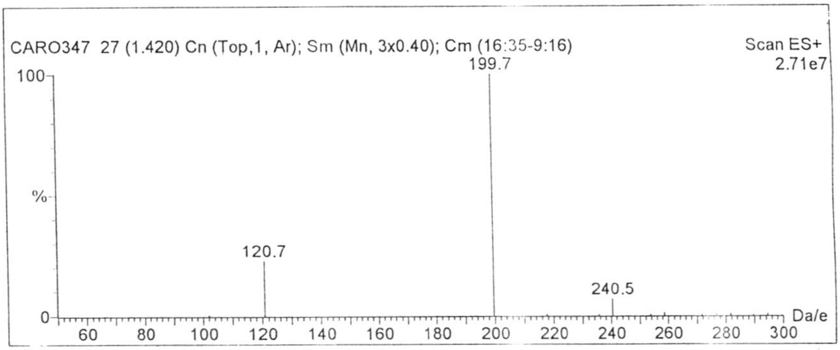

25 V

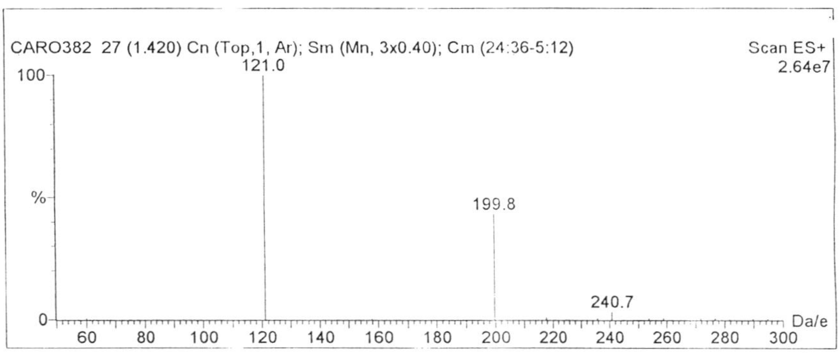

Figure 7. Mass pectra of p-OCH_3 Bz- Pyr at 15V and 25V.

TABLE 3. Summary of the benzylpyridinium salts by increasing order of barrier energy and the values of the survival fraction at different voltages.

Substances	Eb (eV)	N_2 15V	N_2 25V	N_2 35V
p-OCH_3	1.3	0.81	0.37	0.12
p-CH_3	1.6	0.98	0.66	0.13
o-CH_3	1.64	0.99	0.7	0.15
m-CH_3	1.65	0.995	0.76	0.16
m-OCH_3	1.68	1	0.77	0.18
p-Cl	1.73	1	0.77	0.19
p-F	1.8	1	0.75	0.21
p-NO_2	2.1	1	0.98	0.43

The following observations can be made (Table 3 - Fig. 8):

1) As expected the survival rate depends on the applied voltage at the sampling cone. For 15, 25, 35 V the survival rate increases for all the substituent types (Table 3).
2) The energy distribution broadens when the voltage increases. This can be explained by the fact that when the voltage increases the ions acquire more kinetic energy. This kinetic energy is transformed in internal energy upon collision. As the values of the internal energy accessible to the ions increase with the voltage, the energy scale accessible by the ions extends and the internal energy distribution broadens (Fig. 8).
3) The maximum of the curve shifts towards higher energies (Fig. 8). It follows the maximum energy transferable, which corresponds to the centre-of-mass energy and is expressed by the equation:

$$E = 1/2 \frac{m_1 m_2}{m_1 + m_2} v_{rel}^2$$

where $v_{rel} = v_2 - v_1$; m_1 and v_1 represent the mass and speed of the molecular ion; and m_2 and v_2 represent the mass and speed of the gas molecules.

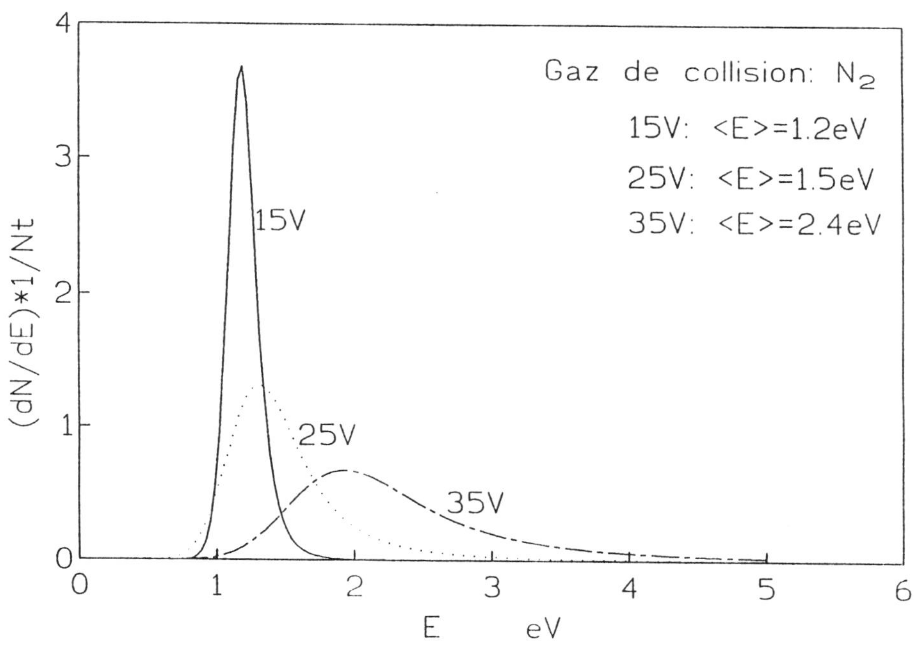

Figure 8. Evolution of the internal energy distribution as a function of the voltage on the sampling cone.

The energy distribution curves do not start at zero. That can be explained by the fact that few experimental points are accessible at low energy, the lowest energy barrier available being 1.3 eV. The mean energy values <*E*> represented in Fig. 8 are obtained by the equation:

$$<E> = \frac{\int f(E)\, E\, \mathrm{d}E}{\int f(E)\, \mathrm{d}E}$$

3.5 COMPARISON BETWEEN COLLISION GASES (N_2, CO_2, AR, SF_6).

ESI spectra were obtained using as collision gases N_2, CO_2, Ar, SF_6. There are two criteria for the choice of the different collision gases: the mass and the presence of vibrational modes of the collision gas. Example spectra are shown in Fig. 9 and the results are summarized in Table 5. The internal energy distributions were calculated at a fixed voltage for the various collision gases and are shown in Fig. 10.

N_2

CO_2

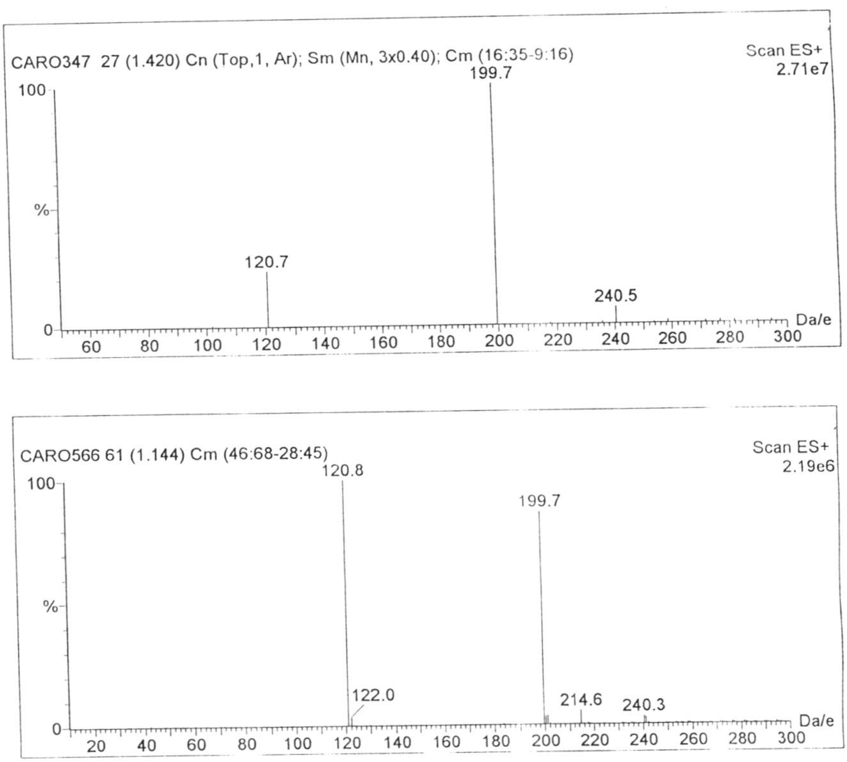

Figure 9. Typical spectra of p-OCH_3 Bz-Pyr compound at 15 volts with N_2, and CO_2, as collision gases.

Ar

SF$_6$

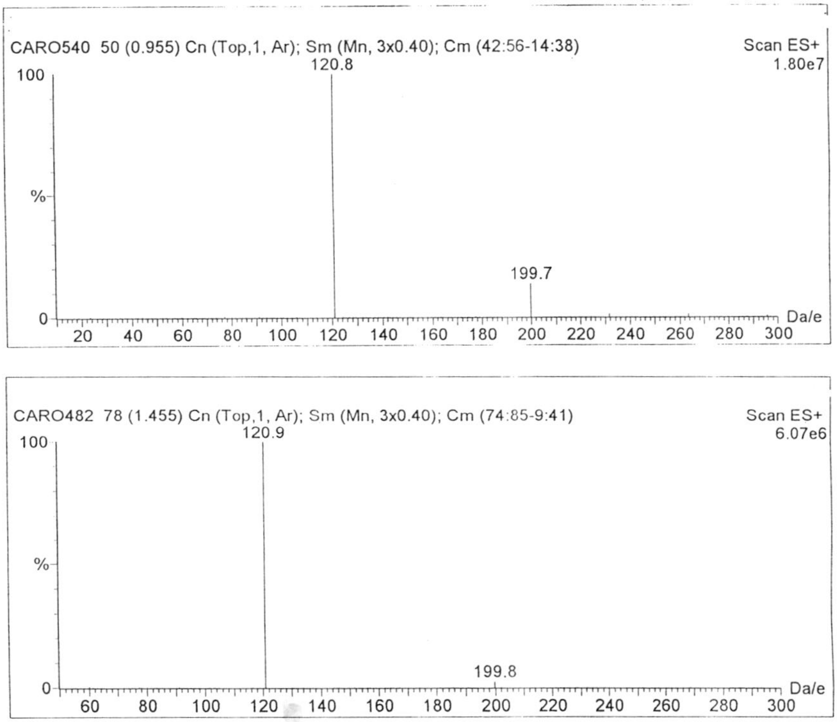

Figure 9 (cont'd). Typical spectrum of p-OCH$_3$ Bz-Pyr compound at 15 volts with Ar and SF$_6$ as collision gases.

TABLE 4. Survival yield as a function of the collision gas at different voltages.

Substances	Eb (eV)	N_2 15V	N_2 25V	N_2 35V	Ar 15V	Ar 25V	CO_2 15V	CO_2 25V	CO_2 35V	SF_6 4V	SF_6 15V
p-OCH$_3$	1.3	0.81	0.37	0.12	0.21	0.12	0.22	0.07	0.04	0.51	0.02
p-CH$_3$	1.6	0.98	0.66	0.13	0.47	0.28	0.71	0.16	0.07	0.9	0.06
o-CH$_3$	1.64	0.99	0.7	0.15	0.57	0.32	0.8	0.19	0.08	0.94	0.08
m-CH$_3$	1.65	0.995	0.76	0.16	0.66	0.33	0.91	0.24	0.1	0.97	0.13
m-OCH$_3$	1.68	1	0.77	0.18	0.63	0.3	0.91	0.23	0.09	0.98	0.13
p-Cl	1.73	1	0.77	0.19	0.61	0.36	0.9	0.28	0.11	0.98	0.14
p-F	1.8	1	0.75	0.21	0.61	0.34	0.87	0.24	0.13	0.96	0.11
p-NO$_2$	2.1	1	0.98	0.43	0.94	0.44	1	0.66	0.3	1	0.4

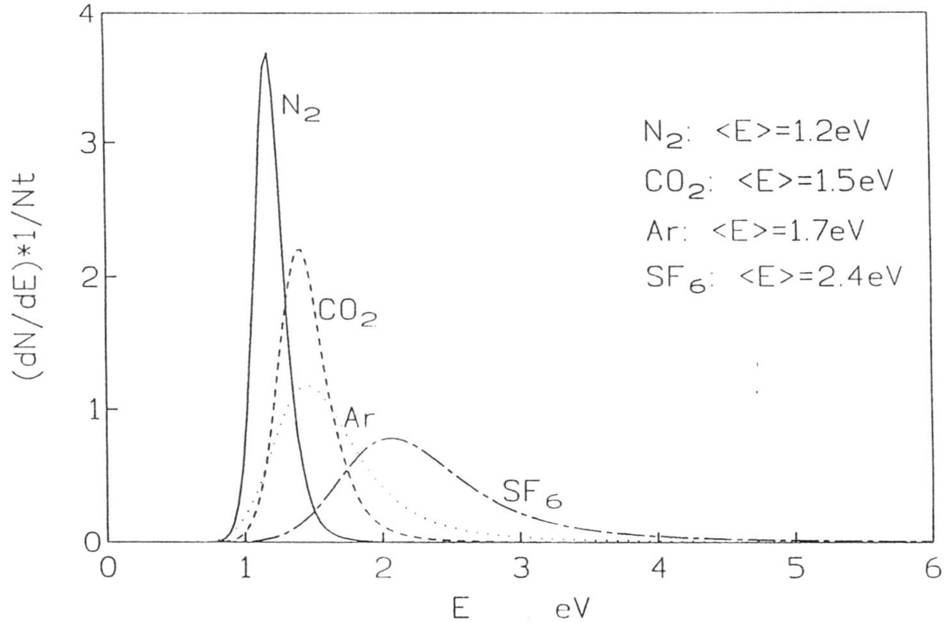

Figure 10. Evolution of the internal energy distribution as a function of collision gas at 15 Volts.

Influence of the mass of the collision gas.

The nature of the collision gas determines the center-of-mass conditions which define the maximum quantity of energy that can be transferred. The program SIMION was used to estimate the acceleration of the ions between the sampling cone and the skimmer cone for the limiting case with no collisions. This gives the maximum speeds of the ions and from these the maximum transferable energies E_{max} may be calculated. The data are summarised in Table 5, which shows that E_{max} increases with the mass of the collision gas. This is consistent with the observed distributions shown in Fig. 10: the distributions shift to higher energy as the mass of the collision gas (N_2, Ar-CO_2, SF_6) increases.

TABLE 5. Mass and maximum transferable energy for different collision gases.

	N_2	Ar	CO_2	SF_6
mass(u)	28	40	44	146
Max energy(eV)	4.08	5.3	5.7	12

The introduction of vibration modes in the collision gas.

A comparison of the energy distributions obtained with two collision gases of almost equal mass (Ar, CO_2) gives an insight on the effect of the presence of vibration modes in the collision gas (Fig. 10). With 15 V on the sampling cone, the curves obtained with the two collision gases (Ar, CO_2), present a maximum at the same energy but the distribution is narrower at the high energy side with CO_2 as compared to Ar. That can be tentatively explained by the fact that in the first case there is a loss of kinetic energy which is transferred to vibration energy of the CO_2.

4. Conclusions

The fragmentation of a set of target molecules (benzylpyridinium salts) has allowed us to establish the internal energy distributions of ions emitted from an electrospray source under various conditions. The correlation of the internal energy with the experimental conditions (voltage on the sampling cone, collision gas) will allow calibration of individual working conditions and therefore the comparison of results obtained on different spectrometers. This calibration of the experimental conditions can be the basis for the development of ESI-spectrum data bases.

5. References

1. Yamaschita M. and Fenn J. B. (1984), Electrospray Ion Source, *J. Phys. Chem.* ,88, 4451.
2. Gatlin C. L. , Turecek F. (1995), Gas Phase Complexes of Amino Acids with Cu (II) and Diimine Ligand, *J. Mass Spectrom.* , 30, 1617.
3. Smith R. D. ,Light-Wahl K. J. (1993), *Biol. Mass Spectrom.* 22, 493.
4. Li Y. T., Hsieh Y. L. , J. Henion D. J. , Senko M. S. , Mac Lafferty F. W. (1993), J. *Am. Soc. Mass Spectrom.* , 4, 557.
5. Canty A. L. , Colton R. , Thomas I. M. (1993), Electrospray Mass Spectrometry of Inert and Labile Metal Alkyl Compounds, *J. Organometallic Chem.* ,455, 283.
6. Kebarle P. and Tang L.(1993), Dependence of Ion Intensity in ESI on the Concentration of the Analytes in the Electrospray Solution, *Anal. Chem.* ,65, 3654.
7. Fenn J. B. (1989),ESI for Mass Spectrometry of Large Biomolecules, *Science*, 246, 64.
8. Kebarle P. and Tang L (1993),From Ions to the Gas Phase: The Mechanism of Electrospray Mass Spectrometry,. *Anal. Chem.* , 65, 972.
9. Whitehouse C. M. , Dreger R. N. ,Yamaschita Y. , Fenn J. B. (1985),Electrospray Interface for Liquid Chromatographs and Mass Spectrometers, *Anal. Chem.* , 57, 675.
10. De Pauw E. , Pelzer G. and Natalis P.(1986),Internal Energy Distribution of Ions Emitted in Secondary Ion Mass Spectrometry, *Ion Formation from Organics Solids, Springer Series in Physics*, Vol. 9, P. 103, Berlin,.
11. Derwa F. and De Pauw E.(1989),Evaluation of Internal Energy of Secondary Ions in LSIMS, *Spectroscopy : Am. Int. Journal*, 7, 227.
12. Katritzky A. R. , Watson C. H. ,Szafran Z. D. and Eyler J. D. (1990), Collisionally Activated Dissociation of N-Alkylpyridinium Cations to Pyridine and Alkyl Cations in the Gas Phase, *J. Am. Chem. Soc.*, 12, 2471.

OBSERVATION OF PROTEIN HEPTAMER FORMATION FOR CHEMICALLY LIGATED EARLY PREGNANCY FACTOR

S.G. LOVE.[1], D.R. ENGLEBRETSEN[1], B. GARNHAM[1],
A.C. CAVANAGH[2], H. MORTON[2], I.V. CHERNUSHEVICH[3],
K.G. STANDING[3] AND P.F. ALEWOOD[1]

[1]*Centre for Drug Design and Development and* †*Department of Surgery, University of Queensland, Brisbane 4072, Australia.*

[2]*Dept of Physics, University Manitoba, Winnipeg, Canada R3T 2N2*

1. Introduction

The ability to use peptide synthesis to produce small synthetic proteins containing approximately 100 amino acids has been demonstrated in the past with the syntheses of Ribonuclease-A, Interleukin-3, HIV-1 Protease, and Ubiquitin [1 - 4]. However these syntheses result in poor yields of the desired product which are difficult to purify because of the heterogeneity of the crude protein [5].

A new approach to the chemical synthesis of proteins, the chemical ligation strategy, has been used to synthesise proteins of very high purity such as HIV-1 PR, chemokine IL-8 and fibronectin type 3 module [6 - 8]. The ligation strategy enables the joining of two peptide segments which contain two mutually reactive chemoselective groups at their N- and C-termini. Reaction of these two segments results in ligation or joining of the two peptides usually via an amide bond surrogate (thioether [9, 10]), or an amide bond isostere (thioester [6]) or formation of a native amide bond [7].

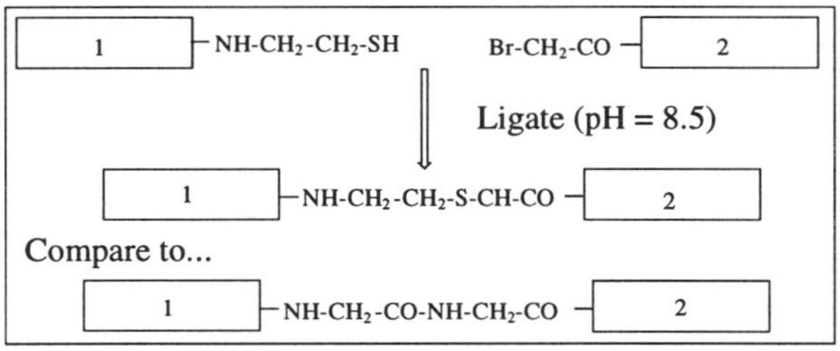

Figure 1. Example of peptide ligation of two peptides to give a thioether peptide bond surrogate.

W. Ens et al. (eds.), New Methods for the Study of Biomolecular Complexes, 171–179.

Here we show that incorporation of multiple thioether ligation sites into the protein early pregnancy factor (EPF) does not affect its activity or structure. This was assessed functionally by the rosette inhibition test and structurally by the ability of the ligated form of early pregnancy factor (*l* EPF), to form complexes of heptamers (Mr 75 kDa) under non-denaturing conditions, as shown by Electrospray Ionisation Time of Flight (ESI-TOF) mass spectrometry. The formation of heptamers is a function of this class of molecule and is consistent with the reported X-Ray crystal structures of two protein homologs, GroES and *Mycobacterium leprae* chaperonin 10 [11, 12].

Early Pregnancy Factor, was described originally as a substance appearing in maternal serum within 24 hours of fertilisation in all mammalian species examined [13]. EPF is indicative of, and necessary for, embryonic well-being and has immunosuppressive properties. EPF is also important as a growth regulator of transformed, neoplastic and normal cells where it is required for cell proliferation; its presence in platelets implies a role in inflammation and wound healing. This molecule possesses an exceptional combination of properties, having great biological significance and has diagnostic and therapeutic potential in many areas of human and veterinary medicine (Figure 2).

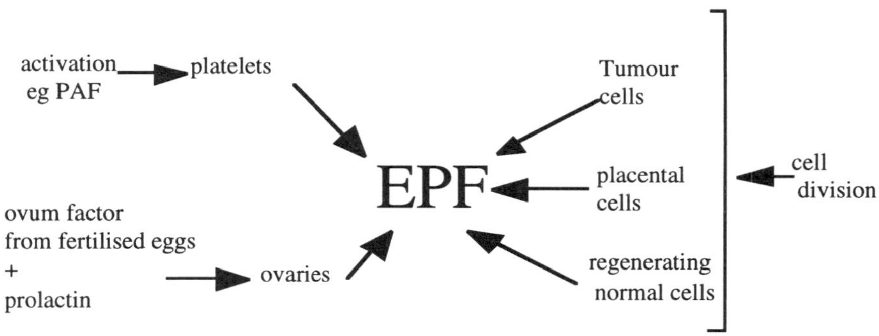

Figure 2. Tissues which produce EPF and stimulants which induce its production. EPF is produced from the sources (indicated on left of figure) in response to a specific releasing factor. For example, platelet activating factor (PAF) acts on platelets while ovum factor from fertilised eggs in cooperation with prolactin, acts on the ovary. For sources indicated on the right of figure, cell division appears to be a direct stimulus.

Recently, EPF was purified from outdated human platelets (45 µg of EPF was obtained from 300 units of platelets isolated from 150 L blood). From DNA cloning experiments EPF was found to be identical to rat liver mitochondrial chaperonin10 (cpn10) with the exception of a single residue [14, 15] (Figure 3). These results were surprising, in light of the known intracellular properties of cpn10. In mitochondria, bacteria and plastids, cpn10 (and its homologs) associate to form heptamers which participate with cpn60 (and its homologs) to form a complex structure which facilitates protein folding/refolding [16, 17].

```
   1         10        20        30        40        50
Ac-AGQAFRKFLPLFDRVLVERSAAETVTKGGIMLPEKSQGKVLQATVVAVGS

         60        70        80        90        101
GSKGKGGEIQPVSVKVGDKVLLPEYGGTKVVLDDKDYFLFRDGDILGKYVD
```

Figure 3. Sequence of EPF showing the position of the ligation sites (underlined).

Chemical Synthesis of EPF

Two forms of EPF were chemically synthesised and their structure and biological activity compared with native EPF.

Stepwise Solid Phase Synthesis of EPF

The first 30 amino acids of EPF were synthesised using manual Boc chemistry with *in-situ* neutralisation [18]. The resin was then transferred to an ABI 430A Peptide Synthesiser and the synthesis continued using standard Boc chemistry with preformed symmetrical anhydride protocols [5]. The synthesis was monitored throughout by ninhydrin analysis and by examination of the products obtained from HF cleavage of fragments 58-101 and 30-101 removed during the course of the synthesis. On completion of the chain assembly the peptide was acetylated at the N-terminal then cleaved by HF. The crude peptide was purified by preparative RP-HPLC then characterised by electrospray mass spectrometry and analytical RP-HPLC to give the final purified protein (Figure 4 and 5a). The reconstructed mass spectrum of EPF shows the presence of a des Gly impurity which co-elutes with the correct amino acid sequence.

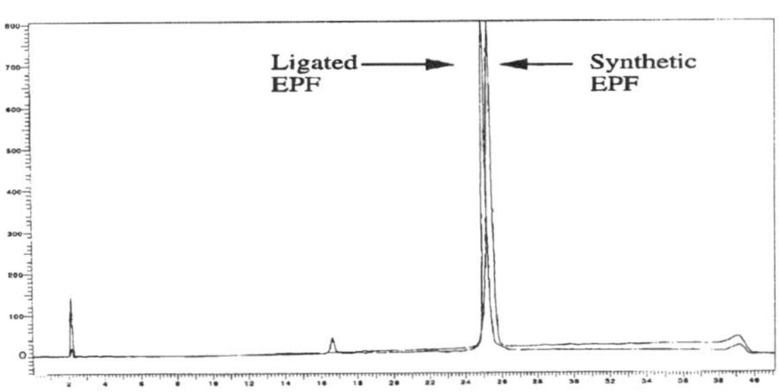

Figure 4. Analytical RPHPLC trace comparing the two synthetic forms of EPF. (10-60% acetonitrile 0.1% TFA 2% per min. Vydac C4 column $A^{214\,nm}$)

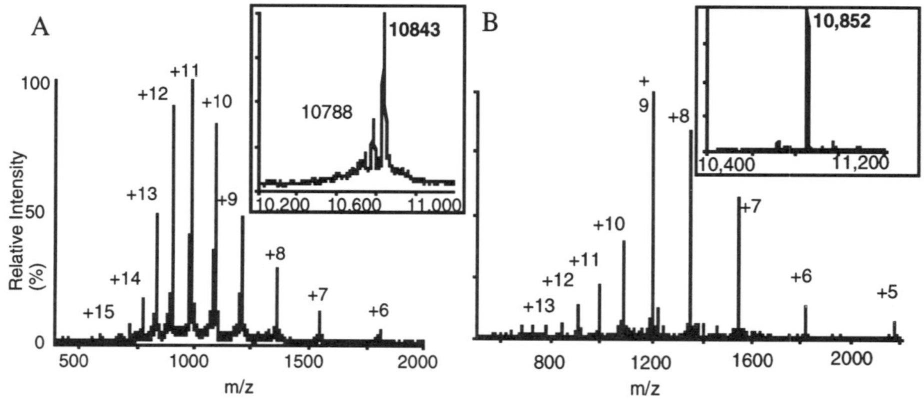

Figure 5. a): ESI mass spectrum showing the charge states for denatured synthetic EPF (calc. mass 10843 Da). b): ESI mass spectrum of denatured ligated EPF (calc. mass 10852 Da).

Synthesis of l EPF by Sequential Fragment Ligation

Four peptide fragments (acetyl-EPF(1-27)-NH-CH_2-CH_2-SH, Cl-acetyl-EPF(30-55)-NH-CH_2-CH_2-SH, Cl-acetyl-EPF(58-75)-NH-CH_2-CH_2-SH and Br-acetyl-EPF(78-101)-OH) encompassing the Gly-Gly ligation sites (Figure 3) were synthesised using manual Boc chemistry using HBTU activation and *in-situ* neutralisation protocols [18]. Fragments with the thiol function at the C-terminus were synthesised using a novel thioether linker resin (Figure 6) [10]. The crude peptides were purified by RP-HPLC then used in the sequential ligation procedure as outlined in Figure 7 to give the final purified protein (Figures 4 and 5b).

The two synthetic forms of EPF (stepwise and ligated) were then assayed for biological activity (the rosette inhibition test) and for their capacity to form heptamers. Heptameric forms of cpn10 in association with cpn-60 are required for protein folding within the cell [16, 17].

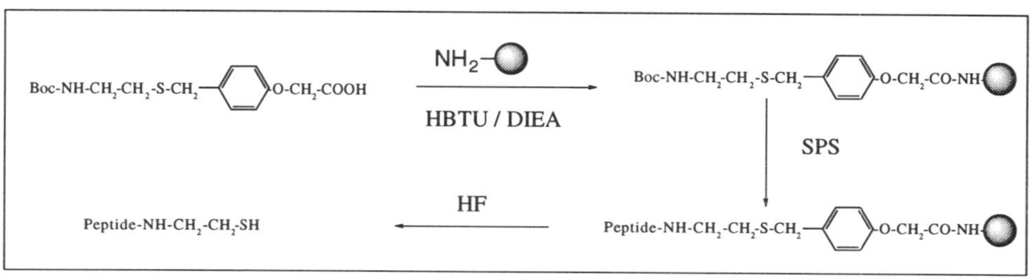

Figure 6. Synthesis of the thioether resin and subsequent peptide synthesis.

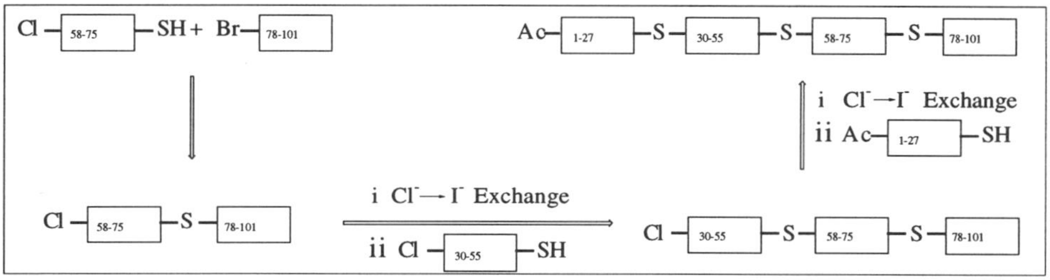

Figure 7. Synthesis of ligated EPF containing multiple thioether ligation sites.

The end point of the rosette inhibition test namely the formation of rosettes between mouse spleen cells and human red blood cells is shown in Figure 8 while the results from testing the various forms of cpn10 and EPF is listed in Table 1. Within the semi quantitative limits of the bioassay the synthetic EPF molecules display identical activity to human platlet isolated EPF and rat liver mitochondrial cpn10.

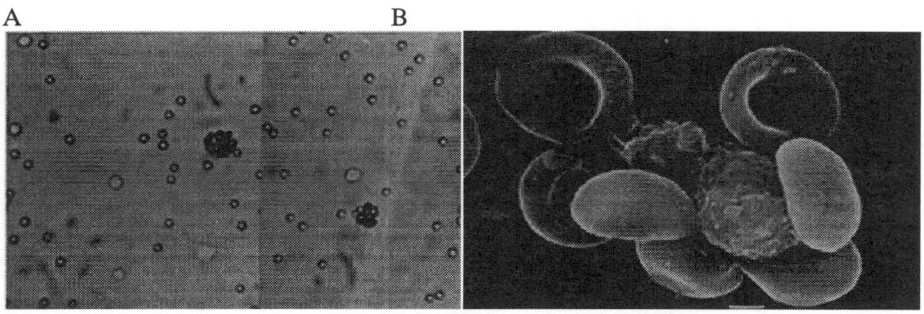

Figure 8. Rosette inhibition test. a) Photo showing the formation of rosettes formed between mouse spleen cells and human red blood cells at 125X magnification b) at 1.0μm.

While the above activities of the synthetic forms of EPF are consistent with them being correctly folded we undertook further experiments. Chemically cross linking followed by SDS PAGE analysis demonstrated that both synthetic EPF molecules have the capacity to form protein oligomers; illustrated by the ligated molecule in Figure 9.

Further analysis with Electrospray Ionisation Time-of-Flight Mass Spectrometry (ESI-TOF) [19, 20] allowed direct observation of heptamer formation of the ligated EPF under native conditions. The use of ESI-TOF at a low skimmer voltage (ΔU=100 V) showed that under non-denaturing conditions (a 10 μM solution in 5 mM ammonium acetate pH 6), the *l* EPF gave primarily heptamers with a mass of 75865.8 kDa (calculated mass 75864)(Figure 10a,b).

TABLE 1 Results from the rosette inhibition test. Samples (50μg/ml) were tested in the rosette inhibition test in 10 fold serial dilutions. Activity is expressed a limiting dose, the highest dilution of sample giving a positive response in the assay.

Sample	**Activity log^{-1}**
Human Platelet derived EPF	13
Rat Liver mitochondrial cpn10	13
Stepwise syn EPF	13
Ligated EPF	13
E.coli cpn10 (GroES)	no activity.

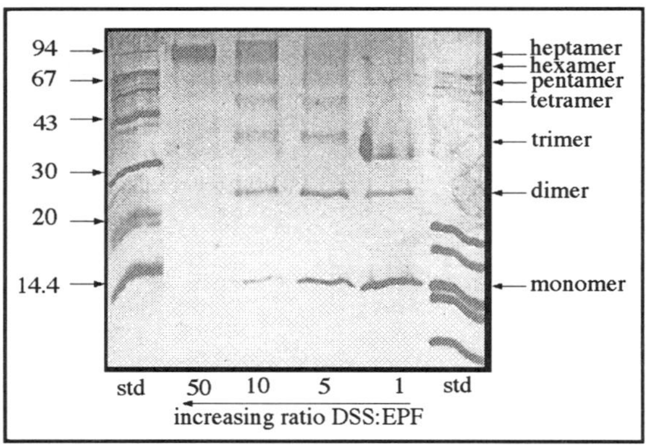

Figure 9. SDS PAGE gel showing the formation of heptamers through cross linking of synthetic ligated EPF using Disuccinimidyl sulphonate (DSS). The appropriate ratio of DSS was added to a solution of 100 μg/ml EPF in 0.1 M $NaHPO_4$ pH 7.4. After 30 min reaction was quenched with 0.5M Tris.HCl. The protein (5 μl)was then loaded on a 13% SDS PAGE gel and silver stained.

By increasing ΔU, fragmentation of the heptamer complex was observed giving rise to the formation of hexamers and monomers (Figure 10c), pentamers and trimers (data not shown). The synthetic non-ligated EPF molecule did not give clear spectra due to heterogeneity resulting from the glycine deletions (data not shown) but did indicate that heptamer formation had also occurred.

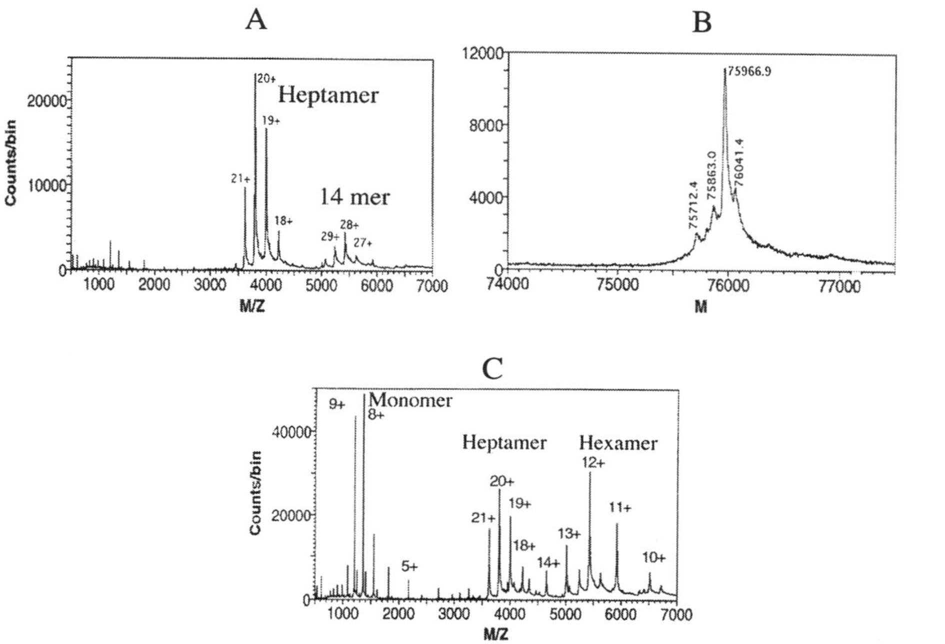

Figure 10. **A.** ESI-TOF mass spectrometry of the ligated form of EPF recorded at a nozzel-skimmer voltage $\Delta U = 100$ V; **B** Deconvolution made from charge states z=18-21 of heptamer shown in A. **C** Same as A but recorded at $\Delta U = 125$ showing monomers at m/z 500-2200, heptamers at 3500-4500 and hexamers at 4500-6500.

Conclusion

The use of chemical ligation has enabled us to rapidly synthesise EPF in sufficent purity for biological and structural studies. ESI-TOF analysis has confirmed that both synthetic forms of EPF have folded correctly. This is the first time a member of the chaperonin 10 class of molecules has been reported to form heptamers through direct observation by ESI-TOF mass spectrometry .

Materials and Methods

Chemicals and Reagents

TFA, DCM, DMF,DIEA and di-tertbutyl pyrocarbonate were from Auspep (Parkville, Melbourne, Australia). HBTU was from Richelieu Biotechnologies (St. Hyacinth, Quebec, Canada). Sigma (St. Louis, MO, USA) supplied hydroxymethyl phenoxyacetic acid. Diisopropylcarbodiimide, 2 aminoethanethiol and bromoacetic acid were from Aldrich (Milwaukee, WI, USA). BDH Chemicals Ltd. (Poole, UK) supplied iodoacetic acid; chloroacetic acid was from Ajax Chemicals (Sydney, Australia). Boc-Asp PAM resins (PAM: phenylacetamidomethyl) was from ABI Inc. (Foster City, CA, USA). Aminomethyl polystyrene resin, 100-200 mesh, 1% DVB crosslinked, amine substitution 0.83 mmole/g and Boc-L-amino acids were from the Peptide Institute (Osaka, Japan) or Nova Biochem (CA.) . Boc-amino acid side chain protecting groups were Arg(Tos), Asn (Xan), Asp and Glu (OChx), Lys(ClZ), Ser and Thr (Bzl), Tyr(2BrZ).

Equipment

Analytical HPLC was carried out using a 4.6 X 150 mm Vydac C4 column (The Separations Group, Hesperia, CA, USA) with a solvent gradient run from 10-70% B over 30 minutes at 1 ml/min (A: 0.1% TFA; B: 0.1% TFA, 10% water, 90% acetonitrile). A Millipore Waters (Milford, MA, USA) HPLC system comprised of a model 600 multisolvent delivery system, 600E controller and model 484 detector was used for the analyses. Preparative HPLC was carried out using a 22 X 250 mm Vydac C4 column with a solvent gradient of 10-70% B over 60 min at 10 ml/min. Mass spectral data were collected using a Perkin Elmer Sciex (Toronto, Canada) API III Biomolecular Mass Analyser ion-spray mass spectrometer equipped with an ABI 140B solvent delivery system. Raw data were analysed using the program MassSpec (Perkin Elmer Sciex). ESI-TOF spectra were recorded at the University of Manitoba using the instrument described elsewhere [19].

Acknowledgements

A.C.C , H.M and DE was supported by NHMRC and CSL Ltd Australia. K.G.S. was supported by NSERC, Canada

References

1 Gutte, B.; Merrifield, R.B., (1971), The Synthesis of Ribonuclease A., *J.Biol.Chem.*, **246,** 1922-1941.

2. Clarke-Lewis, I.; Abersold, R.; Ziltner, H.; Schrader, J.W.; Hood, L.E.; Kent, S.B.H., (1988), Automated Chemical Synthesis of a Protein Growth Factor for Hemopoietic Cells, Interleukin 3., *Science*, **231,** 134-139.

3. Schneider, J.; Kent, S.B.H., (1988), Enzymatic Activity of a synthetic 99 Residue Protein Corresponding to the Putative HIV-1 Protease, *Cell*, **54,** 3128-3134.

4. Ramage, R.; Green, J.; Ojunjobi, O.M., (1987), Solid Phase Synthesis of Ubiquitin, *Tetrahedron Lett.*, **30,** 2149-2153.

5. Kent, S.B.H.; Alewood, D.; Alewood, P.; Baca, M.; Jones, A.; Schnolzer, M., (1992)*Total chemical Synthesis of Proteins: Evolution of Solid Phase Synthetic Methods Illustrated by Total Chemical Synthesis of the HIV-1 Protease.*; in Solid Phase Synthesis: Peptides, Polypeptides and Polynucleotides, Epton. R. Ed.; Intercept Limited: Andover, 1992, pp 1-22.
6. Schnolzer, M.; Kent, S.B.H., (1992), Constructing Proteins by Dovetailing Unprotected Synthetic Peptides: Backbone-Engineered HIV Protease., *Science*, **256,** 221-225.
7. Dawson, P.E.; Muir, T.W.; Clark-Lewis, I.; Kent, S.B.H., (1994), Synthesis of Proteins by Native Chemical Ligation, *Science*, **266,** 776-779.
8. Williams, M.J.; Muir, T.W.; Ginsberg, M.H.; Kent, S.B.H., (1994), Total Chemical Synthesis of a Folded Beta-Sandwich Protein Domain - An Analog of the Tenth Fibronectin Type 3 Module, *J.Amer. Chem.Soc.*, **116,** 10797-10798.
9. Muir, T.W.; Williams, M.J.; Ginsberg, M.H.; Kent, S.B.H., (1994), Design and Chemical Synthesis of a Neoprotein Structural Model for the Cytoplasmic Domain of a Multisubunit Cell-Surface Receptor- Integrin Alpha(IIb)Beta(3) (Platelet GPIIβ-IIIα), *Biochemistry*, **33,** 7701-7708.
10. Englebretsen, D.R.; Garnham, B.G.; Bergman, D.A.; Alewood, P.F., (1995), A Novel Thioether Linker - Chemical Synthesis of a HIV-1 Protease Analogue by Thioether Ligation, *Tetrahedron Lett.*, **36,** 8871-8874.
11. Hunt, J.F.; Weaver, A.J.; Landry, S.J.; Gierasch, L.; Deisenhofer, J., (1996), The Crystal Structure of the GroES Co-Chaperonin At 2.8-Angstrom Resolution, *Nature*, **379,** 37-45.
12. Mande, S.C.; Mehra, V.; Bloom, B.R., (1996), Structure of the Heat Shock Protein Chaperonin-10 of *Mycobacterium Leprae*., *Science*, **271,** 203-207.
13. Morton, H.; Cavanagh, A.C.; Athanasas-Platsis, S.; Quin, K.A.; Rolfe, B.E., (1992), Early Pregnancy Factor has Immunosupressive and Growth Factor Properties., *Reproduction Fertility and Development.*, **4,** 411-422.
14. Cavanagh, A.C., Morton, H., (1994), The Purification of Early-Pregnancy Factor to Homogeneity from Human Platelets and Identification as Chaperonin 10, *Euro.J. Biochem.* **222,** 551-560.
15. Summer, K.M., Murphy, R.M., Webb, G.G., Peters, G.B., Morton, H., Cassady, A.I. Cavanagh, A.C., (1996), *Biomedical and Molecular Medicine*, **58** 52-58.
16. Hartl, F. U.; Martin, J., (1995), Molecular Chaperones in Cellular Protein Folding., *Current Opinion in Structural Biology.*, **5,** 92-102.
17. Saibil, H., (1996), The Lid that Shapes the Pot: Structure and Function of the Chaperonin GroES, *Structure*, **4,** 1-4.
18. Schnolzer, M.; Alewood, P.; Alewood, D.; Jones, A.; Kent, S.B.H., (1992), In situ Neutralisation in Boc-Chemistry Solid Phase Peptide Synthesis., *Int.J. Pep. Prot. Res.*, **40,** 180-193.
19. Verentchikov, A.N., Ens, W., Standing, K.G., (1994), Reflecting Time-of-Flight Mass Spectrometer with an Electrospray Ion Source and Orthogonal Extraction, *Anal. Chem.* **66**, 126-133.
20. Tang, X.; Brewer, C.F.; Saha, S.; Chernushevich, I.; Ens, W.; Standing, K.G., (1994), Investigation of Protein-Protein Noncovalent Interactions in Soyabean Agglutinin by Electrospray Ionization Time-of-Flight Mass Spectrometry, *Rapid Communications in Mass Spectrometry*, **8,** 750-754.

MATRIX-ASSISTED LASER DESORPTION/IONIZATION OF NON-COVALENT COMPLEXES

FRANZ HILLENKAMP
Institute of Medical Physics and Biophysics, University of Münster
Robert-Koch-Straße 31, D-48149 Münster, Germany

1. Introduction

Matrix assisted Laser Desorption/Ionization (MALDI) and Electrospray Ionization (ESI) are considered soft ionization methods and are both the methods of choice for the generation of ions of macromolecules with molecular masses above a few thousand daltons. The exact meaning of "soft" ionization is, however, not well defined and, in fact not well understood in all its aspects. This is in particular true for the desorption of molecular aggregates, the topic of this volume. Whereas relatively consistent results and rules for optimal experimental conditions for the desorption of such aggregates have evolved for ESI over the last few years, such desorption appears to be substantially more difficult with MALDI and no really coherent understanding has been arrived at so far. This contribution attempts to summarize the results obtained in the authors group and published by others and discusses them in view of the mechanisms involved in MALDI. Even though it is not known in detail whether there are systematic differences between the desorption of the different types of molecular aggregates such as the homo- or heterooligomeric quarternary structure of proteins, non-covalent complexes between proteins and prosthetic groups, non-covalent complexes of proteins and ligands such as enzyme-substrate or receptor-ligand complexes and complexes based on charge interaction, they will be discussed separately for clarity of the presentation.

During the early development of MALDI, the intact desorption of proteins retaining their non-covalent quarternary structure appeared straightforward. The spectrum of the homotetramer of glucoseisomerase at mass of 172 kDa was, in fact, one of the first published spectra with a mass in excess of 100 kDa [1]. This spectrum was obtained from an aqueous nicotinic acid solution as a matrix with a laser wavelength of 266 nm. Later spectra of the intact homotrimer of the membrane protein porin from *rhodobacter capsulatus* were obtained under the same desorption conditions [2]. Nicotinic acid, however, has some less desirable features as a matrix, mainly because it

W. Ens et al. (eds.), New Methods for the Study of Biomolecular Complexes, 181–191.

tends to form intense adduct ions of the analyte with matrix derived molecules/ions which sacrifice mass resolution and mass accuracy. In addition, the use of a frequency quadrupled Nd-YAG laser adds substantial cost to the MALDI instrument. Only after the transition was made to Nitrogen lasers with an emission wavelength of 337 nm and matrices absorbing at this wavelength such as 2,5-dihydroxybenzoic acid and various cinnamic acid derivatives did it become obvious that the desorption of proteins with their intact quarternary structure was the exception rather than the rule.

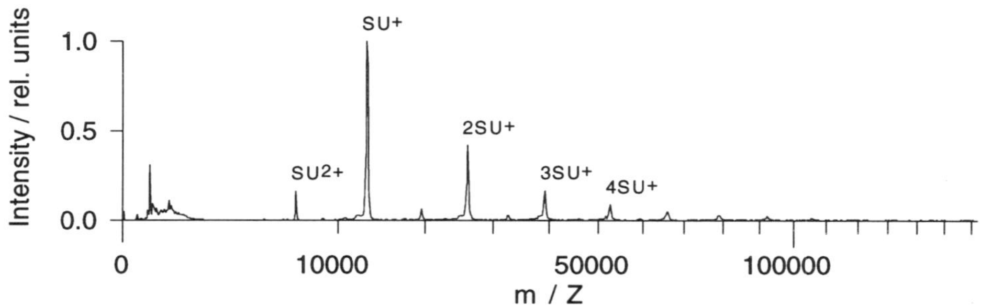

Figure 1. UV-MALDI spectrum of streptavidin. Matrix: DHBs (90% 2,5-dihydroxybenzoic acid, 10% 2-hydroxy-5-methoxybenzoic acid); Wavelength: 337 nm

One of the complications in MALDI-MS is the fact that **non specific** aggregates are frequently observed in the spectra. These may be homooligomers as shown for streptavidin in Figure 1 or heterooligomers out of analyte mixtures as shown in Figure 2. This non-specific oligomer formation is more pronounced for MALDI with laser wavelengths in the infrared (IR-MALDI), but is also frequently observed in UV-MALDI spectra as demonstrated in Figure 2. The signal intensity of such non-specified aggregates decreases approximately exponentially with increasing aggregate order. Though this non-specified aggregate formation usually increases with increasing analyte concentration in the solution, it appears to depend more strongly on not yet identified specific structure aspects of the proteins and also on the ion extraction conditions in the ion source. The latter observation is a strong hint that these aggregates are predominantly formed post desorption in the expanding plume rather than in solution or upon incorporation into the matrix during solvent evaporation. It is as yet not known, whether or not the monomeric units, forming the aggregates, are partially or fully denatured before or upon aggregation. It is clear from these observations that future experimental protocols must be able to clearly differentiate between **specific** complexes and such non-specified aggregates.

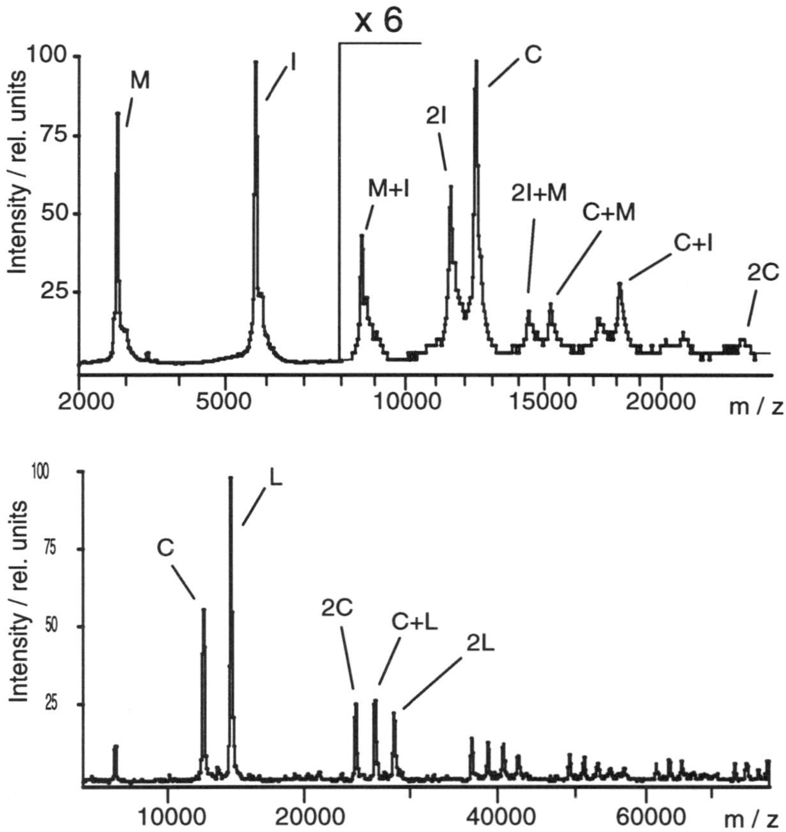

Figure 2. MALDI spectra of non-specific molecular aggregates. Top: UV-MALDI: mixture of Mellitin (M, 2846 Da, 10^{-5} M), bovine insulin (I, 5734 Da, 10^{-5} M) and horse heart cytochrome c (C, 12,360 Da, 10^{-6} M). Matrix: DHB; wavelength: 337 nm. Bottom: IR-MALDI: mixture of horse heart cytochrome c, (C, 12360 Da, 10^{-4} M) and chicken egg lysozyme (L, 14,305 Da, 10^{-4} M). Matrix: Tris/HCl buffer; wavelength: 2.94 μm.

2. The desorption of proteins with intact quarternary structure

There was good reason to assume that the conservation of the non-covalent quarternary structure of proteins was mainly influenced by the matrix rather than the desorption wavelength. A systematic study was, therefore, undertaken in the authors group to identify suitable matrices and possibly relate this specific feature to their chemical structure. To differentiate specific from non-specific aggregates only those spectra were accepted in this study as representing the quarternary protein structure, for which the specific oligomers formed the base peak and a clear deviation from the exponentially decreasing oligomer distribution was found. As a first result of this investigation ferulic acid was identified as one matrix suitable for the 337 nm desorption wavelength which yielded spectra of the intact trimeric structure or porin as well as of the tetrameric streptavidin [3]. It turned out that the matrix solvent also has a strong influence on the results; tetrahydrofuran (THF) was found to give the strongest signals of the quarternary proteins. The most surprising observation, however, was the fact that dominant signals of the specific quarternary protein complexes could only be detected for the first laser exposure of a given sample location; all subsequent exposures of a given location yielded spectra with the monomeric unit as the base peak. This observation was then also shown to hold for the nicotinic acid/266 nm system. For the ferulic acid/THF system the result also depended on the sample crystallization. Signals of the complexes were obtained only at microcrystalline sample locations and only for laser irradiances close to the ion detection threshold. In a follow-up study several other combinations of matrices, solvents and proteins with quarternary structure were identified which resulted in essentially the same behavior, i. e. yielding spectra of the intact oligomers for the firsts exposure of a given sample location and only signals of the subunits for all subsequent exposures [4]. Best results were obtained for 2,6-dihydroxyacetophenone (2,6-DHAP) dissolved either in THF or acetonitrile/TFA (ACN), the largest complex successfully desorbed was that of beef liver catalase of mass 233 kDa. Recently the streptavidin tetramer was also observed as intact ion for the first exposure of a given sample location out of an aqueous solution of DHB [5]. Whereas all the aforementioned spectra were recorded with an reflector time-of-flight instrument (Ref-TOF), a linear instrument (Lin-TOF) was used for the latter.

3. Mechanistic aspects of the MALDI desorption of non-covalent complexes

The generation of ions and their extraction into the mass spectrometer is a sequence of complicated steps, all of which can interfere with the detection non-covalent complexes. The details of the processes during most of the steps are not well understood as yet. The sequence of events is schematically shown in Figure 3. All proteins in the above described experiment were dissolved in physiological aqueous solution with or without buffers and the complexes are assumed to remain stable in this solution. However, most of the UV-matrices are acidic and even if they are soluble in water, the matrix solutions have an pH in the range of 2 - 4. Many matrices such as all of the cinnamic acid derivatives and the acetophenones are only soluble in organic solvents (for a discussion of 6-Aza-2-thiothymine as a matrix, see below). Most of the proteins and other non-covalent complexes are assumed to denature at pH-values below ca. 4 and also in most organic solvents, mostly inferred from their biological activity, though. For the analysis of glucoseisomerase in nicotinic acid/266 nm the addition of 50% EthOH to the aqueous solution did indeed result in a complete dissociation of the protein into its subunits [1]. In view of this common assumption it is certainly surprising that the best results for the detection of intact quarternary proteins were obtained either for matrices dissolved in THF or ACN or the acidic matrices nicotinic acid and DHB in aqueous solutions. Naively a dissociation of the oligomers upon mixing the analyte and matrix solutions would have been expected. It must, however, be kept in mind that only a time of at most a few minutes elapses between the mixing of the two solutions and the complete evaporation of the solvent. Possibly the kinetics of the denaturation are slow on this time scale. It must also be kept in mind that all of the tested organic solvents have a higher vapor pressure than water. A concentration gradient of the mixture of organic solvent, originating from the matrix solution and the water, originating from the sample solution is, therefore, expected to develop during solvent evaporation, even if the two solutions were originally thoroughly mixed. Immediately before solvent evaporation is complete, a predominantly aqueous solution is expected to remain and the quarternary

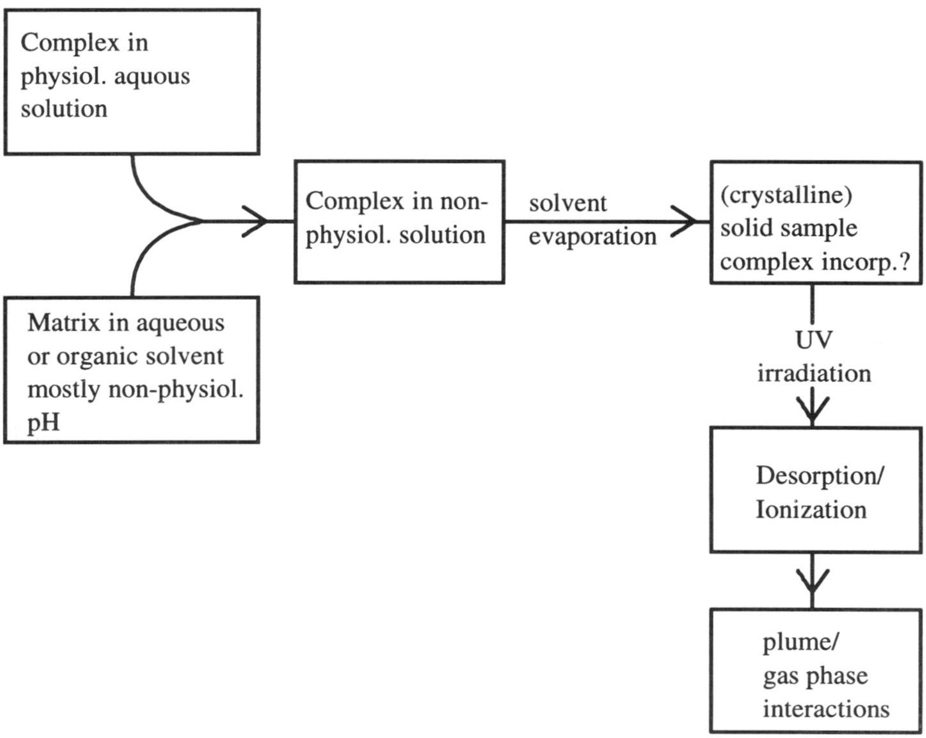

Figure 3. Schematic diagram of the sequence of processes involved in MALDI-MS

proteins, should they have survived the earlier phases of the mixed solution, could precipitate onto the matrix crystals or get only incorporated into the very top layer of the crystals, thus leading to the observation of the intact complex for solely the first shot. Nothing is known, however, about a possibly selective incorporation of complexes versus subunits into the matrix crystals. Even if the experimental results can be partially rationalized by the above arguments, the strong differences between chemically quite similar matrices such as ferulic acid and the other tested cinnamic acid derivatives sinapinic acid and α-cyano-4-hydroxycinnamic acid (4HCCA) are difficult to understand.

It was also established that the detection of only the subunit of the proteins for all but the first exposure did not result from UV-photodissociation by the preceding exposure(s) [3].

The processes during the desorption of material and the jet expansion into the vacuum are the least understood in the whole sequence. It is known that the desorbed material is ejected with a velocity of ca. 500 m/s away from the surface, forming a plume which at least for the first few nanoseconds has a very high density, close to solid state density. Lacking a solvation shield, the molecules are expected to strongly interact with each other through "collisions", leading to charging by proton transfer, but also to cooling and/or heating of internal degrees of freedom. The formation of non-specific complexes is also believed to mainly result from such plume interactions. The main basis for these assumptions is the observation that features such as ion yield, metastable decay and analyte aggregation depend strongly on the parameters of the ion extraction, the electric field strength in front of the sample, in particular. It cannot be excluded that the signals of non-covalent complexes are also modulated during this phase of the desorption process. All these processes take place on a nanometer (thickness of the desorbed layer in UV-MALDI) to micrometer (plume expansion) spatial and on a nanosecond time scale and are, therefore, difficult to investigate and disentangle. Metastable decay of complexes due to their internal excitation in the plume could be another possible process. The comparison of spectra recorded with a Lin-TOF in comparison to those taken with a Ref-TOF did not substantiate this assumption, however. Figure 4 shows two spectra of streptavidin (first shots only) recorded with the same instrument in the Lin-TOF and the Ref-TOF mode of operation. The spectra are identical within the normal experimental fluctuations. The same was true for other matrix/solvent combinations, for which only poor signals of the quarternary complexes were obtained in the Ref-TOF mode. Would the weak signals have been caused by metastable decay, a much improved signal of the complexes should have been observed in the Lin-TOF mode, which was nor observed.

4. Desorption of Heterocomplexes

Recently Woods et al. [6] reported spectra of three different enzymes, incubated with suitable substrates (aminopeptidase I/bovine growth hormone releasing factor (GH), trypsin/rat parathyroid hormone and arylamidase/GH). Small signals of all enzyme/substrate complexes were seen in the spectra of all mixtures for equimolar or near equimolar concentrations of enzyme and substrate, besides strong signals of the single proteins and peptides; the ratio of the complex signal to those of the protein/peptide stayed well below 1:10 in all cases. Sinapinic acid, dissolved in 1:1 1M ammonium citrate-ethanol was used as a matrix,. 1M ammoniumcitrate was also added to all peptide solutions. No complex signals were observed if the matrix was dissolved in an acidic 1:1 ethanol-water solvent without the ammoniumcitrate. Glocker et al. have reported the intact UV-MALDI desorption of the intact RNase S protein/peptide complex as well as signals of the leucine-zipper dimer-peptide. In the latter case

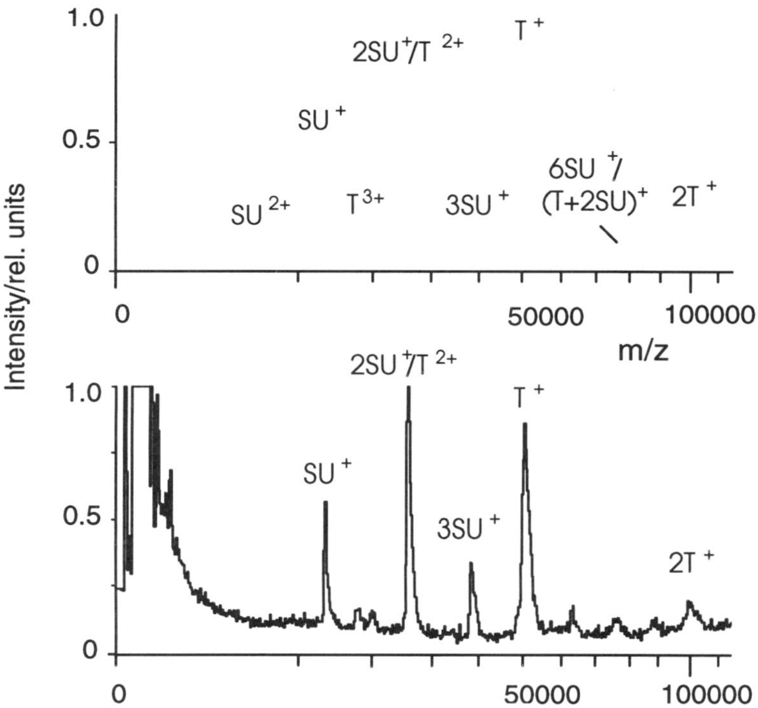

Figure 4. UV-MALDI spectra of streptavidin (first shot only). Matrix: 2,6-dihydroxyacetophenone; wavelength: 337 nm . Top: Reflectron time-of-flight mass spectrometer; Bottom: Linear time-of-flight mass spectrometer. (Please observe that even though the peak height of the T^+-peak is less than that of the $2SU^+/T^{2+}$, the ratio of the peak areas is1.25 in favor of the T^+-signal).

specificity was checked with peptides lacking affinity. A 5 mM ammoniumcitrate solution of 6-Aza-2-thiothymine (pH 6.5) was used as a matrix. Again the complex signals were small compared to those of the single proteins/peptides. The authors of both papers assume that a physiological pH of the analyte and the matrix solution is a prerequisite for the intact desorption of non-covalently bound complexes. Woods et al. have used an organic solvent, however, which should also have denatured the complexes. Glocker has confirmed by personal communication that in their system no difference was observed between the spectra of the first and successive exposures of the same sample location. This is not explicitly mentioned in the paper by Woods at al., but it is most probably safe to assume that the authors would have described such an observation, would they have made it.

The streptavidin/biotin complex is among of the most stable ones known. All attempts in the authors group to desorb the intact complex, or complexes of streptavidin with biotinylated peptides or oligonucleotides by UV-MALDI, using all the matrix/solvent combinations described above, have failed so far. However, if strands of biotinylated oligonucleotides are complexed with streptavidin coated magnetic dynal beads, the oligonucleotides could not be released again by incubation in either a 3-hydroxy-picolinic acid (3-HPA) or DHBs aqueous matrix solutions. This observation was independently confirmed by Tang et al. who saw signals of the hydrolyzed DNA-strand, which had been hybridized to the bound oligonucleotide in the MALDI spectra, but never signals of the strand bound to the beads via the biotin-streptavidin complex [8]. On the contrary, Larsen et al. have observed signals of a variety of peptides and proteins streptavidin/biotin bound to dynal beads in UV-MALDI spectra after matrix incubation in either 4HCCA or DHBs [9]. In all these cases organic solvents were used, however. All protein spectra show additional signals of the streptavidin subunit, suggesting that the streptavidin was denatured under the conditions used, this possibly causing the release of the biotinylated protein.. Similar observations have also been reported by Girault et al [10]. These rather contradictory results suggest that the specific structure of the biotinylated ligand takes part in the interaction at least when the complex is bound to a solid surface.

Another example demonstrating the important influence of the martix has been reported by Rosinke et al. [3]. IR-MALDI (2.94 μm) of cholesterolesterase (partially purified and delipidated) resulted in spectra with only the signal of the protein if urea was used as a matrix, wheres abundant aggregates with lipopolysaccharides were seen in the spectra if succinic acid was used as matrix. This can possibly be attributed to the well known denaturation of proteins by urea in solution.

Proteins carrying a non-covalently bound prosthetic group such as myoglobine are observed as apoproteins in UV-MALDI spectra. However, IR-MALDI of myoglobine using a Tris-buffered physiological solution as a matrix showed clear signals of the hemeprotein besides that of the apoprotein [11]. A similar spectrum was reported by Alimpiev et al. for Surface-Assisted Laser Desorption/Ionization (SALDI) from a frozen water preparation using an IR-wavelength of 3.28 μm for the desorption [12]. A 0.5 J/cm^2 , 7 ns laser pulse was used for the SALDI experiments. This corresponds to an irradiance of $7x10^7$ W/cm^2, certainly not a very "soft" desorption condition.

The publication by Woods et al. [6] also reports spectra of peptide/metal complexes for a zinc finger peptide. The signals were small compared to the pure peptide and Na-adduct ions, but incubation with Co instead of Zn showed the expected decrease in complex signal.

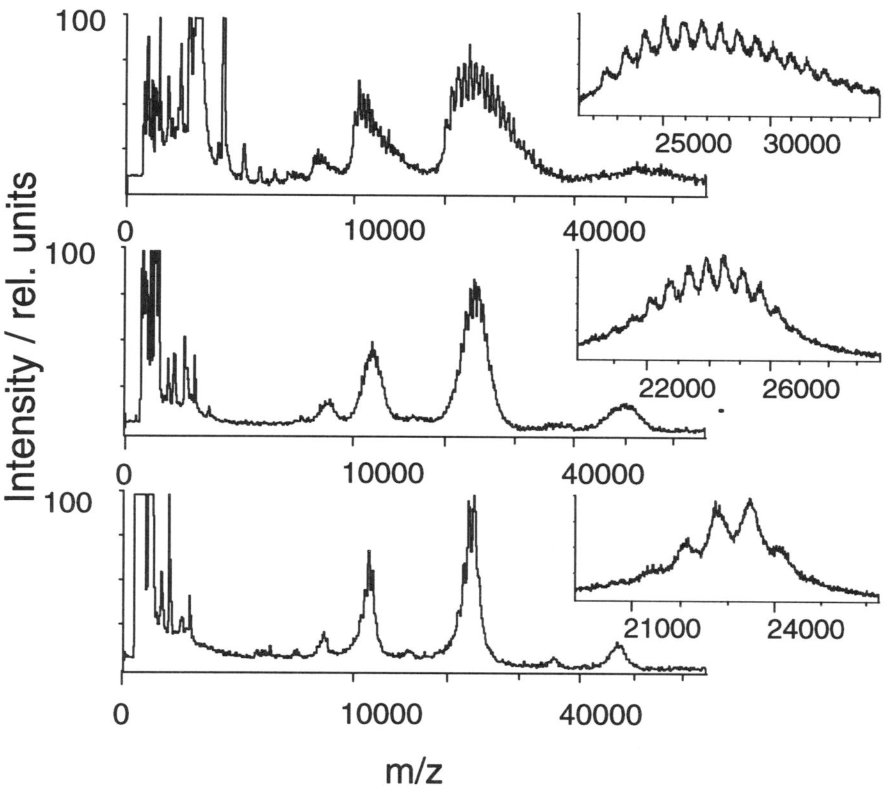

Figure 5. IR-MALDI spectra of trypsin inhibitor (20 kDa), stained with different organic dyes. Direct desorption from blot. Matrix: Succinic acid; wavelength: 2.94 μm. Top: coomassie brilliant blue; Middle: amido black; Bottom: Ponceau S

Ionic ligand/protein complexes are another interesting class of analytes. The intact desorption of complexes of well defined heparin derived oligosaccharides with arginine-rich peptides has been reported by Juhasz and Biemann [13]. Both IR-MALDI at 2.94 μm and UV-MALDI at 337 nm were used in this study. 5-(trifluoromethyl)-

uracil (TFMU) in a 1:1 water-acetonitrile solution was used as the matrix for the IR-MALDI, 3-hydroxypicolinic acid, caffeic acid and sinapinic acid with 10% addition of α-D-fucose, dissolved in water-acetonitrile served as UV-matrices. The complex signals formed the base peak in all spectra. Ionic interaction is also responsible for the dye staining of proteins, electroblotted onto PVDF-membranes after SDS-PAGE separation.

Figure 5 shows IR-MALDI spectra of trypsin inhibitor, stained with different organic dyes on the blot. These spectra were obtained by direct desorption from the blot membrane. Similar spectra can also be obtained for proteins incubated with the dyes in solution and prepared by mixing with a matrix solution followed by drying and desorption. A distribution of dye adducts to the protein is observed in all cases with no measurable signal of the unstained protein. An aqueous succinic acid solution was used as matrix. These complexes are very stable. Even extensive destaining could not remove all of the dye adducts.

5. Conclusion

It is quite obvious that optimal conditions for MALDI of non-covalent complexes have not yet been found. Many of the reported observations are inconsistent or even contradictory. The abundant signals of the intact quarternary protein in first shot spectra for UV-MALDI with a 2,6-DHAP matrix are promising, but the strong dependence of the signals on the details of the preparation and the difficulty to accumulate a sufficient number of first shot spectra make this approach rather impractical. The complex signals obtained by Woods et al. and Glocker at al. also give rise to hope, but the very small complex signals (compared to those of the single constituents) is almost prohibitive for a practical application to unknown samples, not to speak of a quantitation of affinities. A substantially better understanding of the mechanisms influencing the stability of such complexes under MALDI conditions is needed, before a routine application of MALDI-MS to their analysis will be possible. It is hoped that future experiments will solve this problem.

6. References

1. Karas, M. and Bahr, U. (1990) Laser Desorption Ionization Mass Spectrometry of Large Biomolecules *Trends in Anal. Chem.* **9**, 321-325
2. Hillenkamp, F., Karas, M., Beavis, R. C. and Chait, B. T. (1991) Matrix-Assisted Laser Desorption/Ionization Mass Spectrometry of Biopolymers, *Anal. Chem.* **63**, 1193A-1203A
3. Rosinke, B., Strupat, K., Hillenkamp, F., Rosenbusch, J., Dencher, N., U. Krüger and Galla, H.-J. (1995). Matrix-Assisted Laser Desorption/Ionization Mass Spectrometry (MALDI-MS) of Membrane Proteins and Non-Covalent Complexes, *J. Mass Spectrom.* **30**, 1462-1468

4. Cohen, L. R. H., Strupat, K. and Hillenkamp F. Analysis of Quarternary Protein Ensembles by Matrix Assisted Laser Desorption/Ionization, Mass Spectrometry, to be published
5. Karas, M. Institute for Instrumental Analysis, Univ. of Frankfurt/Main, Theodor-Stern-Kai 7, D-60590 Frankfurt/Main, Germany. Personal communication.
6. Woods, A. S., Buchsbaum, J. C., Worall, T. A., Berg, J. M. and Cotter, R. J. (1995) Matrix-Assisted Laser Desorption/Ionization of Noncovalently Bound Compounds, *Anal. Chem.* **67**, 4462-4465
7. Glocker, M. O., Jetschke, M. R., Bauer, S. H. J., Volz, J. and Przybylski, M. (1996) Non-Covalent Protein Interactions and H/D Exchange on Target Characterized by UV-MALDI , Proceedings of the 44th ASMS Conference on Mass Spectrometry and Allied Topics, Portland, Oregon, May 12 - 16, 1996, p. 752. See also Glocker *et al.* in this volume.
8. Tang, K., Fu, D., Kötter, S., Cotter, R. J., Cantor, CH. R. and Köster,H. (1995)Matrix-Assisted laser Desorption/Ionization Mass Spectrometry of Immobilized Duplex DNA Probes. *Nucleic Acid Research*, **23**, 3126-3131
9. Larsen, M. R., Nordhoff, E., Godtfredsen, M., Gerdes, K., Madsen, I., Krisriansen, K. and Roepstorff, P. (1996), MALDI-MS of Peptides and Proteins Affinity-bound to Magnetic Beads Carrying Immobilized Streptavidin. Proceedings of the 44th ASMS Conference on Mass Spectrometry and Allied Topics, Portland, Oregon, May 12 - 16, 1996, p. 1356
10. Hillenkamp, F., Karas, M., and Berkenkamp, St. (1995) MALDI-MS in the Infrared: A Critical Evaluation Proc. 43rd ASMS Conference on Mass Spectrometry and Allied Topics, Atlanta, Georgia, May 21 - 26, 1995, p. 357
11. Girault, S., Chassaing, G., Blais J. C., Brunot, A. and G. Bolbach (1996) Coupling of MALDI-TOF Mass Analysis to the Separation of Biotinylated Peptides by Magnetic Streptavidin beads, *Anal. Chem.* **68,** 2122 2126
12. Alimpiev, S., Chen, Y-Ch., Dratz, E., Kraft, P. and Sunner, J. (1996) Surface-Assisted Laser Desorption Ionization (SALDI) from frozen Water and organic Solutions (1996). Proc. 44th ASMS Conference on Mass Spectrometry and Allied Topics, Portland, Oregon, May 12 - 16, 1996, p. 641.
13. Juhasz, P. and Biemann, K. (1995) Utility of non-Covalent Complexes in the Matrix-Assisted Laser Desorption/Ionization Mass Spectrometry of Heparin-Derived Oligosacharides. *Carbohydrate Res.* **270**, 131-147

CHARACTERIZATION OF TERTIARY STRUCTURE STATES AND SPECIFIC NONCOVALENT COMPLEXES OF PROTEINS BY UV-MATRIX-ASSISTED LASER-DESORPTION / IONIZATION MASS SPECTROMETRY

M. O. GLOCKER, M.R. JETSCHKE, S. H. J. BAUER
AND M. PRZYBYLSKI
Faculty of Chemistry, University of Konstanz, P.O. Box 5560M732, D-78434 Konstanz, Germany

Hydrogen/deuterium exchange reactions (H/D exchange) in solution have been used for probing protein structures and dynamics and have been monitored using a broad spectrum of analytical methods including mass spectrometry. In order to study the effects of Matrix-Assisted Laser-Desorption / Ionization Mass Spectrometry (MALDI-MS) preparation techniques on protein structures, we applied *on-target* H/D exchange reactions. Samples were prepared for MALDI-MS by mixing aqueous protein and peptide solutions with acidic matrix solutions (pH 2) that contained organic cosolvents. H/D exchange for labile protons of peptides and proteins prepared in this manner was found to be > 90%, indicating that proteins were denatured. By subsequent laser irradiation of one single spot on the MALDI-target, depth-profiles of the deposited matrix/sample mixtures were recorded and revealed that *on-target* reactions provide exchange only in the top crystal layers. In deeper crystal layers molecular ions of proteins and peptides with unexchanged hydrogens were observed. No peptides with intermediate H/D exchange were detected. By contrast, the results from H/D-exchange reactions *in-solution*, which were monitored by electrospray/ionization mass spectrometry (ESI-MS), were consistent with folded globular protein structures. Subsequently, conditions for MALDI-MS sample preparations were altered substantially. 6-aza-2-thiothymine (ATT) was used as water soluble matrix and was dissolved in 5mM ammonium citrate (pH 5.5). Proteins and peptides were dissolved in aqueous buffers such as ammonium bicarbonate and were mixed with the matrix solution. Under these conditions non-covalent protein complexes could be analyzed and specificity of complex formation was ascertained by admixture of non-specific peptides which did not yield detectable aggregate ions. Also, trypsin maintained its active structure in the presence of the matrix, as was shown by on-target digests of proteins after UV-MALDI-MS molecular weight analyses.

W. Ens et al. (eds.), New Methods for the Study of Biomolecular Complexes, 193–208.

1. Introduction

The knowledge of higher order structure features of proteins is of importance for understanding structure/function relations and additional characterization methods to X-ray crystallography and nuclear magnetic resonance (NMR) are highly needed. Hydrogen / isotope exchange reactions in solution have been widely used for probing protein structures and dynamics and are monitored using a broad spectrum of analytical methods such as infrared (IR), NMR, and tritium gel filtration [1-4]. Recently, H/D exchange reactions were applied in conjunction with mass spectrometric methods, mostly for investigating higher order structures of proteins in solution [5]. These investigations include ESI mass spectrometric experiments of H/D-exchanged proteins in which conformations of the *intact* proteins were studied [6-8]. Also, the extent of H/D exchange in specific protein-regions was analyzed with FAB-MS of proteolytic fragments [9,10]. Additionally, the use of deuterium incorporation in peptides was combined with MALDI-MS for peptide sequencing by analyzing fragments obtained by post-source-decay (PSD) in time-of-flight (TOF) mass spectrometers [11].

Beyond the characterization of primary structures, the application of MALDI-MS to the characterization of tertiary structures and noncovalent complexes has previously not been considered feasible because of limitations in matrix chemistry and sample preparation methods. Major problems that have hitherto prevented the general analysis of specific noncovalent complexes by MALDI-MS are the harsh, denaturing sample preparation conditions, and the formation of unspecific cluster ions and matrix adducts [12,13]. For the characterization of noncovalent interactions, the mild desorption conditions of IR-MALDI using liquid matrices appear more promising [14,15]; at these conditions, the aquisition of spectra from a single crystal of lysozyme has been reported [16]. One disadvantage of IR-MALDI-MS is the expensive instrumentation that so far has prevented the routine use of this method.

A number of new matrices for UV-MALDI-MS have recently been reported which can be crystallized from buffered solution at a pH range of 2 to 8 [17], but their feasibility to the analysis of noncovalent complexes has not yet been evaluated. Of particular interest has been the recent observation of intact double-stranded oligonucleotides by UV-MALDI-MS with 6-aza-2-thiothymine (ATT) as a matrix, thus avoiding acidic conditions during sample preparation [18].

Our interest in the characterization of intact protein structures and noncovalent complexes by mass spectrometry [5] lead us to investigate the effects of MALDI-MS preparation techniques on protein structures by *on-target* H/D exchange reactions. The results were compared to H/D exchange reactions in solution monitored by ESI-MS. The successful application of an alternative sample preparation method for UV-MALDI-MS using ATT as matrix [19] is also described. At these conditions we have identified specific protein complexes. The novel sample preparation used here has been also successfully applied for enzymatic proteolytic degradation *on-target* in the presence of the ATT matrix.

2. Materials and Methods

H/D exchange reactions in-solution. Deuterated solvents D_2O (99.9 % D), CD_3COOD (98 % D), CH_3OD (99.5% D) and CD_3OD (99.8 % D) were purchased from Aldrich (Steinheim, FRG). H/D exchange *in-solution* was carried out at 20°C in 2% CD_3COOD : CD_3OD (9:1, v/v) and was initiated by dissolving lyophilized proteins to a final protein concentration of approx. 10 pmol/µL. The reaction times were 10 min for HEL (Sigma, St. Louis, USA) and 25 min for RNAse A (Sigma, St. Louis, USA). For both proteins H/D exchanges were also determined after 13 days incubation as well as after 10 min boiling (95°C) in D_2O.

ESI-MS instrumentation and aquisition conditions. ESI mass spectra were obtained with a Vestec 201 single quadrupole mass spectrometer (Vestec Corp., Houston, USA) with a mass range of 2000 amu. The electrospray was generated by applying a voltage of ca. 2kV at the inlet needle, 200V at the nozzle and 20V at the repeller. The temperature in the spraying region was approx. 45°C. Protein solutions were pumped with a Harvard microinfusion pump (Harvard, Franklin, USA) through a fused silica capillary (inside diameter, 75µm) with a flow rate of 2-4 µL/min into the ion source. Spectra were recorded with a scan rate of 7 sec/scan with a mass window of m/z 200-2000 using a Teknivent Vector2 data system (Teknivent, Houston, USA).

On-target H/D exchange reactions. 5 µg of lyophilized RNAse A (365 pmol) or neurotensin (3 nmol; Bachem, Bachendorf, Switzerland) were dissolved in 50 µL 0.1% TFA : CH_3CN (1:2, v/v). 1 µL of this solution and 1 µL HCCA-matrix solution (5 g/L in 0.1% TFA : CH_3CN; 1:2, v/v) were mixed directly on the target. H/D exchange reactions were carried out under N_2-atmosphere by addition of 1 µL, and 3 µL D_2O : CH_3OD (1:1, v/v), respectively. For determining time dependance of depth profiling, H/D exchange reactions were stopped by solvent evaporation using controlled N_2 flow after 8, 23, 42, 65, 157, 180, and 300 sec, respectively. Both, the redissolvation and solent evaporation process were observed using a stereo microscope. The dry target was transfered into the mass spectrometer under N_2-atmosphere.

For on-target deuterium/hydrogen back-exchange (D/H exchange) reactions, completely deuterated matrix was prepared by dissolving 5 g/L HCCA-matrix in 1% CD_3COOD:CH_3CN (1:2, v/v). The sample was twice lyophylized and redissolved in 1% CD_3COOD:CH_3CN (1:2, v/v). Neurotensin (5 µg, 3 nmol) was incubated with 50 µL 1% CD_3COOD:CH_3CN (1:2, v/v) for 20 min. The freshly prepared solutions

(1 μL each) were mixed in dry N_2-atmosphere on the target and dried. For D/H back-exchange 1 μL H_2O : CH_3OH (1:1, v/v) was added and evaporated within 300 sec.

MALDI-MS instrumentation and aquisition conditions. MALDI-MS analyses were carried out with a Bruker Biflex linear time-of-flight spectrometer (Bruker-Franzen, Bremen, FRG), equipped with a UV-nitrogen laser (337 nm) and a dual microchannel plate detector. For the molecular mass determinations of RNAse proteins, the acceleration voltage was set to 25 kV and spectra were calibrated with cytochrome c or myoglobin as internal and external standards. For molecular mass determinations of neurotensin and for peptide mapping experiments, acceleration voltage was set to 10 kV and insulin and neurotensin were used for external mass calibration. Spectra were recorded after evaporation of the solvent and processed using the X-MASS data system. Spectra were typically aquired by accumulation of 10 laser shots for the depth profiling experiments. Approx. 100 laser shots (ca. 20 shots per sample spot) were accumulated for samples in ATT matrix.

Matrix preparation and MALDI-MS analysis of non-covalent complexes. 6-Aza-2-thiothymine (ATT, Aldrich, Steinheim, FRG) was dissolved in 10 mM ammonium acetate and in 10 mM ammonium citrate to yield matrix concentrations of 2, 5, and 10 mg/mL, respectively. RNAse S (Sigma St. Louis, 147 μM) or cytochrome c (Sigma St. Louis, 158 μM) were dissolved in 5 mM ammonium citrate. The ATT solution (0.5 μL) was spotted on the target and 0.5 μL of the protein solution was added. Solvents evaporated slowly (within 30 min) due to high buffer salt concentration, and a thin layer consisting of matrix, buffer salts, and proteins formed. No additional washing steps were carried out. Noncovalent complexes were dissociated *on-target* by addition of a 0.1% TFA solution (1 μL, incubation for 30 min at 20°C). As a control, proteins were prepared for MALDI-MS using standard preparation protocols [20] and α-cyano-4-hydroxycinnamic acid (HCCA) or ATT as matrix. HCCA and ATT were dissolved in 0.1% TFA : CH_3CN (1:2, v/v).

On-target tryptic digestion. After determination of molecular weights, tryptic digestion of proteins was carried out *on-target* in the presence of the ATT-matrix. A freshly prepared trypsin solution (0.5 μL; 0.2 mg/mL trypsin; 50 mM NH_4HCO_3, pH 7.8) was added to the sample / matrix mixtures and incubated for approx. 30 min at 20°C (E:S ratio, 1:10).

3. Results

A principal aim of this study was the investigation of the effects of MALDI-MS preparation techniques on protein structures by *on-target* H/D exchange reactions, using deuterium incorporation as a probe for higher order protein structure. Furthermore, a sample preparation procedure was developed, with which specific noncovalent complexes can be detected by UV-MALDI-MS.

3.1 EFFECTS OF MALDI-MS PREPARATION TECHNIQUES ON PROTEIN STRUCTURES.

For studying the effects of MALDI-MS preparation techniques on protein structures, samples were prepared according to standard preparation protocols (see Materials and Methods) yielding matrix/analyte crystals of 30-50 μm in length which were evenly distributed on the target, approx. 50-80 μm apart from each other. Therefore, the laser beam (approx. 80-100 μm in diameter) can be focused on a single crystal in one experiment. *On-target* H/D exchange reactions were carried out by adding deuterated solvents in dry atmosphere. Both, the redissolvation and solent evaporation process were observed using a stereo microscope. During redissolvation the crystals were entirely covered with solvent. Crystals, or partially dissolved crystals remained visible during the entire process, suggesting that only the top layers were redissolved. Microscopic size measurements revealed that the sizes of the crystals were not altered significantly by the redissolvation / evaporation process. A first set of experiments was carried out with neurotensin, a peptide with 13 amino acids (27 labile protons, 11 backbone amide protons). Complete exchange of all labile protons occured within 300 sec under the conditions used. The resulting molecular ion signal of the deuterated peptide was recorded at m/z 1699, corresponding to 93% exchange of labile protons in the sum of the first ten laser shots (Figure 1a, TABLE 1). Subsequent multiple irradiation of the same crystal on the target resulted in molecular ion-dublets as the only

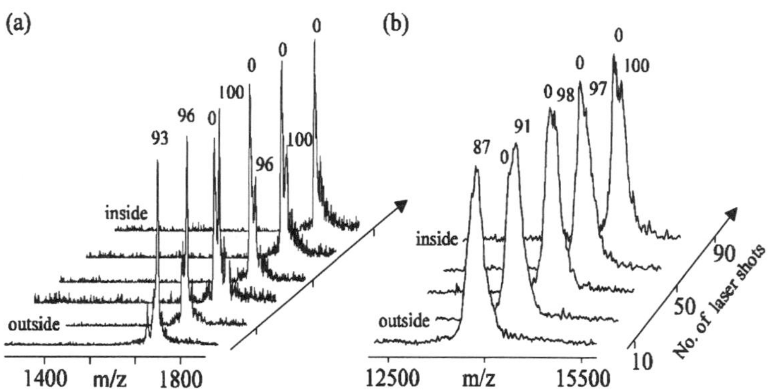

Figure 1. MALDI-MS depth-profiling of *on-target* H/D exchange reactions. *a)* Neurotensin, laser attenuation: 65. *b)* RNAse A, laser attenuation: 45. HCCA, dissolved in CH_3CN / 0.1 % TFA (2:1, v/v, pH 2) was used as matrix and proteins were incubated for 300 sec with D_2O / deuterated methanol (1:1, v/v) prior to data aquisition. Numbers in parentheses represent yield of H/D exchange of labile protons in %.

abundant ion species observed. The molecular ion dublets were consisting of the completely deuterated peptide ($[M+D]^+$ 1701 Da, 100% D) and undeuterated peptide ($[M+H]^+$, 1673 Da, 0% D) with decreasing relative abundance of the molecular ion from the deuterated peptide. Minor peaks, observed in some spectra result from Na^+ and K^+ adducts. As in case of neurotensin the deuterated and non-deuteratd molecular ions were clearly separated, ion intensities were evaluated quantitatively (Tables 1 and 4). The relative abundances of the deuterated molecular ions decreased linearily with number of laser shots, i.e. ablation of crystal surface.

TABLE 1. MALDI-MS depth-profiling of *on-target* H/D exchange with neurotensin.

No. of laser shots	experimental masses [a) b)]		No. of exchanged	% H/D
	$[M+H]^+$	$[M+D]^+$	protons [c)]	exchange [d)]
1-10	- (0)	1699 (100)	26	93
21-30	1672 (30.3)	1700 (69.7)	27	96
41-50	1673 (45.3)	1701 (54.7)	28	100
61-70	1672 (64.9)	1700 (35.1)	27	96
81-90	1673 (73.1)	1701 (26.8)	28	100
101-110	1672 (100)	- (0)	-	-

a) calc. $[M+H]^+$ for neurotensin is 1673 Da and calc. $[M+D]^+$ is 1701 Da for fully deuterated neurotensin.
b) numbers in parentheses denote relative abundances c) refers to calculated $[M+H]^+$. d) refers to labile protons.

The experiment was repeated several times by irradiating different crystals on the same target and displayed the same depth profile for both, the peptide ion signals and for the matrix monomer ($[M+H]^+$: 190 Da; $[M+D]^+$: 193 Da) and dimer ions ($[M+H]^+$: 379 Da; $[M+D]^+$: 384 Da). The somewhat lower molecular weight of the deuterated neurotensin ions, observed in the first 10 shots (TABLE 1) may be due to partial backexchange on the crystal surface during target transfer and evacuation of the source inlet. Similar experiments were carried out with RNAse A, a protein with 124 amino acids (238 labile protons, 122 backbone amide protons) and well characterized tertiary structure. Within 300 sec, nearly complete exchange of the labile protons occured under the conditions used (Figure 1b, TABLE 2).A molecular ion of the deuterated protein was recorded at m/z 13889 (87% exchange). Subsequent multiple irradiation of the same spot on the target resulted in molecular ions consisting of the deuterated protein ($[M+D]^+$ 13922 Da) and undeuterated protein ($[M+H]^+$ 13683 Da) as the predominant protein forms observed, with decreasing relative abundance of the deuterated protein (TABLE 2). Due to higher laser power and the linear time-of-flight mass analyzer used, mass peaks are only partially resolved and depth dependance of ion dublets was less pronounced than for peptides. However, the high extent of exchanged protons (> 90%) indicated that in the exchange experiments RNAse A was present in denatured form.

TABLE 2. MALDI-MS depth-profiling of *on-target* H/D exchange with RNAse A

No. of laser shots	experimental masses [a)]		No. of exchanged protons [b)]	% H/D exchange[c)]
	$[M+H]^+$	$[M+D]^+$		
1-10	-	13889	206	87
21-30	13686	13898	215	91
41-50	13690	13915	232	98
61-70	13678	13912	229	97
81-90	13683	13921	238	100

a) calc. $[M+H]^+$ for RNAse A is 13683 Da and calc. $[M+D]^+$ is 13922 Da for fully deuterated RNAse A.
b) refers to calculated $[M+H]^+$. c) refers to labile protons.

H/D exchange of RNAse A *on-target* was compared to exchange *in-solution*, monitored by ESI-MS. Under ESI-MS conditions, the globular structure of RNAse A, is reflected by the narrow charge distribution that is centered around the $[M+8H]^{8+}$ molecular ion signal (Figure 2a). The intact globular structure of RNAse A is reflected by *in-solution* H/D exchange of only 46% of the labile protons (within 25 min), which increased to 57% by incubation in deuterated solvent for 13 days. This slow H/D exchange is consistent with a globular protein structure. By contrast, RNAse A exchanges 77% of the labile protons in boiling deuterated solvent for 10 min, suggesting partial unfolding (TABLE 3), consistent with NMR studies [21].

Boiled RNAse A resumes its global structure at 20°C as is indicated by the ESI-charge states (Figure 2b). Similar results were also observed for *in-solution* H/D exchange reactions of HEL (TABLE 3). These results show, that H/D exchange *in-solution* analyzed by ESI-MS is suitable to characterize reversible protein denaturation.

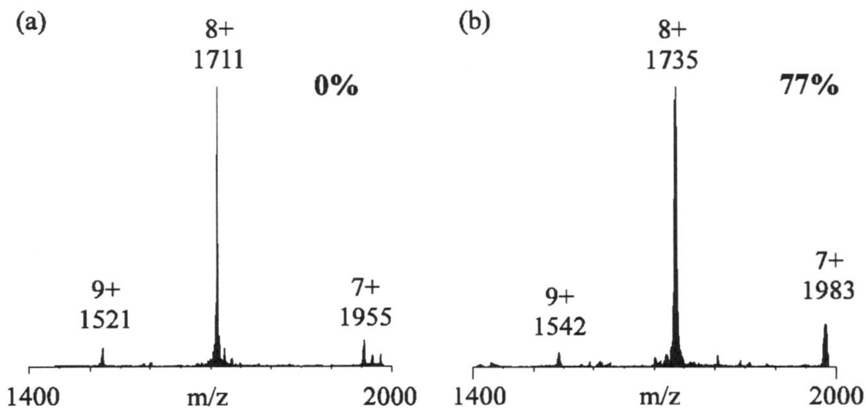

Figure 2. ESI-MS analysis of *in-solution* H/D exchange of RNAse A. *a)* Spectrum in 2% acetic acid / methanol (9:1, v/v), 20°C, pH 3. *b)* Spectrum after H/D exchange in boiling D_2O. Spectrum was recorded from 2% deuterated-acetic acid in D_2O : deuterated-methanol (9:1, v/v) 20°C, pD 3. Yield of H/D exchange is shown in %.

TABLE 3. ESI-MS of *in-solution* H/D exchange with RNAse A and HEL.

protein	exchange time	$[M+D]^{+}$ a)	No. of exchanged protons b)	% of H/D exchange c)
RNAse A	25 min	13791	108	46
	13 days	13819	136	57
	10 min boiling	13864	181	77
HEL	10 min	14433	128	50
	13 days	14480	175	69
	10 min boiling	14534	229	90

a) calc. $[M+H]^{+}$ for RNAse A is 13683 Da and calc. $[M+D]^{+}$ is 13922 Da for fully deuterated RNAse A. calc. $[M+H]^{+}$ for HEL is 14305 Da and calc. $[M+D]^{+}$ is 14510 Da for fully deuterated HEL.
b) refers to calculated $[M+H]^{+}$. c) refers to labile protons.

3.2. TIME DEPENDANCE OF *ON-TARGET* H/D EXCHANGE DEPTH AND MALDI-MS DEPTH PROFILING

On-target H/D-exchange of peptides and proteins enables to distinguish analyte signals from different crystal layers (depth profiling) but exchange yields may be obscured by partial D/H-back-exchange on the crystal surface. In order to avoid surface back-exchange, completely deuterated neurotensin was prepared in solution and mixed with deuterated matrix under dry atmosphere (N_2). After evaporation of the deuterated solvent, crystals of comparable sizes as in the previous experiments were obtained and MALDI-MS molecular weight determination yielded ion signals only for completely deuterated neurotensin. This preparation was applied for on-target back-exchange reactions and showed again only distinct ion doublets for the undeuterated and completely deuterated neurotensin with varying relative abundances (TABLE 4).

TABLE 4. MALDI-MS depth-profiling of *on-target* D/H backexchange with neurotensin.

No. of laser shots	experimental masses a) b)		No. of exchanged	% H/D
	$[M+H]^{+}$	$[M+D]^{+}$	protons c)	exchange d)
1-10	1673 (100)	- (0)	-	-
21-30	1673 (72.4)	1701 (27.6)	28	100
41-50	1674 (46.4)	1702 (53.6)	29	104
61-70	1672 (37.3)	1700 (62.6)	27	96
81-90	1674 (28.6)	1702 (71.4)	29	104
101-110	- (0)	1702 (100)	29	104

a) calc. $[M+H]^{+}$ for neurotensin is 1673 Da and calc. $[M+D]^{+}$ is 1701 Da for fully deuterated neurotensin.
b) numbers in parentheses denote relative abundances c) refers to calculated $[M+H]^{+}$. d) refers to labile protons.

Depth profiling experiments were carried out by varying the H/D exchange time (8 to 300 sec). A linear correlation was observed between *on-target* exchange time and number of shots for which deuterated ion signals were obtained for both, neurotensin

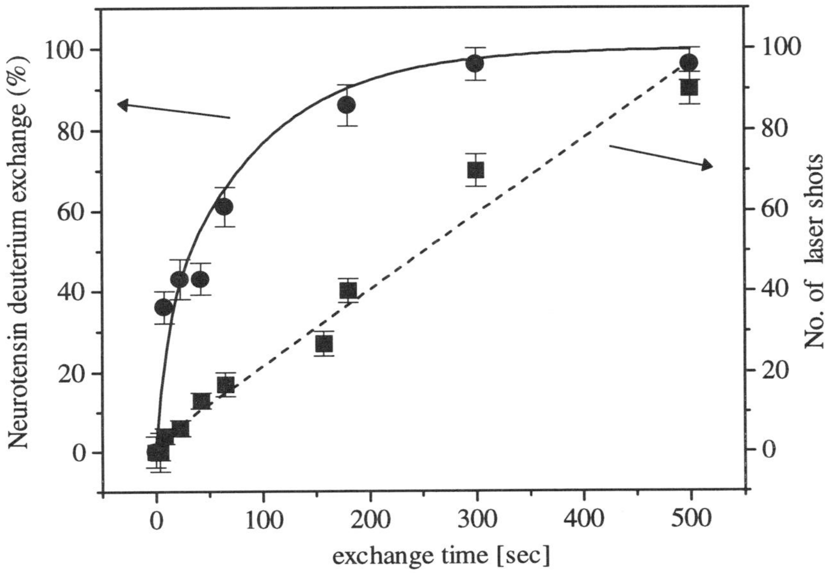

Figure 3. Correlation of time dependance of *on-target* H/D exchange and exchange depth. Exchange depth is linearly dependend to the exchange time, as determined by the number of laser shots for which the deuterated HCCA (■) -ions were recorded. H/D exchange in neurotensin (●) shows a time dependent exponential approximation to the total number of labile protons, comparable to solution exchange behaviour of peptides and proteins.

and matrix (Figure 3). The number of laser shots for which deuterated matrix ion signals were observed before they completely disappeared is plotted against exchange time and showed a linear dependence (Figure 3, dashed line). In addition, H/D exchange in neurotensin shows a time dependent exponential approximation to the total number of labile protons, comparable to results from H/D exchange of peptides and proteins *in-solution* [22]. For short exchange times (< 60 sec) single shot spectra recorded and revealed again distinct ion dublets for neurotensin which were well separated, however deuterium exchange was not complete (Figure 3, solid line). Line broadening, indicating a distribution of different exchange ratios, was not observed. These results are in agreement with recent H/D exchange studies of peptides by vanDorsselear et al. using a sheeth-flow ESI-MS system [23].

We conclude from these results that *on-target* H/D exchange occurs in the superimposed solvent. Furthermore, despite the short on-target H/D exchange times a high deuterium incorporation was observed for RNAse A which is consistent with results from ESI-MS experiments with denatured RNAse A (cf. Figure 2). Generally, the reaction conditions applied here and the sample preparation procedures (acidic pH, organic cosolvents) for MALDI-MS are representing strongly denaturing conditions for proteins. Therefore, an entirely different sample preparation method, avoiding denaturing conditions was required in order to study specific noncovalent protein complexes by UV-MALDI-MS.

3.3 UV-MALDI-MASS SPECTROMETRY OF SPECIFIC NON-COVALENT PROTEIN COMPLEXES

The sample preparation for MALDI-MS in this study uses 6-aza-2-thiothymine (ATT) as matrix without any addition of organic cosolvents. Proteins were dissolved in aqueous buffers such as ammonium bicarbonate, ammonium citrate, and ammonium acetate. First studies were carried out with RNAse S, a protein with well characterized tertiary and quarternary structure. RNAse S is an enzymatic product formed by proteolytic cleavage of RNAse A with subtilisin [24] in which the specific noncovalent complex is composed of the two components, S-peptide (1-19/20) and S-protein (21/22-124). Molecular ions of the intact RNAse S complex have been obtained by ESI-MS at neutral pH in several previous studies [5,25]. Using the ATT-matrix, a singly charged molecular ion of intact RNAse S was clearly detected by MALDI-MS (m/z 13630) when both, protein and matrix were dissolved in aqueous buffer at pH 5.5 (Figure 4). Due to the inhomogenous cleavage by subtilisin and possible buffer salt and matrix adducts which may serve as "solvation spheres", the resulting protein sequences are reflected in the MALDI spectra by broad, unresolved $[M+H]^+$ ions for RNAse S and the S-proteins (m/z 11500), while the $[M+H]^+$ ions for the S-peptides (m/z 2096 and 2167) were resolved. At more acidic, but otherwise identical conditions (pH $\leq$4) the RNAse S complex was completely dissociated and only ions for the separate complex components were observed [19]. This result is well in agreement with (i) the known pH-dependent stability, i.e. instability of RNAse S under acidic conditions, and (ii) with the pH-dependance observed previously in ESI-MS [26,27]. Furthermore, the complexes could also be dissociated by acidification with 0.1% TFA on the sample target, resulting only in ions of dissociated S-protein and S-peptide (data not shown). In addition, the detection of specific complex ions was ascertained by admixture of the oligopeptide angiotensin, unrelated to RNAse S. Both, RNAse S (Figure 4) and angiotensin, were detected by their $[M+H]^+$ molecular ions at m/z 13630 and 1298, respectively, while no ions due to unspecific aggregation of the two polypeptides were observed.

In this study proteins yielded broad molecular ion signals under the sample preparation conditions employed with ATT. Molecular ions showed a steep slope on the low mass side and significant tailing on the high mass side (cf. Figure 5a), which may be explained by buffer salt and matrix adducts. Possibly, buffer salts and/or matrix molecules might act as a solvation sphere, helping to maintain intact structures upon sample preparation, as a prerequisite for the detection of specific noncovalent complexes.[5] The applicability of the present sample preparation procedure with ATT matrix for studying functional protein structures was investigated in further experi-

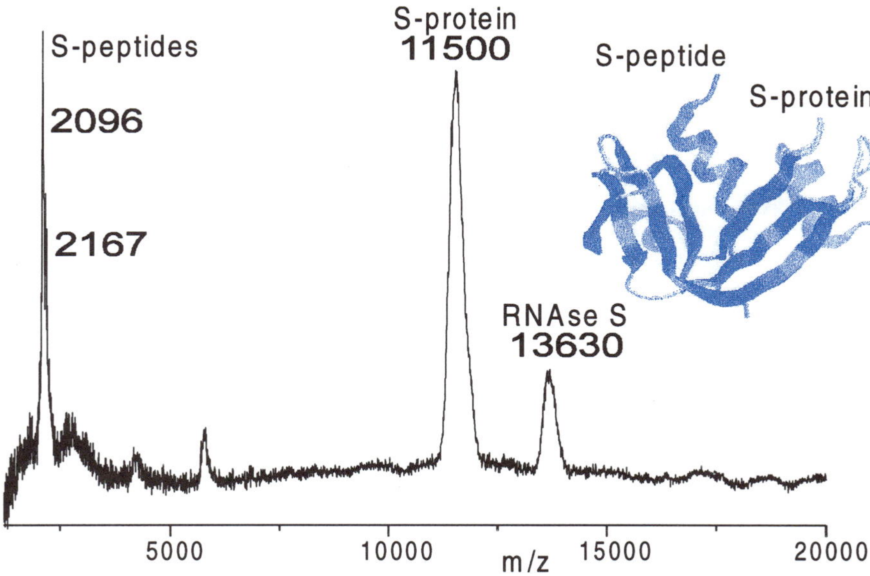

Figure 4. MALDI-MS of RNAse S. An RNAse S solution (147 µM, in 10 mM ammonium citrate) was mixed with ATT solution (5 mg/mL, in 10 mM ammonium acetate). Solutions (0.5 µL each) were directly mixed on the target. Final pH was 5.5.

ments by *on-target* tryptic digestion and MALDI-MS peptide mapping analysis of cytochrome c.

After MALDI-MS analysis of the intact protein as described above, the sample was incubated with buffered trypsin solution for 30 min on the target and then reintroduced into the mass spectrometer. Initial molecular weight determination of intact cytochrome c showed the $[M+H]^+$ ion at m/z 12362 and a small doubly charged $[M+2H]^{2+}$ ion at m/z 6842 (Figure 5a). After tryptic digestion, the molecular ions for cytochrome c had disappeared and ions of the tryptic peptide fragments were detected, yielding almost complete sequence coverage (Figure 5b). Only traces of peptides from tryptic autoproteolysis were detected. The *on-target* tryptic digestion following molecular weight determination was found to be a convenient tool for the sequence characterization of synthetic polypeptides, particularly when the N-terminus is blocked and Edman degradation thus prevented [19].

4. Discussion

In UV-MALDI-MS, solid samples are generally prepared by admixture of a large excess of matrix and in most applications a strongly acidic matrix with suitable absorption properties such as benzoic and cinnamic acid derivatives (e.g. dihydroxy-

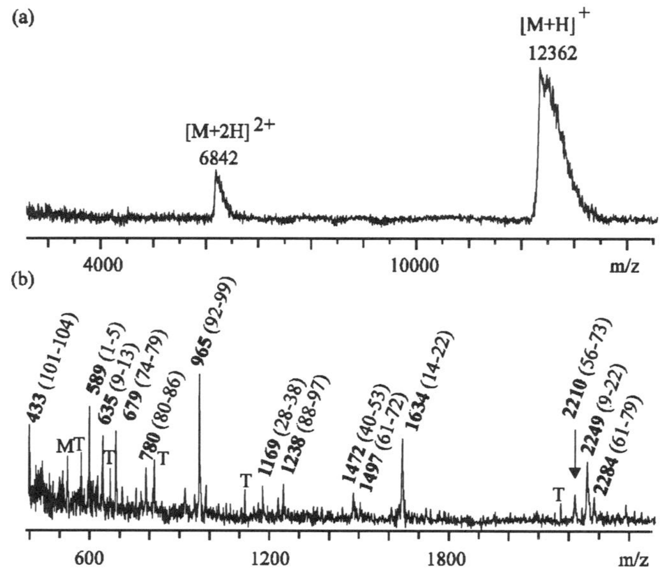

Figure 5. MALDI-MS molecular weight determination and tryptic peptide mapping of cytochrome c. *a)* Spectrum of cytochrome c (158 μM) dissolved in 50 mM ammonium acetate with ATT (5 mg/mL) dissolved in 20 mM ammonium citrate as matrix (pH 5.5). *b)* MALDI-MS peptide mapping after addition of aqueous buffered trypsin solution (pH 7.8) directly on the target. Numbers in parentheses represent partial sequences in cytochrome c. Peaks that derive from trypsin autoproteolysis and matrix peaks are labeled with T and M, respectively. Peptide (14-22) contains the covalently linked heme group with the central iron atom .

benzoic acid, DHB; HCCA) has been applied [28,29]. Using *on-target* H/D exchange reactions, it is possible to study analyte/matrix mixtures, derived from standard sample preparations, such as the "dried droplet method" or the "thin layer" matrix preparation [30,31]. Our results clearly show that these sample preparation methods have denaturing effects on protein tertiary structures. Accordingly, initial attempts to characterize noncovalent complexes did not yield reproducible results or provided only dissociation products [32,33].

Since in the present sample preparation matrix crystals were deposited far apart from each other, laser irradiation results in ablation of material from a single crystal, suggesting that the observed ions derive from proteins that are incorporated in the matrix crystals. This is consistent with model studies in which the proteins were embedded in matrix crystals of millimeter size [34,35]. In these studies, maximum ablation depths of ca. 0.5 μm were estimated resulting from ca. 10 consecutive irradiations and may determine the spatial resolution that can be distinguished in depth by MALDI-MS depth profiling. At the same time, the required layer thickness, i.e. ablation depth, determines the minimum accessible time window for *on-target* chemical reactions. However, *on-target* H/D exchange seems generally applicable in studies of protein conformation and noncovalent protein complexes from samples deposited on the MALDI target.

A first observation of the noncovalent heme-myoglobin complex by MALDI-MS was reported after slow cocrystallization with sinapinic acid in large matrix crystals [36]. MALDI-MS studies have also been performed with ionic complexes between strongly basic polypeptides and polyanions such as oligonucleotides and sulfated oligosaccharides [37,38]. The noncovalent interactions studied in that work have been primarily characterized by ionic complexes without specific molecular recognition. By contrast, examples for specific complexes detected by MALDI-MS are the trimer complex of the ion-channel protein porin, and noncovalent tetramers of streptavidin by using a Nd-YAG laser at 355 nm and hydroxy-benzoic acid derivatives as matrices [33]. However, in these experiments, ions of the noncovalent complexes were only observed upon the first laser shot on a sample spot not previously irradiated.

ATT is readily soluble in aqueous buffers and its solutions can be adjusted from slightly acidic to almost neutral pH. In the present study a pH of 5.5 enabled the detection of the noncovalent RNAse S complex. The pH of the solution appears to be crucial for the detection of specific noncovalent complexes by MALDI-MS. In our experiments, RNAse S was partially dissociated at pH ~5 and completely dissociated at pH $\leq$4 [19], which is consistent with results from NMR studies and from our previous ESI-MS work [5,39]. In the initial studies concerning RNAse S denaturation [40], remaining activity was determined at pH 5 after boiling of RNAse S using various conditions. However, the results obtained by this method are not comparable to our studies as subsequent renaturation before assaying activity can not be ruled out; renaturation was observed for RNAse A (cf. Figure 2) and for RNAse S [40]. Additionally, solvent evaporation during sample preparation may also have an effect on the complex stability and may represent a critical factor for the observation of noncovalent complexes by MALDI-MS, since during evaporation the pH and ionic strength of the solution may be altered, and such effects on the conservation of specific noncovalent associations are not known.

In contrast to the clear solvent effects in the sample preparation no influence of the laser radiation has been found in this study. Spectra of noncovalent complexes were generally obtained by accumulation of ions from single laser shots from the same sample spot. No significant effect of the laser energy, and dependance from the laser penetration depth was observed, within the limitation of the sample/matrix layer employed (s. Materials and Methods).

The proteolytic *on-target* digestion was found to be an efficient tool for the sequence identification of proteins by peptide mapping analyses. The catalytic activity of proteolytic enzymes at the conditions employed here was found comparable to that in solution, and the proteolytic specificity appears to be unchanged. The experimental procedure is simple and convenient and does not require additional sample preparation on the target such as immobilization of the protease [41]. Furthermore, autoproteolysis, albeit detectable did not prevent peptide identification. These results demonstrate that the biological activity of proteases is retained and thus indicate that the functional enzyme structure can be maintained at the sample preparation conditions employed in this study for the detection of noncovalent complexes.

5. Acknowledgements

This work was supported by grants from the Deutsche Forschungsgemeinschaft, Fonds der chemischen Industrie and by the EU-network "Peptide and Protein Structure Elucidation by Mass Spectrometry". We wish to thank Drs. Alain vanDorsselear, Emmanuelle Leize and Noelle Potier for many fruitful discussions on mass spectrometry of supramolecular complexes.

6. References

1. Englander, S.W., Downer, N.W., and Teitelbaum, H. (1972) Hydrogen Exchange, *Annu. Rev. Biochem.*, 903-924.
2. Wagner, G. and Wüthrich, K. (1982) Amide Proton Exchange and Surface Conformation of the Basic Pancreatic Trypsin Inhibitor in Solution, *J. Mol. Biol.*, **160**, 343-361.
3. Miranker, A., Radford, S.E., Karplus, M., and Dobson C.M. (1991) Demonstration by NMR of Folding Domains in Lysozyme, *Nature*, **349**, 633-636.
4 Delepierre, M., Dobson, C.M., Karplus, M., Poulsen, F.M., States, D.J., and Wedin, R.E. (1987) Electrostatic Effects and Hydrogen Exchange Behaviour in Proteins: The pH Dependence of Exchange Rates in Lysozyme, *J. Mol. Biol.*, **197**, 111-130.
5. Przybylski, M. and Glocker, M.O. (1996) Electrospray Mass Spectrometry on Non-covalent Complexes of Biomacromolecules - New Analytical Perspectives for Supramolecular Chemistry and Molecular Recognition Processes, *Angew. Chem. Int. Ed. Engl*, **35,** 806-826.
6. Miranker, A., Robinson, C.V., Radford, S.E., Aplin, R.T., and Dobson, C.M. (1993) Detection of Transient Protein Folding Populations by Mass Spectrometry, *Science*, **262**, 896-900.
7. Wagner, D.S., Melton, L.G., Yan, Y., Erickson, B.W., and Anderegg, R.J. (1994) Deuterium Exchange of a-Helices and b-Sheets as Monitored by Electrospray Ionization Mass Spectrometry, *Prot. Sci.*, **3**, 1305-1314.
8. Katta, V. and Chait, B.T. (1991) Conformational Changes in Proteins Probed by Hydrogen-Exchange Electrospray-Ionization Mass Spectrometry, *Rapid Commun. Mass Spectrom.*, **5**, 214-217.
9. Zhang, Z. and Smith, D.L. (1994) Determination of Amide Hydrogen Exchange by Mass Spectrometry: A New Tool for Protein Structure Elucidation, *Protein Sci.*, **2**, 522-531.
10. Liu, Y. and Smith, D.L. (1994) Probing High Order Structure of Proteins by Fast-Atom Bombardment Mass Spectrometry, *J. Am. Soc. Mass Spectrom.*, **5**, 19-28.
11. Spengler, B., Lützenkirchen, F., and Kaufmann, R. (1993) On-target Deuteration for Peptide Sequencing by Laser Mass Spectrometry, *Org. Mass Spectrom.*, **28**, 1482-1490.
12. Hillenkamp, F., Karas, M., Ingendoh, A., and Stahl, B. (1990) Matrix Assisted UV-Laser Desorption/Ionization: A New Approach to MAss Spectrometry of Large Biopolymers, in A.L. Burlingame and J.A. McCloskey (eds.), *Biological Mass Spectrometry*, Elsevier Science Publishers, Amsterdam, pp. 49-60.

13. Weinmann, W., Parker, C.E., Deterding, L.E., Papac, D., Hoyes, J., Przybylski, M., and Tomer, K.B. (1994) Capillary Electrophoresis-Matrix-Assisted Laser-Desorption Ionization Mass Spectrometry of Proteins, *J. Chromatogr.*, **680,** 353-361.
14. Overberg, A., Karas, M., Bahr, U., Kaufmann, R., and Hillenkamp, F. (1990) Matrix-assisted Infrared-laser (2,94 μm) Desorption/Ionization Mass Spectrometry of Large Biomolecules, *Rapid Commun. Mass Spectrom,* **4,** 293-296.
15. Overberg, A., Karas, M., and Hillenkamp, F. (1991) Matrix-assisted Laser Desorption of Large Biomolecules with a TEA-CO_2-Laser, *Rapid Commun. Mass Spectrom,* **5,** 128-131.
16. Berkenkamp, S., Karas, M., and Hillenkamp, F. (1995) Water as a Matrix in Infrared MALDI-MS, *Proceedings of the 43rd ASMS Conference on Mass Spectrometry and Allied Topics,* 1247.
17. Fitzgerald, M.C., Parr, G.R., and Smith, L.M. (1993) Basic Matrices for the Matrix-Assisted Laser Desorption/Ionization Mass Spectrometry of Proteins and Oligonucleotides, *Anal. Chem,* **65,** 3204-3211.
18. Lecchi, P. and Pannell, L.K. (1995) The Detection of Intact Double-Stranded DNA by MALDI, *J. Am. Soc. Mass Spectrom.,* **6,** 972-975.
19. Glocker, M.O., Bauer, S.H.J., Kast, J., Volz, J., and Przybylski, M. (1996) Characterizaion of Specific Noncovalent Protein Complexes by UV-MALDI-Mass Spectrometry, *J. Mass Spectrom.*, **31,** 1221-1227.
20. Glocker, M.O., Arbogast, B., and Deinzer, M.L. (1995) Characterization of Disulfide Linkages and Disulfide Bond Scrambling in Recombinant Human Macrophage Colony Stimulating Factor by Fast-Atom Bombardment Mass Spectrometry of Enzymatic Digests, *J. Am. Soc. Mass Spectrom.,* **6,** 638-643.
21. Robertson, A.D. and Baldwin, R.L. (1991) Hydrogen Exchange in Thermally Denatured Ribonuclease A, *Biochemistry,* **30,** 9907-9914.
22. Anderegg, R.J., Wagner, D.S., Stevenson, C.L., and Borchardt, R.T. (1994) The Mass Spectrometry of Helical Unfolding in Peptides, J. Am. Soc. Mass Spectrom., 5, 425-433.
23 Potier, N., and vanDorsselear, A. (1995) presented at the 2nd EU network meeting "Peptide and Protein Structure Elucidation by Mass Spectrometry".
24. Kato, I. and Anfinsen, C.B. (1969) On the Stabilization of Ribonuclease S-protein by Ribonuclease S-peptide, *J. Biol. Chem,* **244,** 1004-1007.
25. Loo, J.A., Edmonds, C.G., Smith, R.D., Lacey, M.P., and Keough, T., (1990) Comparison of Electrospray Ionization and Plasma Desorption Mass Spectra of peptides and Proteins, *Biomed. Env. Mass Spectrom,* **19,** 286-294.
26. Ogorzalek Loo, R.R., Goodlett, D.R., Smith, R.D., and Loo, J.A. (1993) Observation of a Noncovalent Ribonuclease A-Protein / S-Peptide Complex by Electrospray Ionization Mass Spectrometry, *J. Am. Chem. Soc,* **115,** 4391-4392.
27. Schreier, A.A. and Baldwin, R.L. (1977) Mechanism of Dissociation of S-Peptide from Ribonuclease S, *Biochemistry,* **16,** 4203-4209.
28. Hillenkamp, F. and Karas, M. (1990) Mass Spectrometry of Peptides and Proteins by Matrix-Assisted Ultraviolet Laser Desorption / Ionization, *Meth. Enzymol,* **193,** 280-305.
29. Smith, D.L. and Zhang, Z. (1994) Probing Noncovalent Structural Features of Proteins by Mass Spectrometry, *Mass Spectrom. Rev.,* **13,** 411-429.
30. Beavis, R.C. and Chait, B.T.(1990) Rapid, Sensitive Analysis of Protein Mixtures by Mass Spectrometry, *Proc. Natl. Acad. Sci. USA,* **87,** 6873-6877.
31. Vorm, O., Roepstorff, P., and Mann, M. (1994) Improved Resolution and Very High Sensitivity in MALDI TOF of Matrix Surface Made by Fast Evaporation, *Anal. Chem.,* **66,** 3281-3287.
32. Hillenkamp, F., Karas, M., Beavis, R.C., and Chait, B.T. (1991) Matrix-Assisted Laser Desorption/Ionization Mass Spectrometry of Biopolymers, *Anal. Chem.,* **63,** 1193A-1203A.
33. Rosinke, B., Strupat, K., Hillenkamp, F., Rosenbusch, J., Dencher, N., Krüger, U., and Galla, H.J. (1995) Matrix-Assisted Laser Desorption/Ionization Mass Spectrometry (MALDI-MS) of Membrane Proteins and Non-covalent Complexes, *J. Mass Spectrom,* **30,** 1462-1468.
34. Strupat, K., Karas, M., and Hillenkamp, F. (1991) 2,5-Dihydroxybenzoic Acid: A New Matrix for Laser Desorption-Ionization Mass Spectrometry, *Int. J. Mass Spectrom. Ion Proc.,* **111,** 89-102.

35. Westman, A., Huth-Fehre, T., Demirev, P., and Sundqvist, B.U.R. (1995) Sample Morphology Effects in Matrix-Assisted Laser Desorption/Ionization Mass Spectrometry of Proteins, *J. Mass Spectrom.*, **30**, 206-211.
36. Xiang, F. and Beavis, R.C. (1993) Growing Protein-Doped Sinapic Acid Crystals for Laser Desorption: An Alternative Preparation Method for Difficult Samples, *Org. Mass Spectrom.*, **28**, 1424-1429.
37. Juhasz, P. and Biemann, K. (1994) Mass Spectrometric Molecular-weight Determination of Highly Acidic Compounds of Biological Significance via their Complexes with Basic Polypeptides, *Proc. Natl. Acad. Sci. USA*, **91**, 4333-4337.
38. Tang, X., Callahan, J.H., Zhou P., and Vertes, A. (1995) Noncovalent Protein-Oligonucleotide Interactions Monitored by Matrix-Assisted Laser Desorption / Ionization Mass Spectrometry, *Anal. Chem*, **67**, 4542-4548.
39. Rosa, J.J. and Richards, F.M. (1981) Hydrogen Exchange from Identified Regions of the S-Protein Component of Ribonuclease as a Function of Temperature, pH and the Binding of S-Peptide, *J. Mol. Biol.*, **145**, 835-851.
40 Richards, F.M., and Vithayathil, P.J. (1959) The Preparation of Subtilisin of Subtilisin-modified Ribonuclease and the Separation of the peptide and Protein Components, *J. Biol. Chem.*, **234,** 1459-1465.
41. Dogruel, D., Williams, P., and Nelson, R.W. (1995) Rapid Tryptic Mapping Using Enzymatically Active Mass Spectrometer Probe Tips, *Anal. Chem*, **67,** 4343-4348.

NON-COVALENT HETERO- AND HOMO-OLIGOMERIC PROTEIN COMPLEXES REASSOCIATE DIFFERENTIALLY DURING MALDI-MS ANALYSIS

B. R. BLOOM[1], C. R. IDEN[2] AND I. A. MASTRANGELO[1]
1 Brookhaven National Laboratory
Biology Department
Upton, New York 11973
2 State Unversity of New York at Stony Brook
Department of Pharmacology
Stony Brook, New York 11794

1. Introduction

MALDI-MS (matrix assisted laser desorption ionization mass spectrometry) may become a powerful tool in biochemistry in view of its accurate measurement of molecular mass of protein up to and above 300,000 Da (1). Noncovalent protein-protein interactions and subunit composition of protein multimers are the basis of molecular recognition and function in the biological world. As a result, detection of complex structure is an important problem. Conventional biochemical techniques used to detect noncovalent protein complexes include nondenaturing gel electrophoresis and size exclusion chromatography, which are useful in many cases, although neither determines molecular weight with accuracy.

Measurement of proteins by MALDI-MS thus far has yielded only a few examples of intact native multimers. In the MALDI ionization process, an acidic organic matrix serves to minimize protein damage from the laser beam by absorbing incident laser energy, resulting in protein and matrix molecules being ejected into the gas phase. Once the ions are formed in the gas phase, they can be electrostatically directed to a mass analyzer. Surviving the MALDI process from sample preparation in low pH solvent/matrix mixtures to mass detection as native oligomers can be attributed to a number of favorable circumstances. The human immunoglobulin IgM, a 980 kDa pentamer observed by Nelson *et al.*, contains covalent disulfide bonds linking subunits into stable associations (2). Farmer and Caprioli were able to detect dimeric glucose-6-phosphate dehydrogenase (104 kDa) and tetrameric avidin (68 kDa) only after chemically crosslinking with gluteraldehyde (3). van Dorsselaer and colleagues observed 300 kDa aerolysin heptamers, exceptionally stable oligomers which retain native quaternary structure when exposed to concentrated guanidinium hydrochloride, pH extremes, SDS, and heat (4).

We investigated the potential of MALDI-MS for studying complex quaternary structures by using standard MALDI preparations on non-covalent hetero- and homo-

W. Ens et al. (eds.), New Methods for the Study of Biomolecular Complexes, 209–216.

oligomeric complexes, of current interest in our laboratory, that have a range of association constants. These proteins include *E. coli* ClpP protease, human tumor suppressor p53, human replication protein A, and the yeast origin recognition complex. ClpP and hp53 were recombinant proteins expressed in *E. coli* and in insect cells, respectively. hRPA and yORC were purified from their native sources. Our results indicate that the extent of dissociation and reassociation of oligomeric subunits during MALDI may reflect subunit interactions present in the native quaternary structures.

2. Materials and Methods

We used a BRUKER (Billerica, MA) ProteinTOF laser desorption time-of-flight mass spectrometer equipped with a 337 nm N_2 laser with a repetition rate of 3 Hz. All data were acquired at 20 kV acceleration potential in the positive ion mode and with the linear detector set to 1.75 kV post-acceleration. Bovine serum albumin, obtained from Sigma Chemicals, served as an external standard for calibration. Sinapinic acid (SA) and protein sequencing grade trifluoracetic acid (TFA) were purchased from Aldrich Chemicals, HPLC grade acetonitrile (ACN) from J.T.Baker Chemicals, and HPLC grade formic acid (FA) from Fluka. The water used had a resistivity of 18.3 M Ω-cm.

E. coli ClpP was stored in 10 mM Hepes-KOH, pH 7.5, 200 mM NaCl, and 0.1 mM EDTA at a concentration of 65 pmol/μL. ClpP was mixed 1:10 (v/v) with a saturated solution of SA in 2:1 (v/v) 0.1% TFA:ACN. The mixture was then quickly vortexed and centrifuged and 1.0 μL was placed on the target, at a final protein concentration of 6 pmol/μL. Final protein concentrations between 6 and 32 pmol/μL gave similar results.

hp53 was stored in 20 mM Tris-HCl, pH 8.0, 100 mM NaCl and 50% glycerol at a concentration of 2.3 pmol/μL. hp53 was mixed 1:4 (v/v) with a saturated solution of SA in 2:1 (v/v) 0.1%TFA:ACN. The mixture was then quickly vortexed and centrifuged and 1.0 μL was placed onto the target at a final protein concentration of 460 fmol/μL. A final protein concentration of 230 fmol/μL yielded similar results.

hRPA was stored in 20 mM Hepes-HCl, pH 7.0, 250 mM NaCl, 1 mM DTT, 0.1 % NP40, and 25% glycerol at a concentration of 1 pmol/μL. hRPA was mixed 1:4(v/v) with a saturated solution of SA in a 2:1 (v/v) 0.1%TFA:ACN. The mixture was then quickly vortexed and centrifuged and 1.0 μL was placed onto the target at a final protein concentration of 250 fmol/μL. Similar results were obtained with a final protein concentration of 850 fmol/μL

yORC was stored in 50 mM Hepes-KOH, pH 7.5, 100 mM KCl, 1 mM EDTA,1 mM EGTA, 5 mM MgOAc, and 5% glycerol at a concentration of 6 pmol/μL. yORC was mixed 1:4 with a saturated solution of SA in a 2:1 (v/v) formic acid:ACN. The mixture was then quickly vortexed and centrifuged and 1.0 μL was placed onto the target, at a final protein concentration of 1.5 pmol/μL. A final protein concentration of 4 pmol/μL gave similar results.

Mass spectra were obtained by summing 100 or more laser shots. Unsmoothed data is shown in all spectra. Molecular weights cited are based on amino acid sequences. By comparison to the sequence mass, the mass measurement accuracy was typically between ± 0.2-0.6 %.

3. Results and Discussion

ClpP is the catalytic subunit of the ATP-dependent caseinolytic protease in E.coli. Scanning transmission electron microscopy (STEM) and small angle X-ray scattering analyses indicate it consists of two heptameric rings of 21,548.7 Da monomer subunits (5). ClpP is a very stable complex with a dissociation constant estimated as less than 1×10^{-12} M (6; J. Flanagan, pers. commun.). The MALDI-TOF spectra of ClpP in Fig. 1 show a multimeric ladder of signals which decrease in amplitude with increasing multiples of the monomer mass. Also visible are small doubly charged peaks of odd multimers extending to the doubly charged 17-mer present at ~183, 000 Da. Doubly charged peaks of even multimers, including the 14-mer, would overlap singly charged peaks of lower multimers and are therefore not visible. These signals, if taken at face value, suggest that the ClpP multimeric state is greater than a tetradecamer. However, other biochemical and physical methods find tetradecamers almost exclusively. While the observed oligomers appear to be consistent with MALDI-induced, non-specific aggregations, the extensive and accurate mass range observed in the ClpP spectra is noteworthy (see Fig. 1, inset). Under the experimental conditions, ClpP monomers remain intact and retain a capacity to form unusually large aggregates.

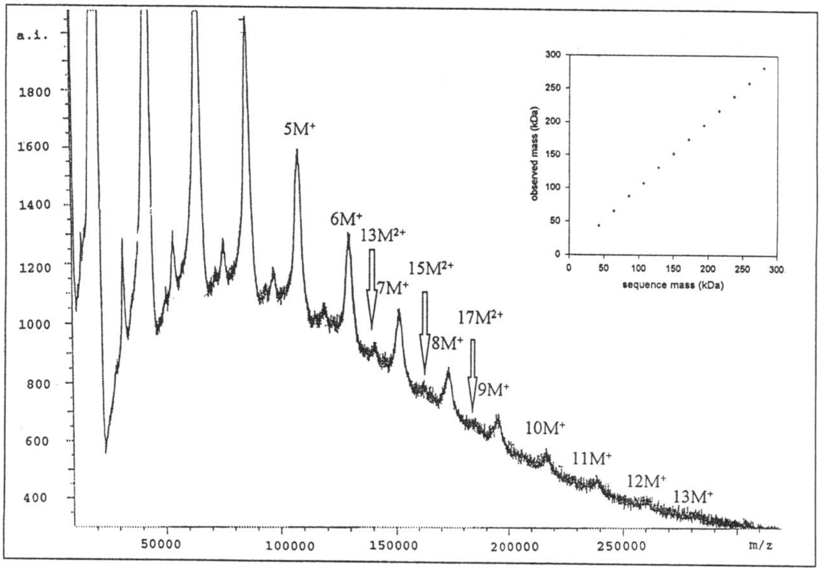

Figure 1. MALDI-TOF positive-ion mass spectrum of *E.coli* ClpP. M for monomer; 2M for dimer, etc. The inset shows the data points which indicate the linearity of the plot of observed mass versus sequence mass (see text).

Human tumor suppressor p53 exists as a stable tetramer composed of 42,653.1 Da monomer subunits. Tetramerization occurs during polypeptide synthesis and subunits can only be interchanged by co-translation (7). The tetrameric structure is important for recognizing specific DNA sequences and for binding at sites of damaged DNA (8,9). X-ray crystallography and NMR spectroscopy reveal the tetramer consists of a dimer of dimers (10-12). MALDI-TOF spectra of p53 are shown in Fig. 2. Distinct singly charged monomer and dimer peaks, and smaller singly charged trimer and tetramer peaks, are visible. In addition, doubly charged monomer and trimer species and triply charged dimer and tetramer peaks are present. Taken together, these dominant peaks might be interpreted as reflections of the underlying tetrameric structure. A weak signal, consistent with a doubly charged pentamer, is also present. Although a minor peak, it indicates that non-specific aggregation has most likely occurred.

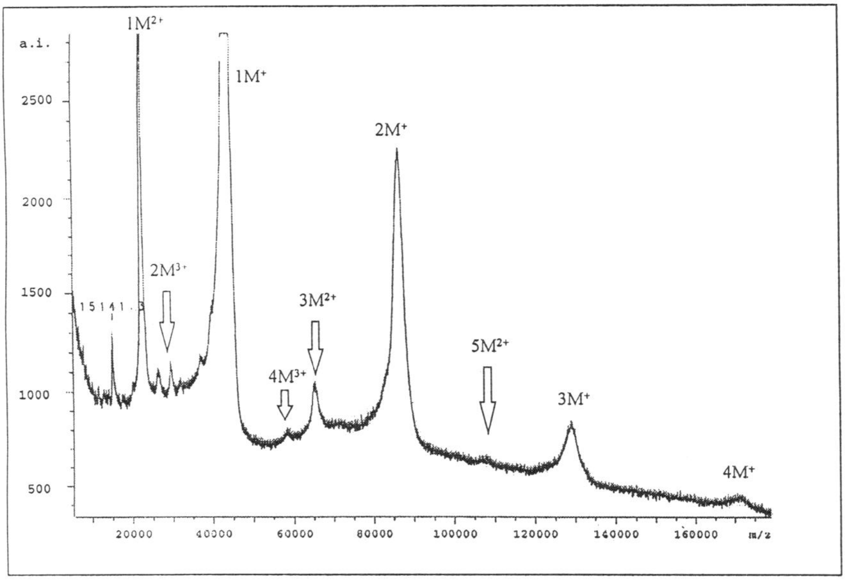

Figure 2. MALDI-TOF positive-ion mass spectrum of hp53.

hRPA is a single-stranded DNA binding protein required to stabilize ssDNA during replication, repair and recombination (13-15). hRPA is a heterotrimer of 68,152.1, 29,246.8 and 13,568.7 Da subunits and all three are essential for viability (16). The hRPA MALDI-TOF spectra are shown in Fig. 3. All three subunits are clearly detected, but the intact heterotrimeric complex was not convincingly seen. Two small peaks observed at 82,320 kD and 97,932 kD indicate pairing of 68/13 kD and 68/29 kD subunits. Other combinations of subunits were not detected. In solution, only the 28 and 14 kDa subunits are known to associate prior to complete complex formation (17,18). The 82 and 97 kDa signals might arise through a combination of laser ablation of the third subunit and non-specific protein interactions in the gas phase. Although sample preparations were almost identical, ClpP and hp53 spectra include multiple oligomeric species compared with hRPA spectra which show minimal oligomeric forms.

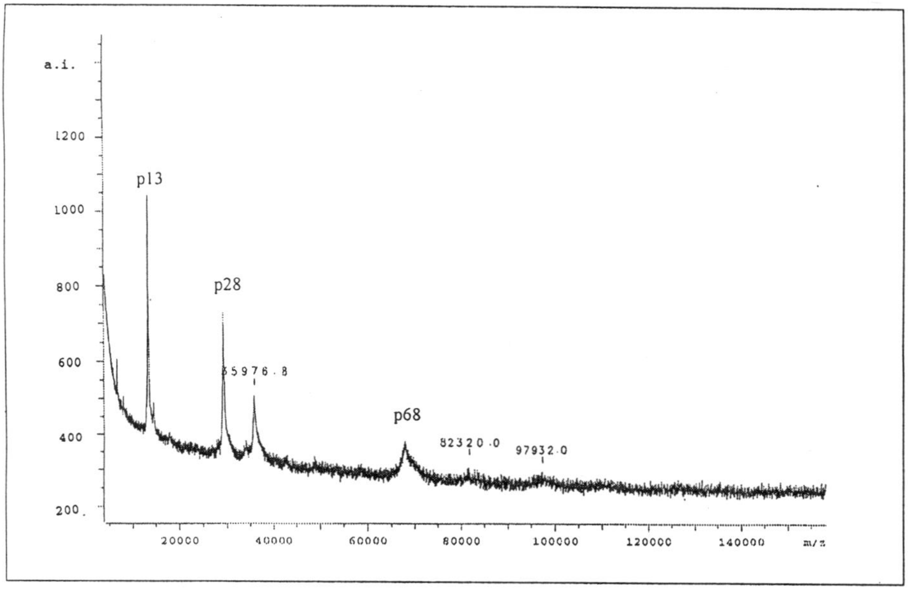

Figure 3. MALDI-TOF positive-ion mass spectrum of hRPA. The strong peak at 36 kDa is observed in different hRPA preparations and may represent a fragmentation product of p68 or a contaminant of purification. We are currently preparing recombinant hRPA which will eliminate the second possibility.

yORC is a six-subunit origin recognition complex recently discovered in yeast and other eukaryotes (18,19). It binds to origins of DNA replication and functions to control DNA synthesis in eukaryotic cells (20). Molecular weights of component peptides are 50,295.5, 55,289.3, 60,705.3, 71,238.0, 72,077.0 and 104,400.1 Da. Fig. 4 shows MALDI-TOF spectra of yORC. Interestingly, no spectra were obtained using the 2:1 (v/v) 0.1% TFA:ACN solvent system with SA. However, substituting a 2:1 (v/v) FA:ACN solvent system allowed all subunits to be detected. The 71.2 kD and 72.1 kD subunits are not resolved. Similar to the minimal aggregates in hRPA spectra, a pronounced 112 kDa peak was the only oligomeric yORC species observed. It could represent paired 50.3 and 60.7 kD subunits or two 55 kD subunits, which have survived laser ablation or reassociated in the gas phase.

To summarize, we have applied MALDI-MS, using a sinapinic acid matrix and a 337 nm laser, to study the multimeric state of non-covalent homo- and hetero-oligomeric protein complexes in the absence of chemical crosslinking. Protein concentrations and storage buffers varied widely. The MALDI spectra for the homo-tetradecamer ClpP extended over an impressive mass range. In the inset in Figure 1, the slope of a least-squares fit to the plotted data equals a monomer mass of 21,618.9 ± 69 Da, accurate to within 0.32% of the sequence mass. Overall, accuracy is best for lower mass oligomers and decreases slightly for higher oligomers. ClpP could be used, to advantage, as a broad-range mass calibration standard. The oligomer ladder observed for ClpP extends several monomer mass units beyond the molecular weight of the native complex in solution. Similarly, the hp53 oligomer ladder reaches beyond the native homo-tetramer to a pentamer mass. In contrast, hRPA and yORC spectra consist almost entirely of a ladder of individual subunits and did not include m/z signals corresponding to the molecular weight of the native complex in solution.

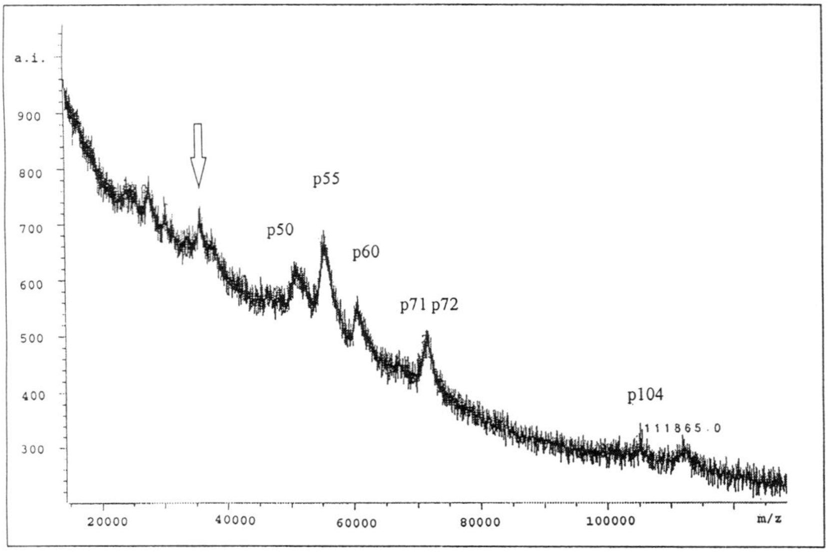

Figure 4. MALDI-TOF positive-ion mass spectrum of yORC. The arrow indicates a doubly charged p71/p72 peak.

The absence of full molecular weight hRPA and yORC complexes, and the virtual absence of oligomers with combined subunit masses, indicate the hRPA and yORC heter-oligomers dissociate and rarely reassociate during the MALDI process. A reasonable speculation for this difference is that homo-oligomer subunits in ClpP and p53 retain some measure of secondary and tertiary structure, including their specialized oligomerization sites These partially denatured regions may play a role in facilitating non-specific reassociation and give rise to oligomeric ladders of aggregates larger than the native protein. Our data indicate that under our conditions homo-oligomer subunits form non-specific protein aggregates far more readily than hetero-oligomer subunits. Structural differences in oligomerization domains and at other sites of interactions between subunits in homo- and hetero-oligomers could contribute to the pronounced reassociation by homo-oligomers.

Acknowledgments

Proteins used in these studies were generously provided by Dr. John Flanagan and James Hartling, Brookhaven National Laboratory (ClpP), Drs. Peter Tegtmeyer and Judith Stenger, State University of New York, Stony Brook (p53), Drs. James Borowiec and Leonard Blackwell, New York University School of Medicine (hRPA), and Dr. John Diffley, Imperial Cancer Research Fund, Clare Hall Laboratories (yORC).

This work was supported by the Department of Energy, Office of Health and Environmental Research (IAM) and Public Health Service grants NIH CA47995 and ES04068 (CRI).

References

1. Siuzdak, G. (1994) The emergence of mass spectrometry in biochemical research, *Proc. Natl. Acad. Sci. USA*, **91**, 11290-11297.

2. Nelson, R.W., Dogruel, D. and Williams, P. (1994) Mass determination of human immunoglobulin IgM using matrix-assisted laser desorption/ionization time-of-flight mass spectrometry, *Rapid Commun. Mass Spectrom.*, **8**, 627-631.

3. Farmer, T.B. and Caprioli, R.M. (1991) Assessing the multimeric states of proteins: Studies using laser desorption mass spectrometry, *Biol. Mass Spectrom.*, **20**, 796-800.

4. Moniatte, M., van der Goot, F.G., Buckely, J.T., Pattus, F. and van Dorsselaer, A. (1996) Characterisation of heptameric pore-forming complex of the *Aeromonas* toxin aerolysin using MALDI-TOF mass spectrometry, *FEBS Letters*, **384**, 269-272.

5. Flanagan, J.M., Wall, J.S., Capel, M.S., Schneider, D.K. and Shanklin, J. (1995) Scanning transmission electron microscopy and small-angle scattering provide evidence that native *Escherichia coli* ClpP is a tetradecamer with an axial pore, *Biochemistry*, **34**, 10910-10917.

6. Maurizi, M.R., Thompson, M.W., Satyendra, K.S. and Seung-Ho, K. (1994) Endopeptidase Clp: ATP-dependant Clp protease from *Escherichia coli*, *Methods in Enzymology*, Academic Press, New York, **244**, 314-331.

7. Milner, J. and Medcalf, E.A. (1991) Cotranslatiion of activated mutant p53 with wild type drives the wild type p53 protein into the mutant conformation, *Cell*, **65,** 765-774.

8. Stenger, J.E., Tegtmeyer, P., Mayr, G.A., Reed, M., Wang, Y., Wang, P., Hough, P.V.C. and Mastrangelo, I.A. (1994) p53 oligomerization and DNA looping are linked with transcriptional activation, *EMBO J.*, **13**, 6011-6020.

9. Lee, S., Elenbaas, B., Levine, A. and Griffith, J. (1995) p53 and its 14kDa c-terminal domain recognize primary DNA in the form of insertion/deletion mismatches, *Cell*, **81**, 1013-1020.

10. Cho, Y., Gorina, S., Jeffery, P.D. and Pavletich, N.K. (1994) Crystal structure of p53 tumor suppressor-DNA complex: understanding tumorigenic mutations, *Science*, **265**, 346-355.

11. Lee, W., Harvey, T.S., Yin, Y., Yau, P, Litchfield, D. and Arrowsmith, C.H. (1994) Solution structure of the tetrameric minimum transforming domain of p53, *Nature Structursl Biology*, **1**, 877-890.

12. Clore, G.M., Omichinski, J.G., Kazuyasu, S., Zambrano, N., Sakamoto, H., Appella, E. and Gronenborn, A.M. (1994) High-resolution structure of the oligomerization domain of p53 by multidimensional NMR, *Science*, **265**, 386-391.

13. Wobbe, C.R., Weissbach, L., Boroweic, J.A., Dean, F.B., Musakami, Y., Bullock, P. and Hurwitz, J. (1987) Replication of simian virus 40 origin-containing DNA in-vitro with purified proteins, *Proc. Natl. Acad. Sci. USA*, **84**, 1834-1838.

14. Heyer, W.D., Rao, M.R.S., Erdile, L.F., Kelly, T.J. and Kolodner, R.D. (1990) An essential *Sacchacomyces cerevisiae* single stranded DNA binding protein is homologous to the large subunit of RP-A, *EMBO J.*, **9**, 2321-2329.

15. Coverly, D., Kenny, M.K., Munn, M., Rupp, W.D., Lane, D.P. and Wood, R.D. (1991) Requirement for the replication protein SSB in human DNA excision repair, *Nature*, **349**, 538-541.

16. Brill. S.J. and Stillman, B. (1991) Replication factor-A from Saccaromyces cerevisiae is encoded by three essential genes and coordinately expxressed at s phase, *Genes Dev.* **5,** 1589-1600.

17. Stigger, E., Dean, F.B., Hurwitz, J. and Lee, S-H. (1994) Reconstitution of functional human single stranded DNA binding protein from individual subunits expressed by baculovirus, *Proc. Natl. Acad. Sci. USA,* **91**, 579-583.

18. Henricksen, L.A., Umbricht, C.B. and Wold, M.S. (1994) Recombinant replication protein a: expression, complex formation, and functional characterization, *J. Biol. Chem.*, **269**, 11121-11132.[

19. Bell, S.P. and Stillman, B. (1992) ATP-dependant recognition of eukaryotic origins of DNA replication by a multiprotein complex, *Nature*, **357**, 128-134.

20. Diffley, J.F.X. and Cocker, J.H. (1992) Protein-DNA interactions at a yeast replication origin, *Nature* **357**, 169-172.

21. Donovan, S. and Diffley, J.F.X. (1996) Replication origins in eukaryotes, *Curr. Opin. Gen. Devel.* **6,** 203-207.

THE DETECTION OF NON-COVALENT INTERACTIONS IN NUCLEIC ACIDS USING MALDI

PAOLO LECCHI and LEWIS K. PANNELL
Laboratory of Bioorganic Chemistry
National Institute of Diabetes and Digestive and Kidney Diseases
National Institutes of Health, Bethesda, MD 20892.

Introduction

The unfriendly conditions for the preparation of samples (an acidic matrix dissolved in organic solvents) has made MALDI (Matrix-Assisted Laser Desorption Ionization) inappropriate for the analysis of fragile complexes of macromolecules. Nevertheless, several authors have proved its feasibility in detecting non-covalent interactions between macromolecules, including: enzyme-substrate complexes (1), protein-polysaccharide interactions (2), peptide-DNA interactions (3), intact quaternary structures of proteins (4), and interactions between complementary oligodeoxynucleotides (5). These experimental results indicate that the ionization process itself might not be the denaturating step in the MALDI analysis. The preparation of the sample seems to play the key role in preserving the structure of the non-covalently bound biopolymers.

In this communication we report further evidence showing the possibility of detecting intact complementary oligonucleotide hybrids by MALDI and we propose a different mass spectrometric approach for the characterization of non-covalent interaction between macromolecules.

Experimental and Results

Mass spectra were obtained on a Kratos MALDI III mass spectrometer (Kratos, Manchester UK) and acquired in negative ion linear mode at an accelerating voltage of 22 kV. Usually, 50 laser shots were collected across the sample surface. The mass spectrometer was externally calibrated using the $[M-H]^-$ peak of oligonucleotides of known molecular weight (MW).

For the usual analysis of oligonucleotides by MALDI, the experimental procedure involves looking for a "hot spot" on the sample target, where the matrix/analyte mixture

W. Ens et al. (eds.), New Methods for the Study of Biomolecular Complexes, 217–224.

yields the highest signal-to-noise ratio when hit by the laser beam. Once the hot spot has been identified, the best results are obtained by averaging spectra from consecutive laser shots (usually 30 to 100). As previously reported (4), fragile non-covalent interactions have been detected in the spectra from just the very first few laser shots. For this reason, we chose a different approach summarized in Fig.1. With this strategy, data were accumulated and analyzed as an array of individual laser shots each of which was acquired at different sites on the sample surface. This produced a three-dimensional plot representing the array of single spectra produced as the laser beam passed over the sample (laser fixed, sample slide scanned).

To facilitate this experimental procedure, it is mandatory to have a smooth surface, such as is obtained using 6-aza-2-thiothymine (ATT) in ammonium citrate buffer (pH 5.5)/CH3CN (50/50) as previously described by our group (6). This protocol proved to give the best results with high signal-to-noise ratio, no formation of alkali adducts to the oligonucleotide samples, no degradation of the oligonucleotides and, most important, higher shot-to-shot reproducibility compared with other matrices.

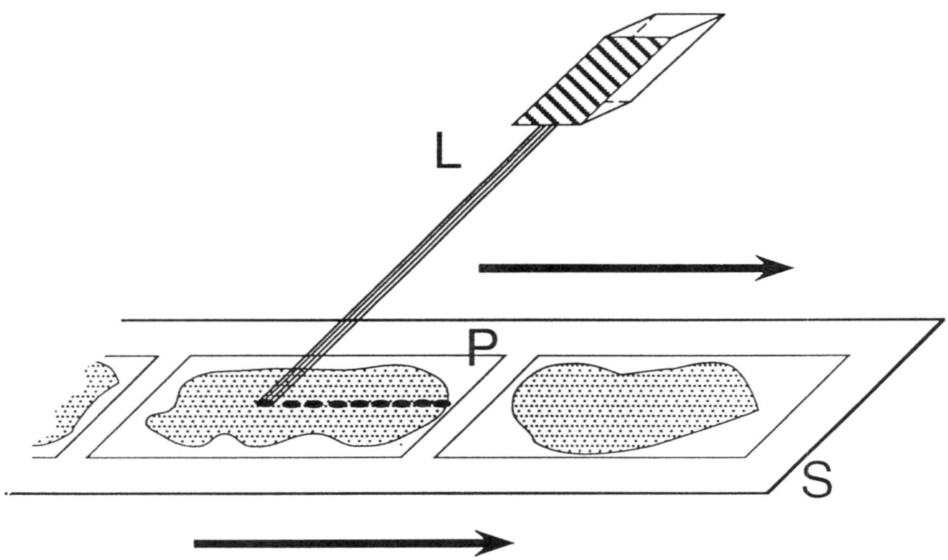

Figure 1. L = Nitrogen laser beam (337 nm wavelength, 3 nsec pulse width and 20 μm "virtual" diameter), P =sample probe (a polished surface of 2x1 mm) located on a steel slide (S) which contains up to 20 small sample areas. During the acquisition of data, S is moved by a stepper motor such that the laser pulses are evenly spread in a line across the sample surface.

In the first example, MALDI was used for the analysis of two complementary oligonucleotides, each 8 bases in length. The spectrum obtained (Fig. 2) shows the two individually resolved dissociated strands, but not the double-stranded (ds) DNA. Failure to detecting ds-DNA was expected as the melting temperature of the 8-mer hybrid in solution is below room temperature (7). The absence of non specific dimer formation indicates that the laser does not cause photo-chemical cross-linking between oligonucleotides.

The use of divalent cations (e.g. calcium and magnesium) has been described as an important factor in increasing the stability of the nucleic acid hybrids in solution (7). We used this strategy to improve the stability of the DNA duplex during the sample preparation. In this experiment, a 16-mer ds-DNA d(pAATATTAGTTAGT TAG)/d(TTATAATCAATCAATCp) was analyzed both with and without calcium. While calcium did increase the ability to detect the ds-DNA, the formation of multiple cation adducts adversely affected the determination of the MW (Fig. 3). The calculated MWs for the two single strands and for the ds oligonucleotides are: 5013.3 Da, 4902.2 Da and 9916.5 Da respectively.

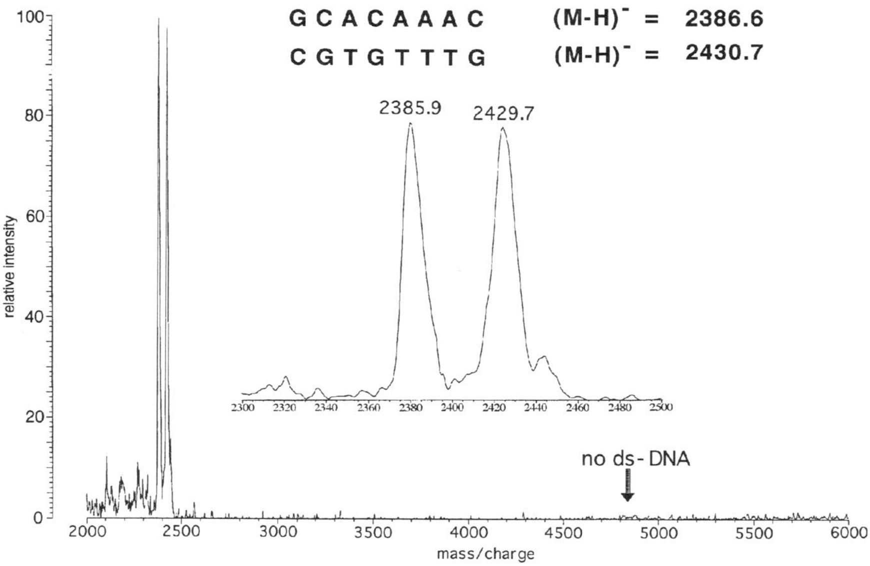

Figure 2. Complementary oligonucleotides 8 bp in length. The inset shows the two completely resolved single strands. No evidence of ds-DNA detection or formation is observed.

To rule out the possibility of non-specific dimer formation and to further refine the experimental conditions, two complementary oligodeoxynucleotides (12-mer and 16-mer in length) were annealed in solution under the following conditions:

i. Equimolar amounts (20μM) of 12-mer and 16-mer oligodeoxynucleotides were mixed either in: a) annealing buffer (100 mM ammonium citrate pH 5.5); b) buffer citrate containing $CaCl_2$ (5 mM); c) buffer citrate containing formamide (20% v/v); or d) buffer citrate containing spermidine tri-HCl 5 mM.
ii. The solutions were heated for 15 min. at 75°C.
iii. The temperature was allowed to decrease gradually from 75°C to 4°C (on the bench top in a cold room).

These should yield a ds-DNA fragment with a protruding 5'-end:

```
5'-AATTCCGTTGCTGTCG-3'        MW = 4863.2
    3'-GGCAACGACAGC-5'        MW = 3664.4
```

The mass spectrum of this compound has a small peak at the MW of the ds-DNA, along with two major peaks at the MWs of the two individual oligonucleotides (not shown). When calcium was added to the solution, as shown before, increased detection of the ds-DNA was observed but a shift in MW occurred due to calcium adduct formation. The addition of formamide (an agent that decreases the stability of the hybrids) caused the loss of any ds-DNA signal. When spermidine (a factor that increases the stability of the hybrids) was added to the solution, a major peak at the MW of the ds-DNA was observed with no detection of non-specific dimers (Fig. 4).

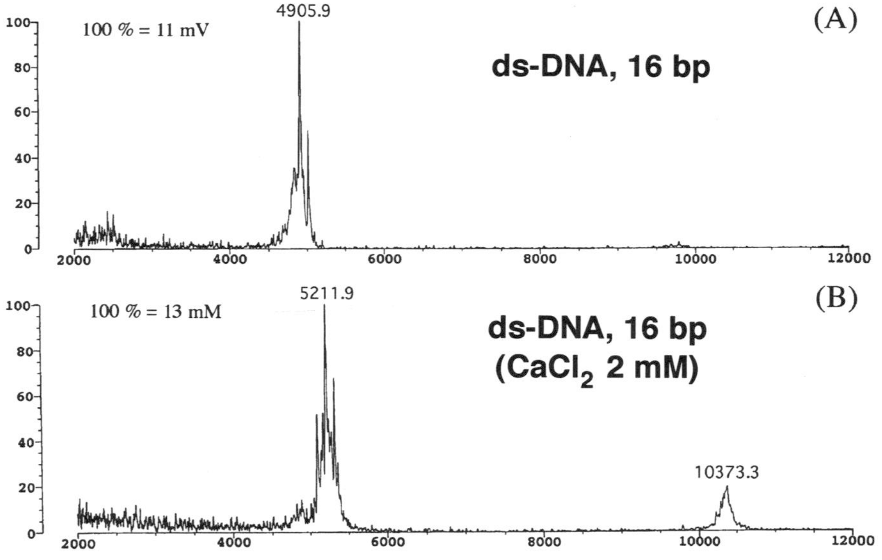

Figure 3. Analysis of a ds-DNA 16 bp in length (Sigma cat # S-9650): (A) without calcium and (B) with calcium (2 mM).

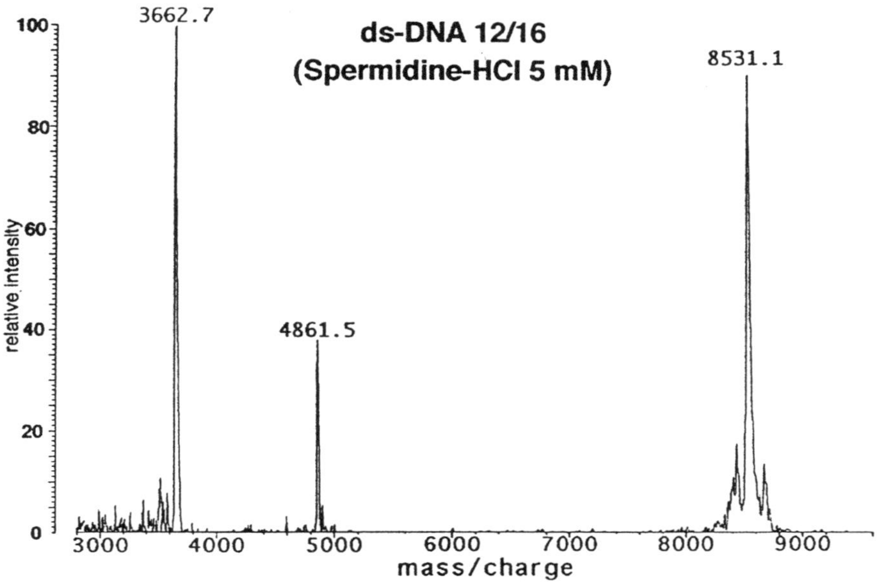

Figure 4. Spectrum of a ds-DNA, 12/16 bp in length, obtained in the presence of spermidine-HCl (5mM).

To exclude the possibility that ds-DNA peaks may arise from the formation of adducts during the desorption process, individual laser shot spectra were examined. Figure 5 shows a progression of 40 laser shots across the sample surface, individually acquired and plotted on a three-dimensional axis. This shows the distribution of the three species (both monomers and the ds-DNA), each of which have different MWs. The ds-DNA in this sample occurs when little or no single-stranded oligonucleotides appear, and vice-versa. This technique is a useful way of distinguishing the desorption of intact ds-DNA from laser-produced artefacts. Desorption-produced dimers would be expected to occur when maximum monomer signal is obtained.

Since RNA/RNA hybrids display higher stability than DNA/DNA or DNA/RNA hybrids, and have consequently higher melting temperatures for a given molecular size, a palindromic 12-mer oligoribonucleotide was tested. The melting temperature of the hybrid is well above room temperature, and a stable dimer is formed (ΔG -27.2 Kcal/mol), and as expected, the ds-RNA was observed in the MALDI spectrum. As before, the spectrum was plotted on a three-dimensional axis to allow the examination of the distribution on the sample spot of single-stranded and the ds-RNA (Fig 6).

Discussion

We have shown the praticability of detecting non-covalent interactions between nucleic acids using MALDI. The use of single shot spectra to distinguish intact double-stranded oligonucleotides from those which could be formed during the mass spectrometry process is critical as a proof of the desorption of the complexes. Most notably, a high level of the single stranded species does not correspond to a high level of the "dimer" peaks.

Some questions are still open:

(1) Is it reasonable to compare the gas phase conditions with the conditions in solution (considering also that the analytes are desorbed from the solid state)?

(2) After the laser shot, is there stabilization (due to an adiabatic expansion and cooling) (8) or is there a degradation of macromolecular interaction due to the increase of the temperature?

(3) Does the laser shot induce photochemical cross-linking between complementary strands? If so, what is the role of the matrix?

The use of an appropriate procedure for sample preparation seems to be the key step in detecting non-covalent interactions between complementary oligonucleotides. ATT was the only matrix among those we tested (e.g.: 4-hydroxy-α-cyano-cinnamic acid, gentisic acid, sinapinic acid, and 5-hydroxy picolinic acid) which allowed the observation of non-covalent interactions.

With some more refining (e.g. detecting oligonucleotides with higher MW), MALDI might become a powerful tool for studying the structure of nucleic acids (e.g. characterization of the secondary structure of DNA and/or RNA) and could be a useful probe in testing hybridization between naturally and chemically modified oligonucleotides (e.g. antisense oligonucleotides) which are of great interest as potential therapeutic agents.

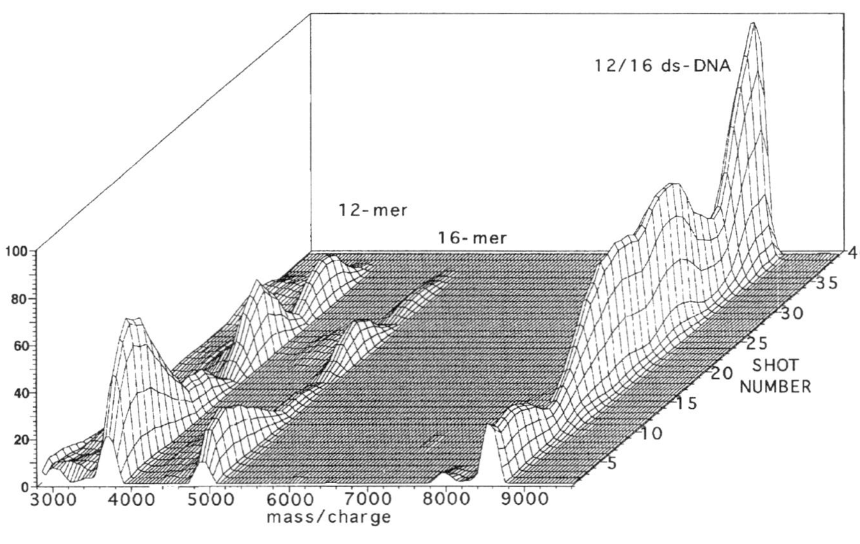

Figure 5. The data presented in Fig 4. has been replotted shot-by-shot (laser pulses) on a 3-D axis.

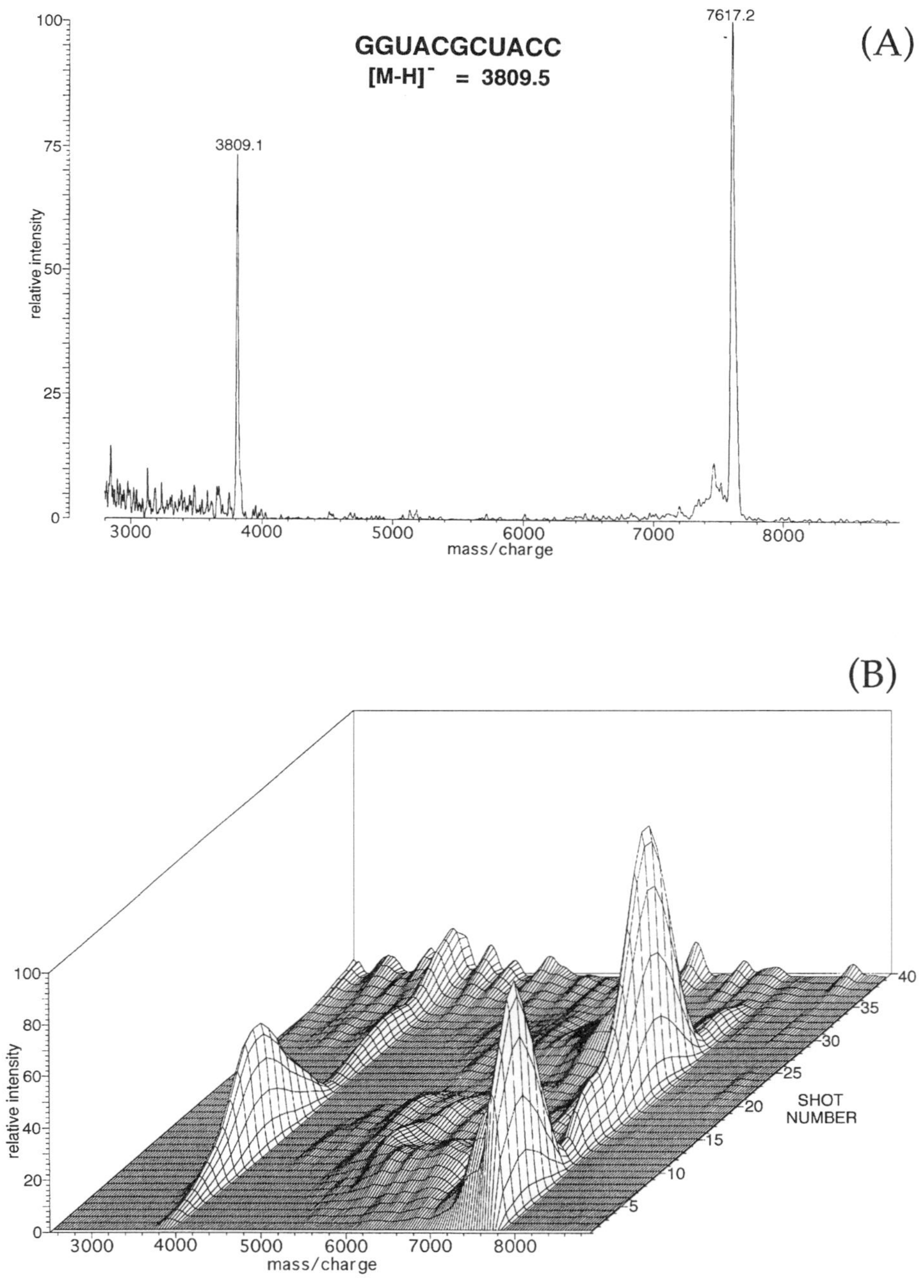

Figure 6. Analysis of a self complementary 12 mer oligoribonucleotide: (A) plotted as a combined spectrum and (B) plotted shot-by-shot on a 3-D axis.

Acknowledgements

We thank the LMB, NIDDK, for the gift of RNA samples. We acknowledge the use of the MALDI Kratos III instrument in the laboratory of Henry Fales (LBC, NHLBI) and our special thanks go to Ralph Nesson for making available an instrument at Shimadzu (Columbia, MD).

References

1. Woods A.S., Buchsbaum J.S., Worrall T.A., Berg J.M and Cotter R.J. (1995) *Anal. Chem.* **67**, 4462-4467.
2. Tang X., Callahan J.H., Zhou P. and Vertes A. (1995) *Anal. Chem.* **67**, 4542-4548.
3. Juhasz P. and Biemann K.(1994) *Proc. Nat. Acad. Sci. USA* **91**, 4333-4337.
4. Rosinke B., Strupat K., Hillenkamp F., Rosenbusch J., Dencher H., Kruger U. and Galla H.J. (1995) *J. Mass Spectrom.* **30**, 1462-1468.
5. Lecchi P. and Pannell L.K. (1995) *J. Am. Soc. Mass Spectrom.* **6**, 972-975.
6. Lecchi P., Le H.M.T. and Pannell L.K. (1995) *Nuc. Acid Res.* **23**, 1276-1277.
7. Saenger W. (1984) *Principles of Nucleic Acid Structure*, 2nd. chapter 6; Springer, New York.
8. Vertes A. and Levine R.D. (1990) *Chem. Phys. Letters* **171(4)**, 284-290.

MASS SPECTROMETRIC METHODS FOR BIOMOLECULAR CHARACTERIZATION

R. W. NELSON, J. R. KRONE, K. A. TUBBS, D. DOGRUEL
Department of Chemistry and Biochemistry
Arizona State University
Tempe, AZ, 85287-1604

1. Introduction

Mass spectrometry has long been viewed to possess the ability to readily contribute to the rapid, sensitive and accurate characterization of biomolecules. However, it has been only within the last decade, with the advent of electrospray ionization (ESI) [1] and matrix-assisted laser desorption/ionization (MALDI) [2] mass spectrometries, that routine biomolecular analyses at physiological levels have become a reality. Indeed, these techniques have opened the door to the regular practice of biological mass spectrometry (at the bench level), and in doing so, have paved the way for a number of mass spectrometric-based methods directed at the intricate investigation of biomolecular structure and function. Generally, these methods revolve around the combination of other analytical techniques or methodologies (*e.g.* chromatography, electrophoresis, enzymatic/chemical treatment of analyte) with mass spectrometry. In order to achieve the highest levels of analytical performance, the integration of techniques must be performed in manners which minimize sample losses, maintain a desired rapidity, and preserve a high level of accuracy.

Over the last few years, we have been involved in the development of such mass spectrometry-based methods. Presented here is a brief overview of our findings on: 1) The integration of micro-scale immunoaffinity isolation with MALDI, 2) The use of *bioreactive* mass spectrometer probes in the structural analysis of proteins, and, 3) The interfacing of surface plasmon resonance (SPR)-based Biomolecular Interaction Analysis (BIA) with MALDI.

2. Mass Spectrometric Immunoassay

A common occurrence during the direct MALDI analysis of complex mixtures is the suppression, in the mass spectrum, of various analyte signals. The causes of this

W. Ens et al. (eds.), New Methods for the Study of Biomolecular Complexes, 225–238.

phenomenon, while not completely understood, can likely be linked to a number of effects including precipitation of analytes due to protein/protein interactions, and the preferential inclusion of certain analytes into the MALDI matrix -- to the exclusion of other species exhibiting less avidity towards the matrix. Another, more fundamental reason for the lack of a given ion signal in the mass spectrum is that the solution concentration of the target species is too low. That is to say, the concentration of the analyte is below that necessary to introduce an adequate amount of sample (for signal generation) into the mass spectrometer when typical MALDI sample preparation protocols are used (preparations generally involving the mixture of low microliter volumes of analyte solution with the equivalent of matrix solution).

An answer to the aforementioned problems is the selective isolation of desired analytes from solution prior to MALDI using an affinity medium. In this manner, unwanted mixture components are discarded from the analysis while targeted species are selectively retained and concentrated (in cases where relatively large volumes of analyte solution are exposed to the affinity medium). The approach has coined a number of acronyms -- surface-enhanced affinity capture (SEAC) [3], affinity-directed mass spectrometry (AMS) [4,5], and, probe affinity mass spectrometry (PAMS) [6], -- all of which generally relate to the same thing; the use of affinity capture in isolating analytes for MALDI.

We have introduced another acronym, mass spectrometric immunoassay (MSIA) [7 - 9]. The general approach is much like the aforementioned methods, however, with subtle yet significant differences meant for the overall improvement of affinity capture for mass spectrometry. First, a high-throughput/surface area affinity-derivatized medium is used for affinity capture (different, for instance, from the two-dimensional media used in PAMS). The incubation steps are performed off-line, flowing, and in manners which allow large volumes (over 10 ml) of solution to be brought into contact with the affinity medium in a very short amount of time (1 - 5 minutes). These steps are generally performed with the affinity medium retained inside a micro-pipettor tip which allows for a final, microliter-volume elution step (generally using the MALDI matrix) directly onto the surface of a mass spectrometer probe. The elution step is a second deviation from the above methods (SEAC), in that a beaded affinity medium is not introduced into the mass spectrometer (a practice that degrades the performance of the mass analysis). Another difference is the covalent linkage of the affinant to the affinity medium. Affinant-derived interferences are thereby eliminated in the mass spectrum -- an important step if second-dimension analyses (*e.g.* enzymatic mapping) are to be performed on the captured ligands. Lastly, when appropriate, internal reference standards are incorporated into the MSIA for use in quantitative MALDI [10,11]. This last step, quantitative detection, is fundamentally the most critical if MSIA, or any form of affinity-capture mass spectrometry, is to be applied for clinical or diagnostic assay.

Although some of the aforementioned steps may seem trivial from a conceptual standpoint, it should be noted that the careful practice of all aspects of MSIA is necessary in order to achieve successful detection of target analytes present in "real" analytical systems. To illustrate, an example of the application of MSIA in the detection of complement protein C1q present in the brain tissue of Alzheimer's disease (AD)-positive patients is given.

2.1. COMPLEMENT PROTEIN C1Q

C1 is a protein of the complement system which lyses microorganisms and infected cells by forming holes in their plasma membrane [12]. The protein consists of a recognition unit *q* and two zymogens, *r* and *s*. C1*q* is an approximately 460 kDa hexamer embodying three kinds of chains (A, B, and C) each about 24 kDa. The three chains compose a total of nine subunits, six of which are disulfide-linked dimers of the A and B chains, and three of which are disulfide-linked dimers of the C chain. Each ABC trimer forms a triple-stranded helix similar to that in collagen with the overall structure resembling a bouquet of six tulips (see Fig. 1). Fig. 2 is a mass spectrum of purified human C1q laser desorbed/ionized from a matrix of sinapinic acid. Although the spectrum looks complicated, all the peaks can be attributed to C1q and its various subunits. As already stated, C1q is composed of A,B, and C chains which exist as 6[AB] disulfide-linked units, and 3[C_2] disulfide-linked units. These two individual units are seen in the mass spectrum at mass-to-charge (m/z) = 53,480, and 48,330 Da, respectively. In the intact C1q molecule, two of the AB units combine with one of the C_2 units to form a nominal 150 kDa subunit (denoted as SU). Ion signals are observed in Fig. 2 for a species at m/z ~ 155 kDa, signifying the presence of 1 SU. The interesting aspect of this species is the non-covalent, collagen-like triple helix responsible for holding the subunit together. Additional signals are observed in the mass spectrum for 2 SU (m/z ~ 310 kDa) and 3 SU (the intact C1q molecule,

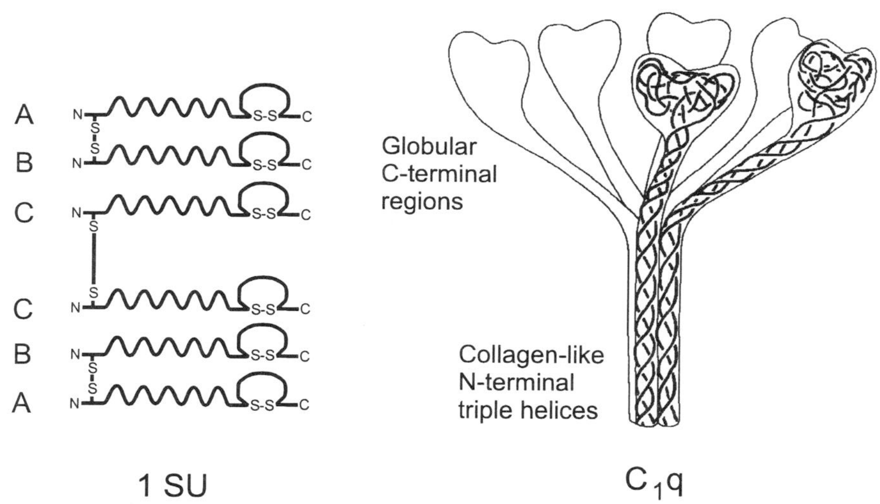

Figure 1. Illustration of complement C1q. Covalent dimers, 2(AB) and C_2 **(Left)**, combine to form a single subunit of the three-subunit collagen-like triple-helix structure **(Right)**.

m/z ~ 465 kDa) and various multiple charge states. Also observed are 1 SU plus additional AB or C_2 units (marked by asterisks at m/z ~ 210 kDa and 260 kDa). Fitting within the context of these proceedings is perhaps the most interesting feature of the C1q spectrum; an apparent higher degree of stability for the fully-closed assembly than the next closest intermediate. This conclusion is drawn from the lack of ion signals for the 2 SU plus (either) AB or C_2 (or combination thereof) species. The absence of these signals (which should be observed at m/z ~ 360 kDa and 410 kDa) while still observing the intact C1q, suggests a relatively low degree of stability for the intermediates, and supports the notion of an organized non-covalent trimeric structure (3 SU) for the C1q. It is worth noting that non-covalent complexes are typically not observed in sinapinic acid MALDI spectra [13], illustrating a need for further investigations into the intricacies of the C1q (and other triple-stranded helix systems).

2.2. C1Q MASS SPECTROMETRIC IMMUNOASSAY

As interesting as the observations in Fig. 2 may be, the purpose of the mass spectrum is primarily as a signature for an MSIA directed at detecting C1q in the brain tissue of AD-positive patients (post-mortem).

Depicted in Fig. 1 are the six, c-terminal globular-like regions of the C1q molecule. Each of these globular-like regions contains a binding site for the Fc region

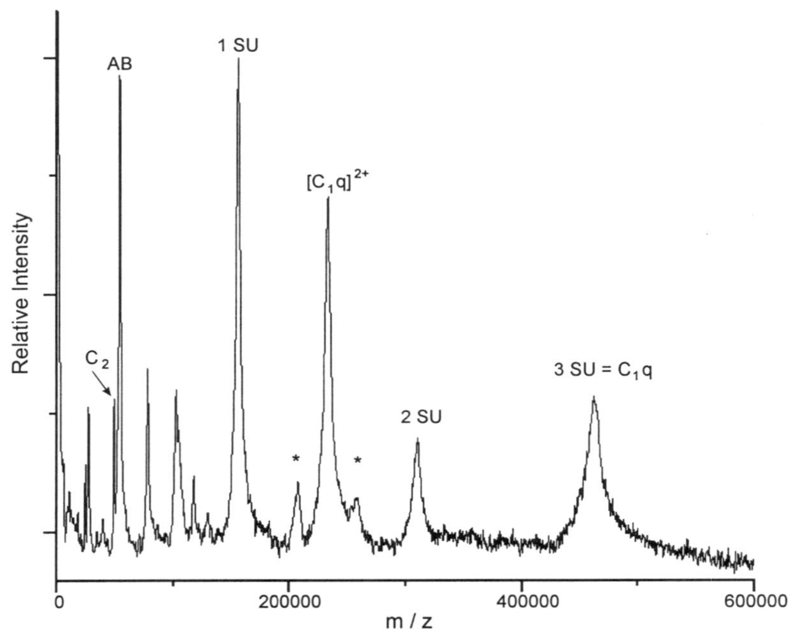

Figure 2. MALDI mass spectrum of complement C1q. Covalent (AB, C_2) and non-covalent subunits (SU) leading to the fully-assemble C1q molecule (m/z = 465,0000 Da) are observed. Intermediate components (denoted by *) are not observed between 2 SU and the C_1q assembly.

of an antibody molecule. Binding of antibodies to the region triggers a complement cascade crucial in the nonspecific humoral defense against microbial pathogens. There exist two different complement pathways, of which components of only the "classical pathway" -- including C1q -- have been found in the brains of AD-positive patients. It is speculated that the complement cascade components of the classical pathway are associated with the formation of senile plaques and activation within Alzheimer's disease tissue resulting in bystander cell lysis which may contribute to AD-positive neuropathology [14].

Fig. 3 shows the results of screening AD-positive brain homogenate for the presence of C1q. Fig. 3A is a MALDI spectrum of whole brain homogenate. The few signals observed in the mass spectrum are of low mass species (< 50 kDa), and give no indication of the presence of C1q. Results of the MSIA analysis of the same brain homogenate using a polyclonal anti-C1q IgG affinity reagent are shown in Fig. 3B. The presence of C1q is indicated by the signature subunit series (including the intact C1q at m/z ~ 465 kDa), as well as the basic building blocks: AB and C_2. Also readily observed in the spectrum is the presence of an additional species observed at m/z ~ 150 kDa species. This species is presumably brain-borne IgG retained by the Fc-binding regions of the C1q molecule. Control spectra of the affinity reagent indicated no "leakage" of IgG from the media (due to the covalent binding of the IgG to the media). Blank MSIA spectra of AD negative brain homogenate gave no indication of either IgG or C1q.

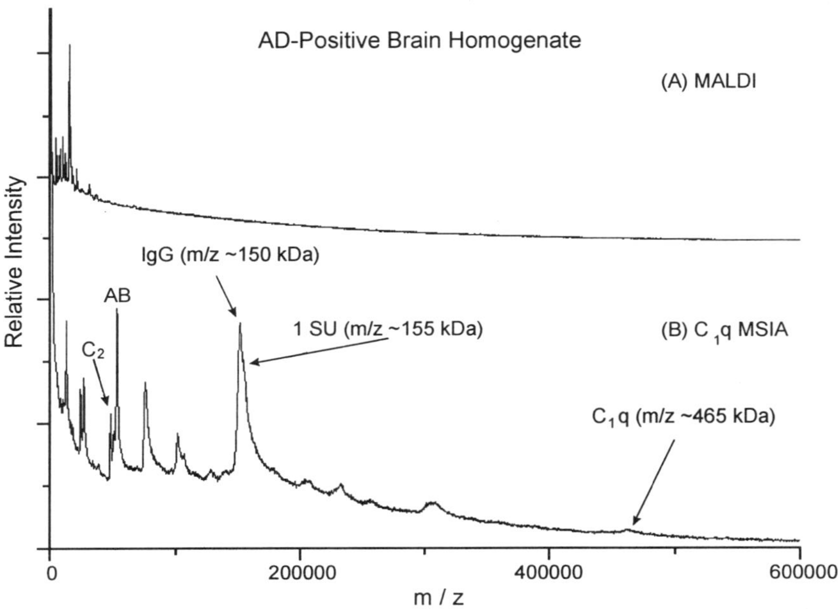

Figure 3. **(A)** MALDI of AD-positive brain homogenate. **(B)** Anti-human C1q MSIA of AD-positive brain homogenate. Components of the C1q, and retained brain-borne IgG, are observed during MSIA. Further details of MSIA procedure are given in References 7-9.

Although this example demonstrates the successful implementation of MSIA for use in the determination of (suspected) AD positive markers, it does little as far as providing a viable diagnostic for Alzheimer's disease. The single underlying reason is that the target media was brain tissue, the sampling of which is generally not undertaken on a regular basis. In order to be more useful, the C1q MSIA needs to be applied to a fluid system (serum or CSF) and performed in a rigorously quantitative manner. Development is still underway.

3. *Bioreactive* Mass Spectrometer Probe Tips

To date, the principle use of biological mass spectrometry has been, in some form or another, related to the analysis of the primary structure (sequence) of peptides and proteins. Analyses range from verification of sequence by absolute molecular weight determination [15], to partial sequence elucidation using tandem mass spectrometry [16]. Other analyses move more into the identification of unknowns using enzymatic/chemical mass mapping techniques in combination with genome/proteome database searches [17]. Moreover, mapping approaches can be employed with mass spectrometry to readily identify variants of molecules differing in mass by as little as a single mass unit [18] (due to residue miscall or point defects). There exists, however, the potential of using mapping techniques beyond sequence verification and in the investigation of the higher-order (tertiary and quaternary) structure of proteins.

A few issues are worth considering before using mass mapping in the investigation of protein higher-order structure. The first of these is that the digestion needs to be performed in a manner free of artifacts introduced by the digesting agent. Autolysis products, for instance, pose a hindrance to analysis (by interfering with analyte signals), and should be kept to a minimum when possible. A second issue is that of analytical sensitivity/precision. Although structural analyses are typically performed on analytes either (recombinantly) expressed, synthesized, or readily isolated/purified -- so there is ample intact reagent -- there is a practical issue of losing digestive fragments in transfer to and from reaction vessels, and in sample preparation. That is to say, certain fragments will exhibit affinities for vessel and pipettor tip walls and be lost in transfer. With the loss of digestive fragments comes a corresponding loss in the precision of the analysis. Another issue is the speed at which the digestive steps can be performed. Here, obviously, the faster the better, as short duration digestions (< 1 minute) could be performed on analytes taken from one environment to another, denaturing-to-renaturing for instance, allowing the possibility of catching molecules (as a function of time) at various stages during folding. Alternatively, rapid digestions can be performed as a function of denaturant concentration to "landmark" various stages of molecular conformation (tertiary structure) and/or assembly (quaternary structure). We have begun using mass mapping in the study of protein conformation and assembly, and find an ideal way of performing the digestions is with devices termed *bioreactive* mass spectrometer probes [19 - 22]. These are conceptually and practically simple devices constructed for the purposes stated above: the rapid, sensitive and accurate mass mapping of proteins without interference from the digestive reagents. In essence, mass spectrometer probes are converted into the digesting agent through the covalent immobilization of enzymatic or chemical reagents. Digestion is initiated upon application of an analyte, and terminated with the addition of a MALDI matrix. Sample losses are kept to a minimum as there is no transfer or handling beyond the initial application, and digestion times are short because high enzyme-to-substrate ratios can be used without introduction of autolytic interferences. Overall, the *bioreactive* devices have been found to be a highly effective means for analytical mapping of proteins.

3.1. HEMOGLOBIN MASS MAPPING

Bovine hemoglobin exists as a naturally occurring non-covalent tetramer comprised of two alpha chains (Hbα) and two beta chains (Hbβ). The molecule also contains four prosthetic heme groups to facilitate oxygen transport. Hemoglobin is the major oxygen transport protein among invertebrates, exhibiting an overall (quaternary) conformational change (*R* to *T*) when exposed to O_2. Moreover, the molecule can be made to traverse various stages of unfolding upon exposure to denaturants. Mass mapping is an ideal means of tracking these changes as each stage of the unfolding will exhibit different digestion patterns (dependent on the steric hindrance of the cleavage sites).

Shown in Fig. 4 are mapping spectra taken of bovine hemoglobin exposed to four different concentrations of the denaturant n-octyl glucoside (NOG). Trypsin *bioreactive* probes were used in the mapping. Anywhere from five to approximately forty ion signals are observed in the mass spectra. The signals can be, with few exceptions, assigned (by accurate molecular weight determination) unambiguously to cleavage fragments of either Hbα or Hbβ. Coverage maps corresponding to the spectra are shown in Fig. 5. The data are interpreted as the hemoglobin going through three stages of unfolding with the increasing denaturant concentration. The first stage of unfolding is actually disruption of the tetrameric assembly. At low (< 5 mM) NOG concentrations, little (digestive) coverage of the hemoglobin is observed. This observation is consistent with an overall steric shielding of both chains due to the tetrameric structure of the molecule being fully intact. Between a NOG concentration of 5 mM and 10 mM, the extent of digestion is observed to increase as regions of the Hbα, previously shielded by the tetrameric structure, become accessible. Changing the denaturant concentration even further (to 20 mM) results in a dramatic difference in the number of cleavage fragments and corresponding sites observed for the Hbα (19 fragments; 11 sites). Indicated within this is an unfolding of the initial globular state of the Hbα. A similar transition is observed for the Hbβ chain upon further increase of the denaturant concentration (to 30 mM). Signals consistent with cleavage of the Hbβ chain at 13 sites, producing 20 cleavage fragments, are observed in Figs. 4D and 5D.

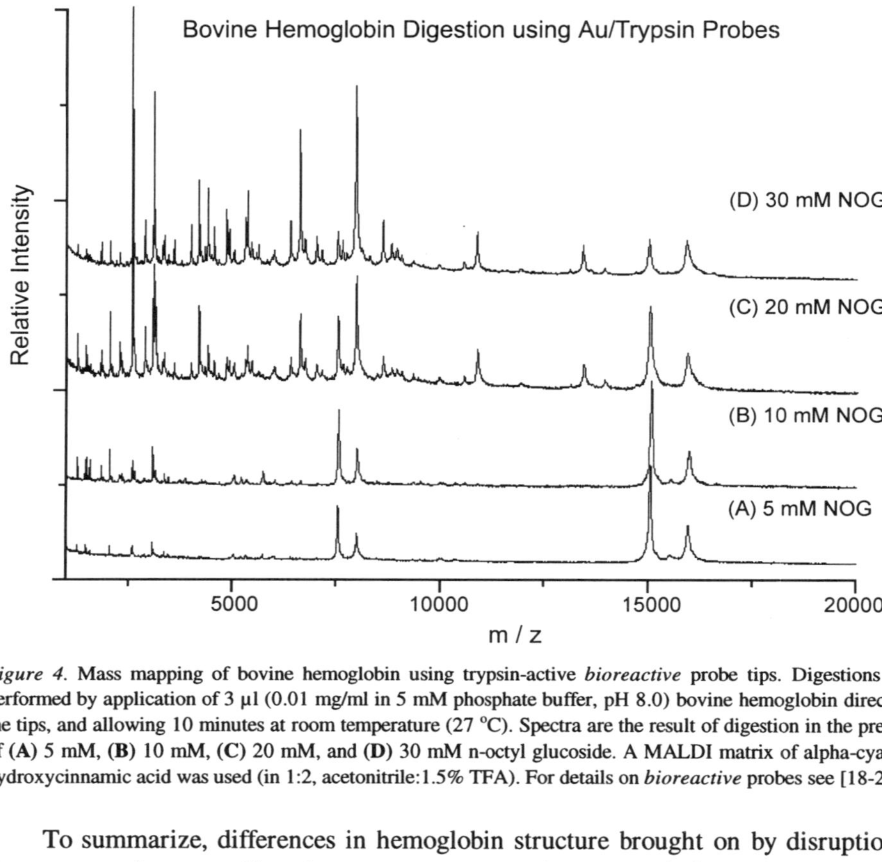

Figure 4. Mass mapping of bovine hemoglobin using trypsin-active *bioreactive* probe tips. Digestions were performed by application of 3 μl (0.01 mg/ml in 5 mM phosphate buffer, pH 8.0) bovine hemoglobin directly to the tips, and allowing 10 minutes at room temperature (27 °C). Spectra are the result of digestion in the presence of **(A)** 5 mM, **(B)** 10 mM, **(C)** 20 mM, and **(D)** 30 mM n-octyl glucoside. A MALDI matrix of alpha-cyano-4-hydroxycinnamic acid was used (in 1:2, acetonitrile:1.5% TFA). For details on *bioreactive* probes see [18-21].

To summarize, differences in hemoglobin structure brought on by disruption of the tetrameric assembly (quaternary structure), or variations in constituent conformation (tertiary structure), are readily recognized during mass mapping. *Bioreactive* mass spectrometer probe tips were used as a means of studying such conformational differences free of artifacts, and with high speed and sensitivity. The future direction of this research is the design and implementation of *bioreactive* probes having much higher enzymatic activities (and hence, rates of digestion) for the tracking of the dynamics (as a function of time) involved in protein folding.

4. Biomolecular Interaction Analysis Mass Spectrometry (BIA/MS)

An assay that seems ideally suited for interfacing with MALDI is biosensor analysis based on the non-destructive detection principle of surface plasmon resonance (SPR). This detection principle relies on the variation of the SPR angle with the refractive

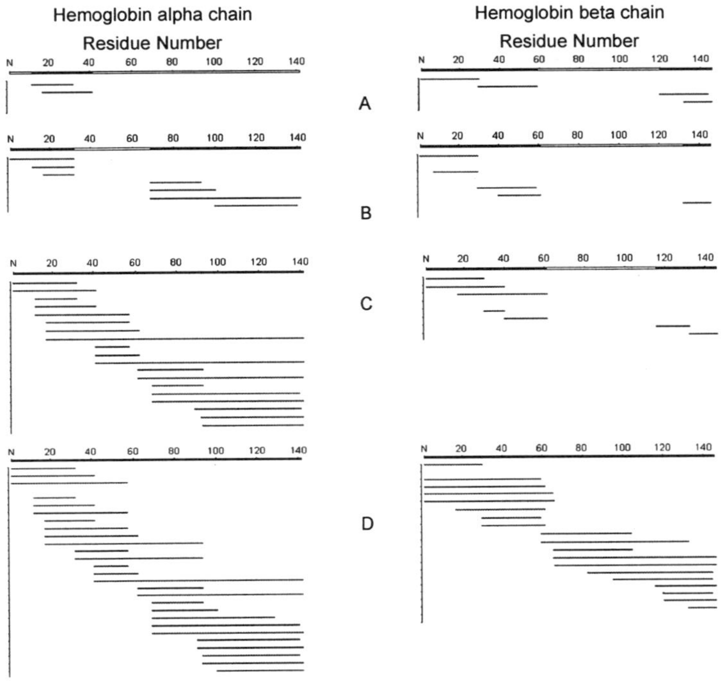

Figure 5. Coverage maps of bovine hemoglobin digested using trypsin-active *bioreactive* probes; **(Left)** Hbα coverage, **(Right)** Hbβ coverage. Intact hemoglobin was exposed to NOG concentrations of **(A)** 5 mM, **(B)** 10 mM, **(C)** 20 mM, and **(D)** 30 mM.

index changes of a surface due to adsorption of molecules. A technique that utilizes SPR detection in the investigation of structure-function relationships is biomolecular interaction analysis (BIA). BIA is a biosensor technology used in monitoring binding of an analyte (in solution) to a surface-immobilized binding partner. The sensor (chip) surface used in this method consists of a carboxylated-dextran hydrogel layer coupled to a thin gold surface. Affinity ligands (antibodies, receptors, streptavidin) are bound within the dextran yielding a quasi-three-dimensional affinity matrix. Any change in surface coverage resulting from ligand-affinant interaction is detected as a change in the SPR signal. The resulting sensorgrams report the mass quantity of analyte bound to the chip surface (in RU with 1000 RU = 1 ng/mm^2) as a function of time. Sensorgram data can then be used to determine kinetic parameters (on- and off-rates), and the molar absorptivity constants of the interaction.

Although BIA yields information pertinent to ligand binding and kinetics, SPR detection is non-specific. As a result, the identity of retained ligands may not always be certain. This situation can hold particularly true in complex systems where there exists the potential of binding multiple or unknown ligands. Such unknown binding is of great consequence during the curve fitting of rate laws to data which is assumed to be derived from a known system. By mass spectrometrically analyzing ligands retained during BIA, it is possible to unambiguously verify the retention of the desired ligands. Alternatively, the presence of non-targeted ligands is as easily determined and possibly corrected. In recognition of the complementary nature of the two techniques, we have investigated the interfacing of BIA with MALDI mass spectrometry [22 - 24]. To avoid transfer losses and achieve the best sensitivity, ligands were sampled directly from the sensor chip surface.

4.1. HUMAN MYOGLOBIN BIA/MS

A biomolecular recognition system of polyclonal anti-human myoglobin IgG/human myoglobin was investigated using BIA/MS. The antibody was used as the immobilized affinant to retrieve trace human myoglobin from a stock solution of human serum albumin (HSA) in HEPES buffered saline (HBS). SPR-based BIA analyses were performed during both the immobilization and affinity capture procedures.

Sensorgrams of the antibody immobilization, and myoglobin binding are shown in Fig. 6. Fig. 6A shows a sensorgram obtained for a single flow cell during the coupling of anti-myoglobin IgG to the surface of the CM5 sensor chip. Anti-human IgG (~ 2 mg/mL in HBS) was flow incubated over the chip surface for ~ 7 minutes before a ~ 2 minute rinse with HBS, followed by another ~ 7 minute incubation with the IgG. An estimate of the binding capacity of the flow cell can be made from the final difference in the sensorgram reading. A change in response of ~ 15,000 RU translates to ~ 15 ng of IgG (~ 100 fmole) covalently linked to the surface of the 1 mm^2 area of the flow cell. Considering two binding sites per antibody molecule, a myoglobin binding capacity of 200 fmole is estimated for the flow cell. All four flow cells present on the sensor chip were derivatized using identical conditions and resulted in virtually the same amount of bound antibody. Fig. 6B shows sensorgrams obtained during the incubation of the anti-myoglobin-derivatized flow cells with human myoglobin. Sensorgrams for two flow cells, FC 3 and FC 4, are shown. FC 3 was exposed to human myoglobin by flow injecting an aliquot of myoglobin (400 ng/mL in 40 mg/mL HSA) over the flow cell for approximately 1.5 minutes (flow rate = 10 µl/minute). The observed difference in the FC 3 sensorgram signal of ~ 125 RU translates to approximately 0.125 ng of myoglobin retained within the confines of the flow cell (~ 10 fmole, assuming the total response is due to myoglobin). FC 4 served as a blank, and was used exposed only to the HSA-containing buffer (40 mg/mL, for ~ 1 minute at a flow rate of 10 µl/minute). A response difference of ~ 20 RU is observed, indicating the retention of ~ 0.020 ng of material (within the confines of the flow cell).

Fig. 7 shows mass spectra obtained from the direct MALDI-TOF analysis of FC 3 and FC 4. Fig. 7A is a mass spectrum obtained within FC 3. The presence of human myoglobin is verified by observance of ion signal at m/z = 17,200 Da (corresponding to the singly-charged ion of the human myoglobin). The measured molecular mass is

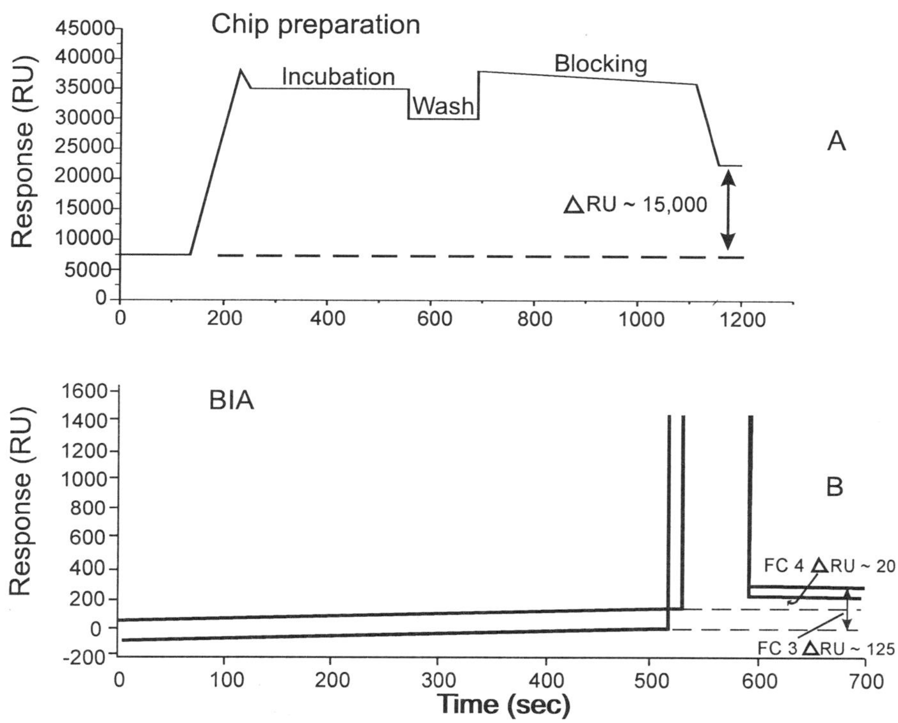

Figure 6. Sensorgrams of: (**A**) Sensor chip preparation with polyclonal anti-human myoglobin IgG, and, (**B**) BIA analysis of active (**FC 3**) and blank (**FC 4**) flow cells. See References 21-23 for further details on BIA/MS experimental procedures.

approximately 150 Da higher than that calculated. The shift to higher mass is due to a chemical stabilization process applied to the myoglobin [21]. The stabilization process introduces a degree of heterogeneity into the sample as well, thereby accounting for the overly broad signals. Also observed in the spectrum are additional signals (marked with *) indicating the retention of non-targeted species. Fig. 7B is a mass spectrum obtained within FC 4 (blank of HSA). The spectrum lacks any indication of retained myoglobin, however, the majority of prominent non-targeted species observed in FC 3 are also observed to be retained in FC 4. By all appearance, the non-targeted species are due to background species present in the HSA carrier.

A few comments on BIA/MS are warranted. First, interfacing of the techniques in the manner demonstrated here (analysis of retained ligands by direct MALDI analysis of the sensor chip) is easy, requiring little more than application of matrix to the sensor chip and introduction into an appropriately configured mass spectrometer. Additionally, the direct analysis approach affords quite remarkable sensitivity. It should be noted that there is, at most, ~ 4 fmole of any one of the background species retained within FC 4. This estimate is true only if all of the retained material (20

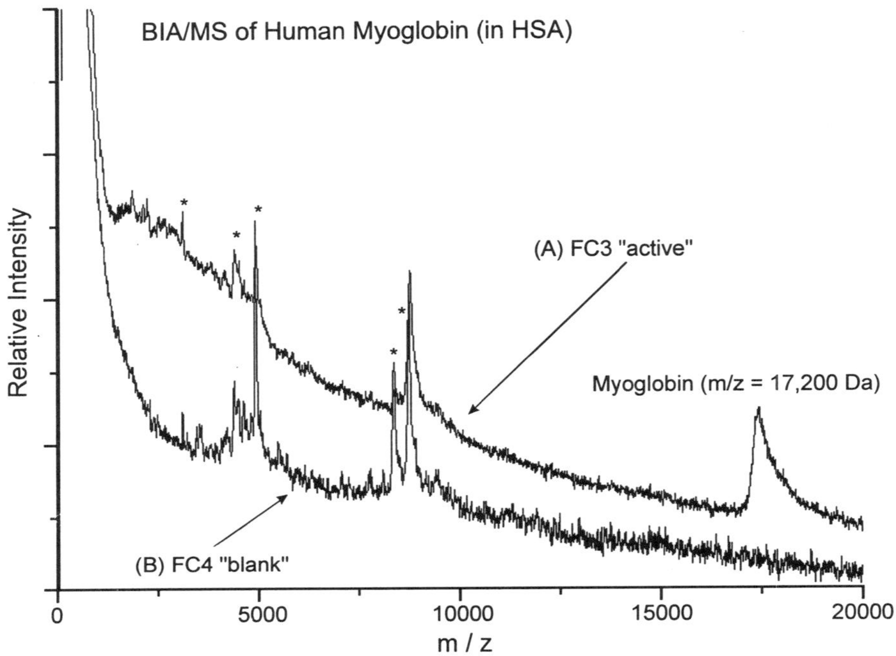

Figure 7. Species retained during the BIA analyses shown in Fig. 6: **(A)** FC 3, **(B)** FC 4. Mass spectra were produced by the direct MALDI analysis of the flow cells. Species indicated by * appear to be the result of non-specific binding of buffer components.

picograms) can be accounted for by only the lowest molecular weight species in Fig 7B, an assumption that is obviously not true. Along theses lines, the quantitative estimate for FC 3 is the retention of 125 picograms of myoglobin. Considering ~ 20 picograms of the retained material to be the background species observed in FC 4, simple quantitative background subtraction estimates an amount closer to 100 picograms. The amount of myoglobin retained within FC 3 is therefore ~ 5 fmole of myoglobin; an amount typically lost in transfer and handling (it is worth remembering that only (this) 5 fmole of myoglobin was available for analysis). Finally, BIA/MS is found to be highly complimentary with SPR. SPR detection allows for the monitoring of steps involved in the preparation of affinity media, as well as the real-time viewing of the binding of ligands during BIA. MALDI analysis then compliments the BIA by revealing the nature of the binding event.

Considering the comparable sensitivities, as well as the complementary nature of SPR and MALDI detection, BIA/MS appears to be an ideal combination. Future improvements (*e.g.* combination with immobilized enzymologies and second dimension mass spectrometry) should provide the means with which to selectively isolate and characterize biomolecules and associated interactions.

5. Concluding Remarks

With the advent of techniques capable of generating macromolecular vapor-phase ions, mass spectrometry has entered into the mainstream of biomolecular analysis. The combination of routine mass analysis with traditionally employed bioanalytical and preparatory methods affords new and powerful approaches for biomolecular study. Presented here have been ideas on the sensible combination of affinity capture, proteolytic digestion and biomolecular interaction analysis with MALDI mass spectrometry to achieve the highest degrees of analytical performance (speed, sensitivity and accuracy) in the characterization of biomolecules and biomolecular interactions.

6. Acknowledgments

Financial support from Pharmacia Biosensor, AB, and an Arizona State University Biomedical Planning Award are greatly appreciated.

7. References

1. Fenn, J. B., Mann, M., Meng, C. K. Wong, S. F. and Whitehouse, C. M. (1989) Electrospray Ionization for Mass Spectrometry of Large Biomolecules, *Science* **246**, 64-71.
2. Karas, M. and Hillenkamp, F. (1988) Laser Desorption/Ionization of Proteins with Molecular Mass Exceeding 10,000 Da, *Anal. Chem.* **60**, 2288-2301.
3. Hutchens, T. W. and Yip, T. T. (1993) New Desorption Strategies for the Mass Spectrometric Analysis of Macromolecules, *Rapid Commun. Mass Spectrom.* **7**, 576-580.
4. Zhao, Y., and Chait, B. T. (1994) Protein Epitope Mapping by Mass Spectrometry. *Anal. Chem.* **66**, 3723-3726.
5. Zhao, Y., Muir, T. W., Kent, S. B. H., Tischer, E., Scardina, J. M., and Chait, B. T. (1996) Mapping Protein-Protein Interactions by Affinity-Directed Mass Spectrometry, *Proc. Natl. Acad. Sci. USA* **93**, 4020-4024.
6. Brockman, A. H., and Orlando, R. (1995) Probe-Immobilized Affinity Chromatography/ Mass Spectrometry, *Anal. Chem.* **67**, 4581-4585.
7. Nelson, R. W., Krone, J. R., Bieber, A. L., and Williams, P. (1995) Mass Spectrometric Immunoassay, *Anal. Chem.* **67**, 1153-1158.
8. Krone, J. R., Nelson, R. W. and Williams P. (1996) Mass Spectrometric Immunoassay, *Proceedings of Ultasensitive Biochemical Diagnostics, SPIE* **2680**, 415-421.
9. Krone, J. R. (1996) Mass Spectrometric Methods for Application in Immunoassay and Biomolecular Characterization, Ph.D. Thesis.
10. Nelson, R.W., McLean, M.A., and Hutchens, T.W. (1994) Quantitative Determination of Proteins in Solution by Matrix-Assisted Laser Desorption/Ionization Time-of-Flight Mass Spectrometry, *Anal. Chem.* **66**, 1408-1415.
11. Muddiman, D. C., Gusev, A. I., and Hercules, D. M. (1996) Application of Secondary Ion and Matrix-Assisted Laser Desorption/Ionization Time-of-Flight Mass Spectrometry for the Quantitative Analysis of Biological Molecules, *Mass Spectrom. Rev.* In press.
12. Voet, D. and Voet, J. G. (1995) *Biochemistry*, John Wiley and Sons, Inc., New York, NY.
13. Hillenkamp, F. (1996) in these proceedings.

14. McGeer, P. L., Akiyama, H., Itagaki, S. and McGeer, E.,G. (1989) Activation of the Classical Complement Pathway in Brain Tissue of Alzheimer Patients, *Neurosci. Lett.* **107**, 341-346.
15. Li, Y., McIver, R. T. and Hunter, R. L. (1994) High-Accuracy Molecular Mass Determination for Peptides and Proteins by Fourier Transform Mass Spectrometry, *Anal. Chem.* **66**, 2077-83.
16. Figeys, D., Vanoostveen, I., Ducret, A. and Aebersold R. (1996) Protein Identification by Capillary Zone Electrophoresis/Microelectrospray Ionization-Tandem Mass Spectrometry at the Subfemtomole Level, *Anal. Chem.* **68**, 1822-1828.
17. Henzel, W. J., Billeci, T. M., Stults, J. T., Wong, S. C., Grimley, C. and Watanabe, C. (1993) Identifying Proteins from Two-Dimensional Gels by Molecular Mass Searching of Peptide Fragments in Protein Sequence Databases, *Proc. Natl. Acad. Sci. USA* **90**, 5011-5015.
18. Zaia J., Annan R. S. and Biemann K. (1992) The Correct Molecular Weight of Myoglobin, A Common Calibrant for Mass Spectrometry, *Rapid Comm. in Mass Spectrom.* **6**, 32-36.
19. Dogruel, D., Williams, P. and Nelson, R. W. (1995) Rapid Tryptic Mapping using Enzymatically-Active Mass Spectrometer Probe Tips, *Anal. Chem.* **67**, 4343-4348.
20. Nelson, R. W., Dogruel, D., Krone, J.R. and Williams, P. (1995) Peptide Characterization Using *Bioreactive* Mass Spectrometer Probe Tips, *Rapid Comm. Mass Spectrom.* **9**, 1380-1385.
21. Dogruel, D., Nelson, R. W., Krone, J. R., Lewis, J. K. and Williams, P. (1995) Rapid Protein Characterization Using *Bioreactive* Mass Spectrometer Probe Tips, *Proceedings of Ultasensitive Biochemical Diagnostics, SPIE* **2680**, 390-396.
22. Nelson, R. W., Krone, J. R., Dogruel, D., Tubbs, K. A., Granzow, R. and Jansson, Ö. (1997) Interfacing Biomolecular Interaction Analysis with Mass Spectrometry and the Use of *Bioreactive* Mass Spectrometer Probe Tips in Protein Characterization, in D. Marshak (ed.), *Techniques in Protein Chemistry VIII*, Academic Press, San Diego. In press.
23. Krone, J. R., Nelson, R. W., Dogruel, D., Williams, P. and Granzow, R. (1996) Interfacing Mass Spectrometric Immunoassays with BIA, *BIAjournal* **3**, 16-17.
24. Krone, J. R., Nelson, R. W., Dogruel, D., Granzow, R. and Williams, P. (1997) BIA/MS: Interfacing Biomolecular Interaction Analysis with Mass Spectrometry, *Anal. Biochem.* **244,** 124-132.

X-RAY CRYSTALLOGRAPHIC AND MASS SPECTROMETRIC STRUCTURE DETERMINATION AND FUNCTIONAL CHARACTERISATION OF SUCCINYLATED PORIN FROM *RHODOBACTER CAPSULATUS*: IMPLICATIONS FOR ION SELECTIVITY AND SINGLE-CHANNEL CONDUCTANCE†

WOLFRAM WELTE[1], KAY DIEDERICHS[1], MICHAEL PRZYBYLSKI[2], MICHAEL O. GLOCKER[2], ROLAND BENZ[3] AND JASON BREED[1]

[1] *Fakultät für Biologie,* [2]*Fakultät für Chemie, Universität Konstanz, P.O. Box 5560 M656, 78434 Konstanz, Germany*

[3]*Universität Würzburg, Lehrstuhl für Biotechnologie, Theodor-Boveri-Institut (Biozentrum), Am Hubland, 97074 Würzburg, Germany*

Abstract

The role of charges near the pore mouth has been discussed in theoretical work about ion channels. To introduce new negative charges in a channel protein, amino groups of porin from *Rhodobacter capsulatus* 37b4 were succinylated with succinic anhydride, and the precise extent and sites of succinylations and structures of the succinyl-porins determined by mass spectrometry and x-ray crystallography. Molecular weight and peptide mapping analyses using matrix-assisted laser desorption-ionisation mass spectrometry identified selective succinylation of three lysine-ε-amino groups (Lys-46, Lys-298, Lys-300) and the N-terminal α-amino group. The structure of a tetra-succinylated porin was determined to 2.4 Å and was generally found unchanged in comparison to native porin, to form a trimeric complex. All succinylated amino groups found in a mono/di- and a tetra-succinyl-porin are localised at the inner channel surface and are solvent-accessible: Lys-46 is located at the channel constriction site, whereas Lys-298, Lys-300 and the N-terminus are all near the periplasmic entrance of the channel. The Lys-46 residue at the central constriction loop was modelled as succinyl-Lysine from the electron density data and shown to bend towards the periplasmic pore mouth. The electrical properties of the mono- and tetra-succinylated porin were determined by

† Dedicated gratefully to the memory of Prof. G. Adam (*1933, †1996), who deceased unexpectedly on Sept. 8. He has taught many of us in biophysics.

W. Ens et al. (eds.), New Methods for the Study of Biomolecular Complexes, 239–276.

reconstitution into black lipid membranes, and showed a negative charge effect on ion transport and an increased cation selectivity through the porin channel. The properties of a typical general diffusion porin changed into those of a channel that contains point charges near the pore mouth. The single-channel conductance was no longer a linear function of the bulk aqueous salt concentration. The substantially higher cation selectivity of the succinylated porins compared to the native protein is consistent with the increase of negatively charged groups introduced. These results show tertiary structure-selective modification of charged residues as an efficient approach in the structure-function evaluation of ion channels, and x-ray crystallography and mass spectrometry as complementary analytical tools for precisely defining the chemically modified structures. The experimental part of the work has recently been published elsewhere [1]. In this presentation theoretical aspects have been much extended, especially the possibility of asymmetric solute fluxes through porins.

Abbreviations:
BrCN, Cyanogen bromide; BLM, Black Lipid Membrane, MS-porin, mono/di-succinylated porin; TS-porin, tetra-succinylated porin; ESI-MS, electrospray mass spectrometry; HCCA, α-cyano-4-hydroxycinnamic acid; MALDI-MS, matrix-assisted laser desorption ionisation mass spectrometry; nS, nano Siemens; OM, outer membrane; PDMS, Plasma desorption mass spectrometry;SDS-PAGE, sodiumdodecylsulfate polyacrylamide gel electrophoresis; TFA, trifluoro-acetic acid; TPCK, Tosyl-phenyl-alanyl-chlormethyl-keton;

1. Introduction.

Gram-negative bacteria protect themselves against bactericidal substances by the outer membrane (OM) which envelops entirely the vulnerable plasma membrane, creating an additional aqueous compartment, the periplasmic space. The OM is asymmetrically composed of a lipopolysaccharide monolayer towards the medium and a phospholipid monolayer towards the periplasmic space [2]. Among the gram-negative bacteria there is considerable variation of the lipopolysaccharide component.

The plasma membrane contains highly specific active transport systems for all essential small foodstuff molecules like ions, sugars, nucleotides, organic acids. Consequently, the concentration of these molecules in the periplasmic space is often as low as 1 - 50 μM, while in the external medium the concentrations are much higher. For the survival and for proliferation of the cell it is essential that the foodstuff flux across the OM is not growth limiting, i.e. that the OM poses a low resistance barrier for diffusion of these foodstuff molecules down this concentration gradient.

As the lipid bilayer of the OM is impermeable to this class of molecules, the membrane has to be studded with transmembrane proteins which form aqueous channels that allow for the passive permeation of small polar molecules. These „channel proteins“, called porins, possess a certain exclusion limit of typically 600 daltons (for reviews see [3, 4]).

1.1 FUNCTIONAL CHARACTERISATION OF PORINS

The permeability P, according to Fick's law, determines the diffusional flux φ when a type of molecules is present at a concentration c_e outside of the cell and c_p in the periplasmic space:

$$\varphi = (c_e - c_p) \cdot P \tag{1}$$

The "enzyme activity" for porins thus is the channel permeability P. As the quantitative determination of very small amounts of uncharged substances requires use of radioactively labelled molecules, simpler methods are often used to deduce information on permeability. One example is the "liposome swelling assay". Porins are reconstituted into liposomes in presence of an impermeable polymer like e.g. dextran. If an added test substance is capable of permeating through the porin channel, water will follow to keep the interior isoosmolar with the exterior solution and the liposome will swell. The accompanying increase in light scattering can be monitored in a time dependent manner [5]. In order to analyse the kinetic data on absolute scale, the density of channels in the liposomes has to be determined. Moreover, in most cases porins insert in random orientation into the lipid bilayer, so that a superposition of an influx and an efflux is monitored, whereas only the influx appeares to be relevant for the biological function of porin.

Much easier is the measurement of an ion flux, as even very small currents through single channels can be determined with great accuracy. For such experiments two chambers filled with electrolyte solution are separated from each other by a black lipid film ("BLM"), which ideally should be just a lipid bilayer (see [6] for a detailed description).The black colour results from destructive interference of the two reflected beams seen when the film is observed in reflected light. When detergent-solubilised porin trimers are added to one side, the proteins insert usually with random orientation into the bilayer. With each insertion a stepwise increase of current between the two chambers is observed which can be ascribed to one porin trimer in one of the two orientations. The steps allow for calculation of the single channel conductivity σ from the voltage $\Phi_r - \Phi_l$ applied to the black lipid membrane:

$$I = (\Phi_r - \Phi_l)\ \sigma \tag{2}$$

σ can be determined for ions of different size and varying concentration, at different pH values.

The similarity of equations (2) and (1) is evident, as a difference in electric potential on the two solutions adjacent to the membrane and a difference in concentration both represent a difference in chemical potential or free energy per mole of the ions (see e.g. Ch. 6.4.2 in [7]):

$$\mu = \mu_o + RT \ln [f(c)c] + z F \Phi \tag{3}$$

We will approximate the activity coefficient f(c) by 1 which is justified at low concentrations. By introducing formula (3) for $\Phi = 0$ into (1) we obtain [8]:

$$\varphi = \exp[-\mu_0/RT] (\exp[\mu_e/RT] - \exp[\mu_p/RT]) \cdot P$$

Using $\langle\mu\rangle = (\mu_e + \mu_p)/2$ and placing a factor $\langle c\rangle = \exp[\langle\mu\rangle/RT]$ outside the brackets, we find:

$$\varphi = \langle c\rangle (\exp[(\mu_e - \mu_p)/2RT] - \exp[-(\mu_e - \mu_p)/2RT]) \cdot P$$

For $\Delta\mu$ not too large, this can be approximated by

$$\varphi = \langle c\rangle/RT \cdot (\mu_e - \mu_p) \cdot P \tag{4}$$

If we also substitute electric potentials for chemical potentials by inserting (3) into (2) and use $I = \varphi z$, where z is the valency of the ion, we find:

$$\varphi = (\mu_r - \mu_l) \cdot \sigma/zF \tag{5}$$

From comparison an approximate equation is obtained which relates the permeability with the conductivity:

$$P \cdot \langle c\rangle/RT = \sigma/zF. \tag{6}$$

There are, however, two differences to the native situation: Firstly, in presence of a potential difference across the membrane the anion flux is oppositely directed to the cation flux and both fluxes may be different. Except for the case that the channels are predominantly anion- or cation specific or that we use an electrolyte with one impermeable ion we will measure always a superposition of oppositely oriented particle flows. In vivo, however, anions and cations will permeate both in pairs from the external phase to the periplasmic space, so that no net current will flow. Secondly, the electric field applied to measure σ will add a free-energy ramp to the channel which is absent in the measurement of P. All this is ignored in relation (6). It is thus an assumption that our experimental scores for evaluating and comparing porins coincide with the ones acting in evolution and we do not know how well this assumption holds.

In order to get a parallel flow of anions and cations, we can apply different electrolyte concentrations to the two chambers separated by the BLM. This set-up allows to determine the ratio of parallel flows of anions and cations through the channels. The voltage necessary to suppress any current flowing through the channel allows to calculate the ratio of permeability of cations versus anions (P_{cation}/P_{anion}) using the Goldmann-Hodgkin-Katz equation. These measurements are, however, usually done with many channels so that they yield information on a superposition of randomly oriented channels.

1.2. STRUCTURE OF PORINS

The structures of five bacterial porins (four general diffusion porins and two specific porins) have been determined by X-ray crystallography: *Rhodobacter capsulatus* [9, 10], *E. coli* OmpF and PhoE [11], *Rhodopseudomonas blastica* [12], maltoporin from *E.coli* [13] and sucrose porin from *E.coli* (Forst et al., in preparation). All form trimers and consist of a 16-18 stranded β-barrel of similar overall shape and tilt of the β-strands. The connections are all between neighbouring strands. All possess sharp β-hairpin loops on their periplasmic surface. The opposite barrel rim is rough due to an irregular termination of the β-strands and to loops of varying length connecting the neighbouring strands on the external side. While the porin structures are similar in the periplasmic half of the OM, they differ from each other in the medium-apposed half. In all porins the third external loop is exceptionally large (some 45 amino acid residues). This segment folds in an extended conformation from the end of the fifth strand of the barrel rim along the inside barrel wall and back to the sixth strand. The free cross-section of the β-barrel cylinder of *Rhodobacter capsulatus* porin which is about 30 by 30 Å, is constricted to an opening of about 8 by 10 Å at the position of the third loop inside the barrel which is roughly halfway through the channel (Figure 1). The permeation of hydrated ions through this constriction loop may require transient stripping of part of their hydration shells [10].

The amino acid residues coating this central "constriction zone" or "eyelet" show a remarkably asymmetric charge distribution. While the respective segments of the third loop are carrying acidic amino acids, the opposite barrel wall (near the centre of the trimers) is coated with basic amino acid residues (Figure 2). Reconstitution experiments with porin from *Rhodobacter capsulatus* suggest that there is a link between the rate of permeation of cations through the porin and the number of acidic residues as well as a similar link between anion permeability and basic residues [10, 11, 14].

We have chemically modified porin from *Rhodobacter capsulatus* 37b4 by introducing negative charges through succinylation with succinic anhydride. The effect of the succinylation on single channel conductance and ion selectivity was studied by electrical measurements after reconstitution into lipid bilayer membranes. The molecular structures of the succinylated porins were characterised by direct molecular

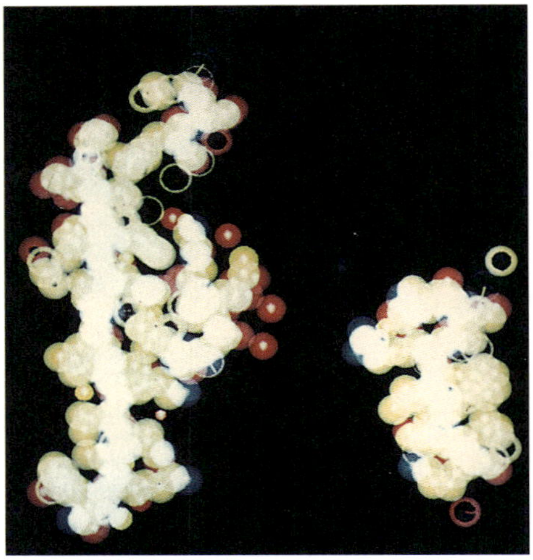

1 a

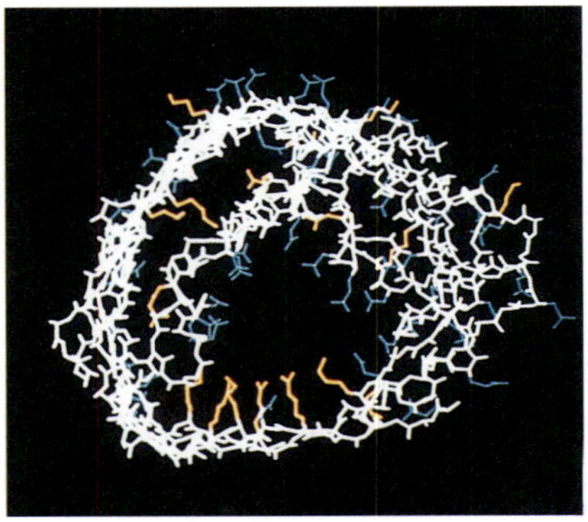

1b

Figure 1. a) Slab through a porin monomer along a plane spanned by the threefold axis of the trimer and a vector pointing from this axis to the centre of the constriction site. The third loop, attached to the peripheral barrel wall, creates the constriction site. Colours used are white for peptide backbone atoms; red for side-chain oxygen; blue for side-chain nitrogen. b) View of the porin monomer along the threefold axis from the periplasmic side. The scheme shows the backbone atomic bonds in white, the acidic residues in blue and the basic residues in orange, illustrating the strong charge asymmetry across the constriction site.

mass determination using matrix-assisted laser desorption-ionisation mass spectrometry (MALDI-MS) which showed the selective introduction of up to four succinate groups; their localisation (three lysine residues and the N-terminal α-NH_2-group) was identified by mass spectrometric peptide mapping analyses following proteolytic degradation, and by sequence identification of proteolytic peptides isolated by HPLC. The modified protein crystallised isomorphically to the native protein so that the crystal structure and electron density map allowed the comparison of the structures by difference maps. The changes in the electrical properties are discussed in this study in view of the structures of the succinylated porins.

1	11	21	31	41
EV*K*LSGDAR**M**	GV**M**YNGDDWN	FSSRSRVLFT	**M**SGTTDSGLE	FGASF*K*AHES
51	61	71	81	91
VGAETGEDGT	VFLSGAFG*K*I	E**M**GDALGASE	ALFGDLYEVG	YTDLDDRGGN
101	111	121	131	141
DIPYLTGDER	LTAEDNPVLL	YTYSAGAFSV	AAS**M**SDG*K*VG	ETSEDDAQE**M**
151	161	171	181	191
AVAAAYTFGN	YTVGLGYE*K*I	DSPDTAL**M**AD	**M**EQLELAAIA	*K*FGATNV*K*AY
201	211	221	231	214
YADGELDRDF	ARAVFDLTPV	AAAATAVDH*K*	AYGLSVDSTF	GATTVGGYVQ
251	261	271	281	291
VLDIDTIDDV	TYYGLGASYD	LGGGASIVGG	IADNDLPNSD	**M**VADLGV*K*F*K*F

Figure 2. Amino acid sequence of porin from *Rhodobacter capsulatus*. Lysine residues are boxed; the sequence comprising the third channel constriction loop is shaded.

2. Results

2.1. MASS SPECTROMETRIC STRUCTURE IDENTIFICATION OF SUCCINYLATED PORINS

Native porin was subjected to succinylation with an approximately 15- and 100-fold molar excess of succinic anhydride for 60 minutes at 25°C, respectively (ca pH 6.5), at conditions that were previously established to provide selective N-acylation at lysine-ε-amino groups and the N-terminus [15, 16]. The modified proteins were dialysed against crystallisation buffer and purified by FPLC on Q-sepharose. The succinylation did not lead to dissociation of the native trimer complex of porin, in accordance with the tertiary structure-selective acylation conditions of surface-accessible lysine residues [17], and the protein derivatives showed a slight increase in the apparent mass by SDS-PAGE.

Precise molecular mass determinations of two succinylated porins were obtained by MALDI-MS which, together with other "soft"-ionisation methods such as electrospray (ESI-MS) and Cf-252-plasma desorption (PDMS), have recently provided a breakthrough in the direct characterisation of proteins and other biopolymers [18, 19, 20]. MALDI mass spectra of the unmodified and two succinylated porins, obtained under acidic conditions in the solid crystalline matrix α-cyano-4-hydroxycinnamic acid (leading to dissociation of the trimer complex), are compared in Figure 3 a-c. All three proteins yielded a consistent series of singly charged (protonated) molecular ions together with the doubly and triply charged ions (M^{2+}, M^{3+}). A molecular mass of 31,541 Da in close agreement with the sequence molecular weight (M_r, 31,537 Da) of the monomer was determined for unmodified porin. The porin derivative modified with an 15-fold excess succinic anhydride yielded a molecular mass of 31,684 Da

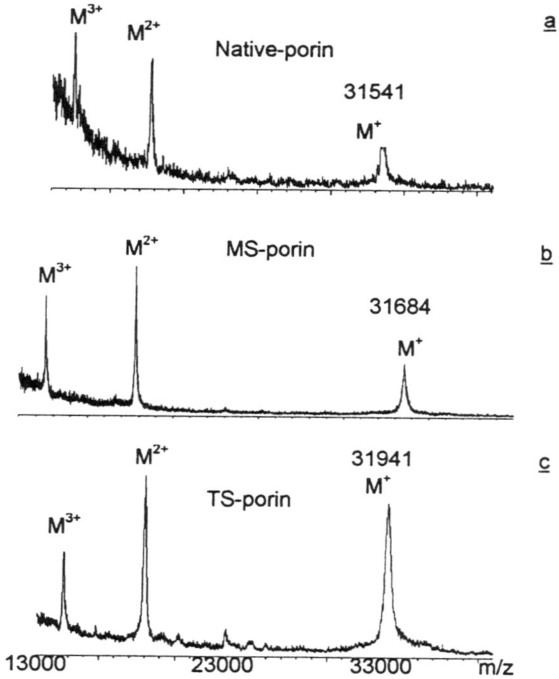

Figure 3. MALDI-MS analysis of (a) unmodified (Native-porin) (b) mono/di-succinylated porin (MS-porin), and (c) tetra-succinylated porin (TS-porin). Sample preparations for mass spectrometric determinations were carried out by adding 1 µl of a 5-µg/µl solution of protein in crystallisation buffer (s. Materials and Methods) to 19µl of a matrix solution of α-cyano-hydroxycinnamic acid in acetonitrile: 0.1% TFA (2:1). Mass calibration was perfumed by using the precise average-isotope singly charged molecular ion signals of hen egg-white lysozyme, cytochrome C, trypsin and BSA as internal standards.

which (by calibration with the precisely defined molecular ions of trypsin, insulin and BSA as internal standards) corresponds to a modification degree of 1.4 and hence to a mono- to disuccinylated porin; the higher modified porin providing a molecular mass of 31,941 Da corresponding to the presence of four succinyl groups exactly (ΔM, 400 amu; Figure 3 b,c).*

Furthermore, the molecular ion peak widths for the succinylated porins compared to that of the unmodified porin (representing average-isotopes due to the time-of-flight analyzer employed), indicated a relatively high homogeneity and selective introduction of up to 4 succinyl-groups. Thus, the narrow molecular ions in the spectrum of the native-like Ts-porin suggested the presence of only very small amounts of tri- and

* For simplifying the discussion, the mono-/di-succinylated and tetra-succinylated porins are assigned as MS-porin and TS-porin in the following text.

penta-succinylated protein (Fig. 3 c). This result was in contrast to the much broader molecular ion signals observed for an approximately octa-succinylated, denatured protein prepared at more drastic reaction conditions with a 500-fold reagent excess (not shown). The selective modification in the MS- and TS-porin was ascertained by the subsequent determination of specific succinylation sites and unsuccinylated lysine residues.

The complete structural characterisation and identification of the succinylation sites of the MS-porin and TS-porin was obtained by (i) mass spectrometric peptide mapping of peptide mixtures after proteolytic degradation with BrCN and trypsin, (ii) mass spectrometric analysis of peptide fragments isolated by HPLC, and (iii) Edman sequence determination of the intact succinylated porins and of relevant isolated proteolytic peptides. Edman sequence analyses showed that the TS-porin was completely blocked, and in the case of the MS-porin provided a partial N-terminal sequence (EVKLS-) only with low sensitivity beyond background in comparison to native porin, indicating partial to complete (TS-porin) N_α-succinylation. Direct MALDI-MS peptide mapping analysis of BrCN fragments from MS-porin and TS-porin are compared in Figure 4 (see also Table 1). Molecular ions of three BrCN peptides were found which showed an increase in molecular mass by 100 amu each compared to the corresponding fragments from unmodified porin. Two of these peptides, (32-72) and (14-72) identified complete mono-succinylation at Lys-46 for TS-porin and partial Lys-46-succinylation in MS-porin (a modification at Lys-69 was excluded by analysis of tryptic peptides as shown below). Furthermore, mono-succinylation in the N-terminal part was confirmed by the peptide (1-31) in TS-porin (m/z 3645), in contrast to the presence of some unmodified (1-31) in MSporin. The succinyl-(1-31) peptide from TS-porin isolated by HPLC was amenable to Edman sequencing only at a very low background level which indicated only a trace of alternative succinylation at Lys-3. This result was in agreement with the mass spectral identification of the HPLC-isolated BrCN peptide succinyl-(1-10) which was entirely blocked at the N-terminus (Fig. 6 a). Further possible succinylation sites at Lys-169 were ruled out by BrCN peptides (135-178) and (73-181) which were found unmodified both in the MS- and in TS-porin. A large BrCN polypeptide fragment (182-291; M_r ca 12,500 Da) was not found by direct MALDI-MS analysis and could not be isolated by HPLC, probably due to its low solubility. However, the remaining two succinylation sites at Lys-298 and Lys-300 were identified by HPLC isolation, MALDI-MS and partial sequence analysis of the C-terminal peptide (292-301) which was unmodified in native porin and MS-porin but was shifted by $\Delta M = 200$ amu in TS-porin (m/z 1324; Table 1).

Peptide fragments obtained by tryptic digestion provided complete agreement with the BrCN degradation in the determination of the four succinylation sites. Untreated trypsin containing additional chymotryptic activity was used in these experiments, as the intact porin structure was found difficult to digest with trypsin even

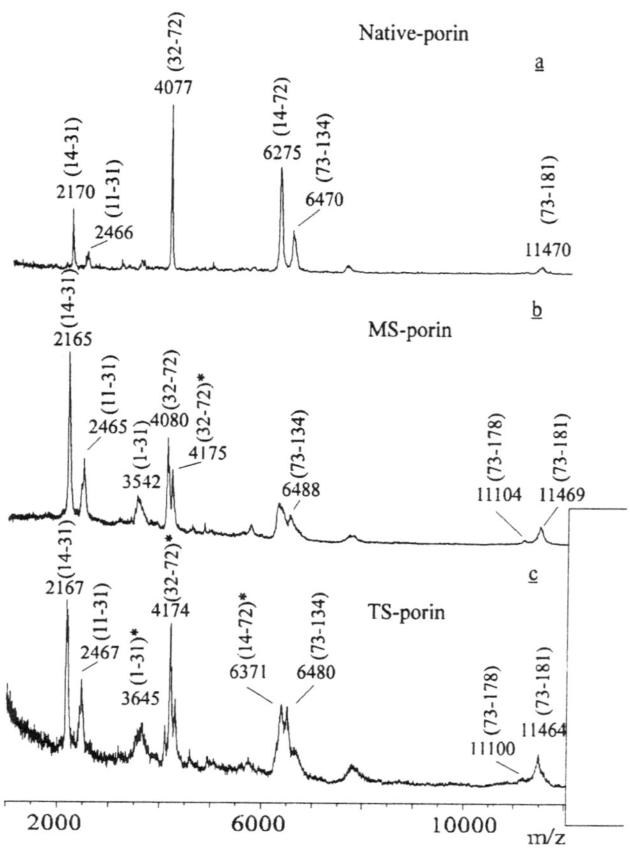

Figure 4. MALDI-MS peptide mapping analysis of BrCN-peptides from (a) Native-porin, (b) MS-porin and (c) TS-porin. A 1 µl aliquot of the BrCN reaction mixture was added to 19 µl of the α-cyano-4-hydroxycinnamic acid matrix solution (s. Materials and Methods). Masses given represent the homoserine-carboxylate forms of BrCN-peptides (see Table 1 and text). Partial sequences are given in parentheses and are marked by an asterisk for succinylated peptides.

after denaturation with urea [21] (however, results were confirmed by comparison with peptide mapping after digestion with TPCK-treated trypsin). HPLC separations of tryptic peptides from native porin are compared in Figure 5, and relevant lysine-containing peptide fragments (and cleavage sites at Lys residues) summarised in Table 1. Despite the multiplicity of proteolytic peptides, MALDI-MS of isolated fractions and partial sequence determinations provided complete primary structure coverage of succinylated porins. In addition to the blocked N-terminal peptide N_α-succinyl-(1-9) in TS-porin (m/z 1034), the succinylation at Lys-46 was ascertained by the modified peptide (27-69) and by the chymotryptic fragment (42-62) which provided a molecular mass due to mono-succinylation and the correct N-terminal sequence

Table 1. Identification of proteolytic fragments and succinylation sites from unmodified and succinylated porins by BrCN and trypsin degradation.

	BrCN fragments[a]					
	MH^+ ions, m/z[c]					
sequence/Lys residue[b]	m.w. calcul.[c]	native porin	MS-porin	TS-porin	sequence analysis[d]	succinylated Lys residue
(1-10)/3	1075	1060[e]	1159[e]	1159[e]	-	a-NH_2
	2464[e]	2466	2465	2467		
(14-31)/-	2165	2170	2165	2167		
(1-31)/3	3541	3542	3542	3645	EVKL[f]	a-NH_2, K-3[f]
(32-72)/46,69	4076	4077	4080/4175	4174	SGT	K-46[g]
(14-72)/46,69	6274	6275	6275/6370	6371		K-46[g]
(73-134)/-	6477	6470	6488	6480		
(73-178)/138,169	11097	11102	11104	11100		
(73-181)/138,169	11462	11470	11469	11464		
(292-301)/298,300	1124	1123	1125/1224	1324	VAD	K-298, K-300
			Tryptic fragments[h]			
(1-9)/3	934	-	935/1034	1034	-[i]	a-NH_2
(1-26)/3	2979	-	3079	3079	-[i]	a-NH_2, K-3[f]
(4-9)/3	619	621	620	621		
(10-24)/-	1779	1779	1779	1780		
(10-46)/46	4100	4102	-	-		
(27-46/46	2095	2095	-	-		
(27-69)/46	4344	4345	4445	4446		K-46
(42-62)/46[j]	2097	-	2199	2198	GASF[j]	K-46
(70-97)/69	3037	3037	3036	3039		
(98-110)/-	1406	1407	1407	1409		
(129-156)/138[j]	2788	2790	2791	2790	SVA[j]	
(170-191)/169,191	2348	2347	2349	2349		
(192-198)/191,198	737	737	738	738		
sequence/Lys residue[b]	m.w. calcul.[c]	native porin	MS-porin	TS-porin	sequence analysis[d]	succinylated Lys residue
(199-208)/198	1173	1174	1172	1174		
(199-212)/198	1662	1663	1662	1665		
(213-230)/230	1797	1798	1799	1798		
(213-232)/230[j]	2030	2031	2030	2032		
(231-298)/230,298	6821	6824	-	-		
(270-301)/298,300[j]	3189	-	-	3390	DLGG[j,k]	K-298, K-300

[a] BrCN fragments identified from MALDI-MS peptide mapping, except for peptides (1-10) and (1-31) which were isolated by HPLC.
[b] Lys residues and N-terminus contained in proteolytic peptide, or involved in cleaved peptide bond.
[c] Average isotopic masses of BrCN peptides, homoserine-carboxylate unless otherwise noted.
[d] Edman sequence determination of HPLC-isolated peptide.
[e] BrCN peptide, homoserine-lactone. (Fontana & Gross, 1986)
[f] Low amount of K-3 succinylation derived from sequenator background.
[g] K-69 succinylation excluded from tryptic cleavage at K-69/I-70 bond.
[h] From HPLC-isolated fragments obtained by digestion with TPCK-untreated trypsin
[i] Peptide blocked at N-terminus.
[j] Peptide resulting from chymotryptic cleavage; other chymotryptic fragments: (199-210); (202-210); (92-104).
[k] See Fig. 6(c).

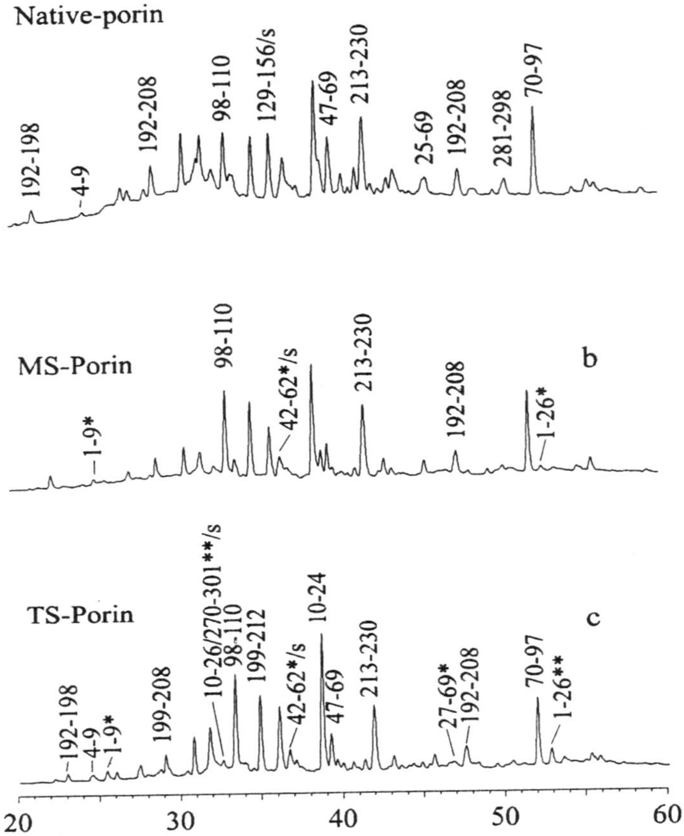

Figure 5. HPLC separation of tryptic peptides from (a) Native-porin, (b) MS-porin and (c) TS-porin. Proteins were digested with untreated trypsin, reaction mixtures injected on a 25 × 0.4 cm nucleosil-C8-column and eluted with a linear TFA/acetonitrile gradient, as described in Materials and Methods. Succinylated peptides identified by MALDI-MS of isolated fractions, and by Edman sequence analysis are marked by asterisks "*" and by /s, respectively.

as well (GASF; see Fig.6 b). Also, Lys-298 and Lys-300 in TS-porin were found inaccessible to tryptic cleavage. These succinylation sites were identified by MALDI-MS and sequence analysis of the bis-succinylated chymotryptic peptide (270-301), despite the presence of an additional peptide (10-26) in the HPLC fraction (Fig. 6 c). By contrast, other Lys-containing peptide fragments were found unmodified by MALDI-MS, e.g. (70-97), (129-157), (170-191), (199-208), and (213-230) (Table 1). In summary, these results ascertained the predominant succinylation at 4 specific amino groups in TS-porin (Lys-46, Lys-298, Lys-300, N-terminus) and partial N-ter-

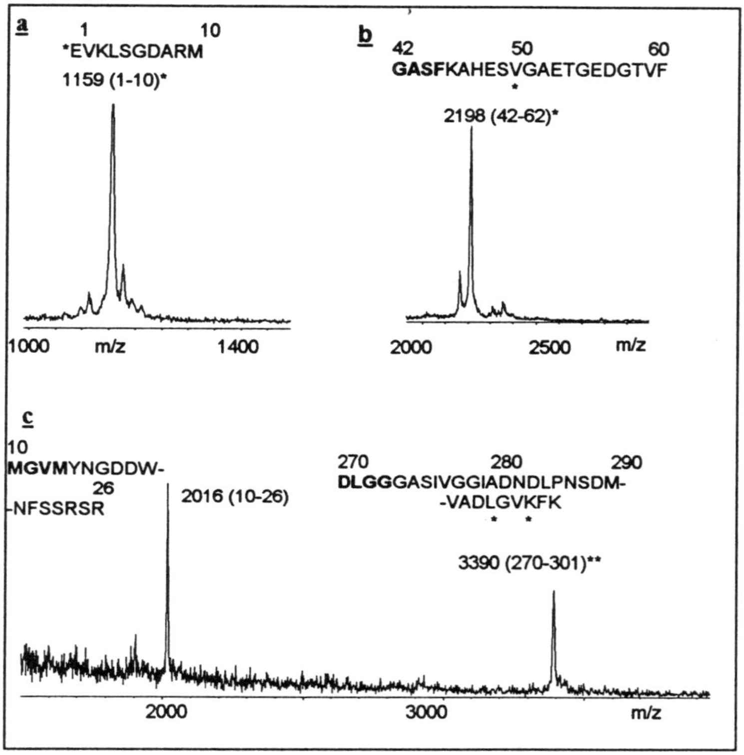

Figure 6. MALDI mass spectra of succinylated peptide fragments isolated by HPLC after BrCN and trypsin digestion of TS-porin. (a), spectrum of BrCN peptide, N_α-succinyl(1-10), lactone form; (b), Lys-46-ε-succinyl-(42-62); (c), mixture of peptides (10-26) and (Lys-298,300)-bis-succinyl-(270-301) (s. Fig. 5 and Table 1). N-Terminal residues determined by automated Edman degradation are indicated by bold letters.

minal and Lys-46 succinylation in MS-porin. Furthermore, other lysine modifications were ruled out except for a low extent of succinylation at Lys-3 (and trace at Lys-69).

2.2 SINGLE CHANNEL CONDUCTANCE OF SUCCINYLATED PORINS

Single-channel experiments revealed that the MS- and TS-porin still had channel-forming activity, i.e. the channel-forming properties of the porin trimers remained intact upon the succinylation. The native and the succinylated porins had approximately the same channel-forming activity indicating that if any, only a minor structural change of the porin trimers had occurred. Figure 7 b (lower trace) shows a single-channel recording in the presence of TS-porin. The protein was added to a black diphytanoyl phos-

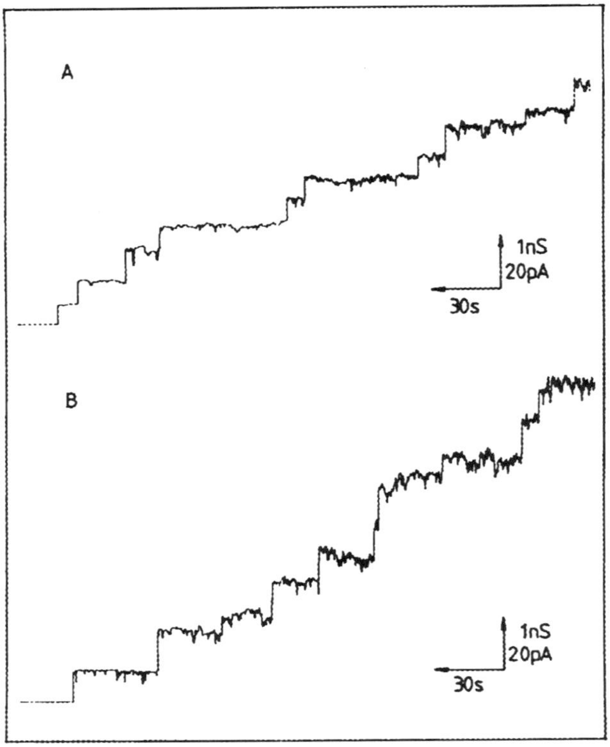

Figure 7. Single-channel recording of diphytanoyl phosphatidylcholine/n-decane membranes in the presence of 20 ng/ml native (A) and TS-porin (B) from *Rhodobacter capsulatus*. The aqueous phase contained in both cases 0.3 M KCl (pH 6). The applied membrane potential was 20 mV; T = 20°C.

phatidylcholine/n-decane membrane at a concentration of about 20 ng/ml. A stepwise increase of the membrane conductance was observed after a delay of several minutes, probably caused by slow aqueous diffusion, reflecting the insertion of the TS-porin trimer into the membrane. Similar to the native porin trimers, the channels had a long lifetime (mean lifetime at least 5 min). The upper trace of Figure 7 shows a single-channel recording observed with native porin under otherwise identical conditions (0.3 M KCl). The comparison of the single-channel records indicated that the single-channel conductance was higher for the MS- and the TS-porin than for native porin.

A histogram of the conductance fluctuations observed with native and TS-porin is shown in Figure 8 under the same conditions as in Figure 7. (0.3 M KCl). The TS-porin showed a single-channel conductance of 1.45 nS, while the native porin had only a conductance of 1.0 nS. Apart from this difference, the histograms for native and TS-porin were very similar as demonstrated in Figure 8.

Single-channel conductance experiments were performed with native, MS- and

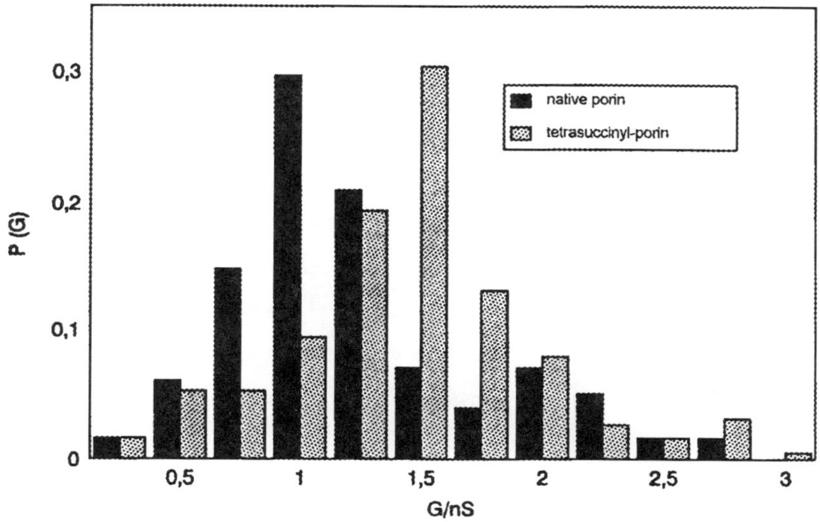

Figure 8. Histogram of the probability P(G) for the occurrence of a given conductivity unit observed with membranes formed of diphytanoyl phosphatidylcholine/n-decane in the presence of 20 ng/ml native porin and TS-porin. P(G) is the probability that a given conductance increment G is observed in the single-channel experiments. It was calculated by dividing the number of fluctuations with a given conductance increment by the total number of conductance fluctuations. The aqueous phase contained 0.3 M KCl. The applied membrane potential was 20 mV; T = 20°C. The average single-channel conductance was 1 nS for 252 single-channel events (native porin) and 1.45 nS for 258 single events (TS- porin).

TS-porin in the presence of a variety of KCl-concentrations. Surprisingly, the single-channel conductance of TS-porin was not a linear function of the bulk aqueous salt concentration as has been found previously for the native porin [14] and still is approximately the case for the MS-porin. Instead, we observed that the conductance followed the square root of the concentration with an approximate slope of 0.5 on a double logarithmic scale (Table 2 and Figure 9). This result suggested negative surface charge effects on ion transport through the TS-porin channel. Similar negative surface charge effects have previously been found for the hemolysine channels of *Escherichia coli* and other related bacteria [22, 23, 24] and for porins from *Pelobacter venetianus* [25] and *Acidovorax delafildii* [26]. Single-channel experiments were also performed with salts other than KCl to obtain information on the selectivity of the TS-porin. These results (see Table 2) showed that replacement of chloride by the less mobile acetate had only little influence on the conductance of the TS-porin. The influence of the cations on the single channel conductance was more pronounced (Table 2) suggesting that the TS-porin is cation selective as has been observed for the native porin [14].

TABLE 2. Average single-channel conductance, G, of native MS-porin and TS-porin in different salt solutions. The membranes were formed of diphytanoyl phosphatidylcholine dissolved in n-decane. The aqueous solutions were unbuffered and had a pH of 6 unless otherwise indicated. The applied voltage was 20 mV, and the temperature was 20 °C. G was calculated from at least 80 single events. c is the concentration of the aqueous salt solutions. See Fig. 8 for standard deviations at 0.3 M KCl.

		G(nS)		
Salt	c(M)	native porin	MS-porin	TS-porin
LiCl	1	1.3±0.11	1.3±0.10	1.0±0.10
KCl	0.01	0.030±0.003	0.090±0.001	0.30±0.05
	0.03	0.095±0.01	0.24±0.003	0.53±0.062
	0.10	0.32±0.035	0.50±0.045	0.90±0.085
	0.30	1.00±0.01	1.10±0.13	1.45±0.25
	1.0	2.90±0.31	2.60±0.27	3.30±0.30
	3.0	7.4±0.73	6.0±0.55	6.5±0.58
KCH_3COO(pH 7)	1.0	2.0±0.22	2.2±0.30	2.7±0.22

2.3. SELECTIVITY OF THE SUCCINYLATED PORIN

Zero-current membrane potential measurements were performed to obtain further information on the ion permeabilities of the TS-porin. Membranes of diphytanoyl phosphatidylcholine/n-decane were formed in 10 mM salt solutions and MS-porin and TS-porin were added to the aqueous phase when the membranes were in the black state. After incorporation of 100 to 1000 channels into a membrane, ten-fold salt gradients were established by addition of small amounts of concentrated salt solution to one side of the membrane. For all salts (KCl, LiCl, and KCH_3COO) tested in these experiments, the more diluted side of the membrane became positive which indicated preferential movement of cations through the MS- and TS-porin channels, i.e. the channels are cation-selective as suggested from the single-channel data (see Table 3). The zero-current membrane potential was 44 and 49 mV with KCl for MS-porin and TS-porin, respectively, compared to 39 mV for the unmodified porin.

Analysis of the zero-current membrane potential in Table 3 using the Goldman-Hodgkin-Katz equation [27] suggested that anions also could have a certain permeability through the TS-porin channels because the ratios of the permeabilities P_{cation} and P_{anions} were between 21 and 23. On the other hand, the asymmetry potentials were very similar for the three salts composed of anions and cations of different mobilities which

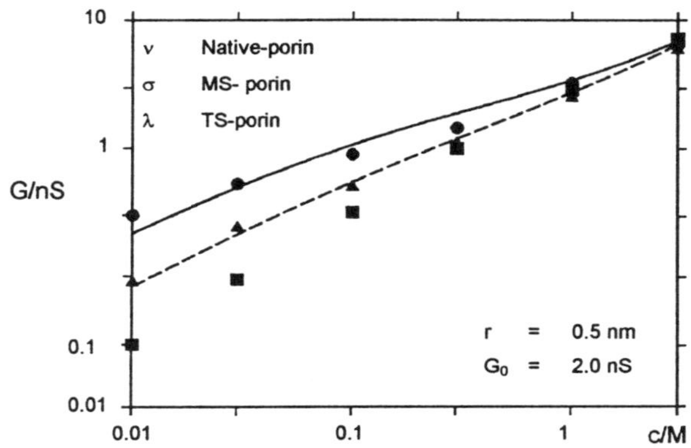

Figure 9. Single channel conductance of native (filled squares), MS-porin (filled triangles) and TS-porin (filled circles) as a function of the KCl-concentration in the aqueous phase. The solid line represents the fit of the single channel conductance with a combination of eqs (1) to (3) by assuming a radius of the succinylated porin channel of 0.5 nm and charge -1.0 e (e = 1.6 • 10^{-19} C) localised at the channel mouth. The broken line represents a similar fit of the single-channel conductance by assuming the same radius and a charge of -0.6 e localised at the channel mouth. The single-channel conductance of the native poring is a linear function of the bulk aqueous concentration. c, concentration of the KCl-solution in M (molar); G, average single channel conductance in nS.

TABLE 3. Zero-current membrane potentials V_m of diphytanoyl phophatidylcholine/n-decane membranes in the presence of native, porin, MS-porin, and TS-porin of *Rhodobacter capsulatus* measured for a 10-fold gradient of different salts. V_m is defined as the difference between the potential at the it the dilute side (10 mM) and at the concentrated side (100 mM). The pH of the aqueous salt solutions was 6 unless otherwise indicated; T = 20C. The permeaility ratio P_{cation}/P_{anion} was calculated with the Goldman-Hodgkin-Katz equation [27] on the basis of at least 3 individual experiments.

	V_m (mV)		P_{cation}/P_{anion}
		native porin	
KCl	39±2.0		9.0±1.0
LiCl	36±1.6		7.0±0.5
K-acetate (pH 7)	45±2.2		15±1.2
		MS-porin	
KCl	44±1.4		13±1.4
		TS-porin	
KCl	49±1.5		23±1.5
LiCl	48±2.0		21±2.0
Kacetate (pH 7)	49±1.3		22±1.8

is not expected for a general diffusion pore [27]. The ratios of permeabilities P_{cation}, P_{anion} were between 7 and 15 for the native porin channel and significantly increased for the three different salts upon succinylation (see Table 3). This suggested that the MS-porin and TS-porin had become substantially more selective for cations because of point charges attached to the channel (see below and Discussion). This change in selectivity must be due to the amino-succinylated residues identified, i.e. conversion of

Lys-46 and the N-terminus into negatively charged groups in the case of the mono/di-succinylated porin, and of Lys-46, Lys-298, Lys-300 and the N-terminus for the tetra-succinylated porin.

2.4. STRUCTURE OF THE TS-PORIN

The tetra-succinylated porin from *Rhodobacter capsulatus* 37b4 crystallised in space group R3 under similar conditions as native porin from this strain. The unit cell constants of native and TS-porin crystals were $a = b = 92.0$ Å and 91.7 Å and $c = 145.9$ Å and 145.7 Å, respectively. The structure was analysed with difference Fourier electron density maps using the known structure of the native porin (see Materials and Methods and Table 4). The crystals of the TS-porin diffracted to 2.4 Å; data were evaluated to 2.7 Å.

Positive electron density, as revealed by 2 F_{obs}-F_{calc} density maps, was found near 4 ε–amino groups of the 10 lysine residues: Lys-46, -69, -230, and -300. Out of these residues, only Lys-46 and Lys-300 (both located on the inner channel surface) were identified by mass spectrometry as succinylation sites. By contrast, Lys-69 albeit on the inner channel surface was found unsuccinylated (or modified only to a very low extent) which can be explained by the considerable shielding of this ε–amino group by the adjacent Phe-62 and Glu-71 residues [28]. Contiguous density of sufficient volume to

Table 4. Refined structure of TS-porin from Rhodobacter capsulatus 37 b4.

Resolution 10 - 2.7 Å		
Number of reflections		12,289
Completeness		97.2 %
Internal R-value on Intensities		8.8 %
Crystallographic free R-value		20.8 %
Crystallographic working R-value		19.9 %
R.m.s.		
	bond lengths (Å)	0.009
	bond angles (deg.)	1.6
	dihedral angles (deg.)	27.9
	improper angles (deg.)	1.2
Number of non-hydrogen atoms		
Polypeptide		2219
	Water	259
	Detergent	42
	Calcium	4
Average B-factors ($Å^2$):		
	All atoms	23
	Polypeptide atoms	22
	Main chain atoms	18
	Side chain atoms	25
	Water molecules	38
	Detergent molecules	58

harbour a succinyl group was found only near Lys-46. Refinement was done after introducing the N-ε–(succinyl)-Lys-46 (see details in Table 4), yielding a final R-factor of 19.9%. The 2Fobs-Fcalc density of succinyl-lysine-46 is shown in Fig. 10.

Stereo views of the Cα-chain of TS-porin, the succinylated residues Lys-46, Lys-300, Lys-298, Glu-1(α-NH_2) and the unmodified inner channel and outer surface Lys residues are shown in Figure 11 a,b. The structural comparison with native porin revealed that all succinylated lysine residues are located at the inner channel surface and are solvent-accessible, i.e. not shielded by neighbouring groups: Lys-46 is located at the channel constriction site whereas the other succinylated residues Lys-300, Lys-298, and the N-terminus are all near the periplasmic pore mouth. Of the further Lys residues at the inner surface (Lys-3, -69, -191, -198), a low to trace extent of modification was found only for Lys-3 and Lys-69. Lys-69 is shielded by the adjacent Phe-62 and Glu-71 residues, Lys-191 by Asn-196, and Lys-198 is shielded within 4 Å by the Phe-83 and Asp-85 residues. These results are in agreement with the previously established predominant reactivity of free surface accessible lysine residues towards amino acylation [17]. Three other lysine residues (Lys-138, -169, -230) on the outer surface of the barrel are not succinylated which can be explained by shielding due to surrounding detergent molecules [10]. The low reactivity of the Lys-3 residue at the inner channel surface towards succinylation can be explained by shielding due to the Ser-5 and Thr-30 residues, and by a local conformational change that may occur upon succinylation of the adjacent N-terminal (Glu-1)-amino group [15, 17].

The side chain of Lys-46 originally pointed almost horizontally towards the centre of the opposite third external loop. Upon succinylation, it bends towards the periplasmic surface. The attached 3-carboxypropionyl group bends further in this direction and to the right when seen along the membrane normal towards the external phase. The average angle of the succinyl-lysine side chain with the membrane is approximately 35 degrees. The terminal carboxyl group shows no polar interactions with protein residues. As no electron density was observed for the succinylated Lys-300, Lys-298 and the N-terminal amino groups in the difference map, a considerable disorder must be assumed for these succinyl groups, at the periplasmic entrance of the channel.

3. Discussion

The electrical properties of porin from *R. capsulatus* were substantially modified upon the succinylation. The observed changes are in complete agreement with the specific succinylation sites identified: all of the succinylated amino groups are solvent-accessible and are located at critical positions for the passage of ions (see Figure 11). In contrast to the other, native charged side chains in this region, the structure of succinyl-lysine-46 as modelled from the electron density map does not point in extended

Following page:
Figure 10. 2Fobs-Fcalc density and model of succinylated lysine-46 residue. The density is plotted at a 1 σ level.
Figure 11. Stereo views of Cα-chain in ribbon representation and distribution of lysine residues in the tetra-succinylated porin channel. All lysine residues are depicted in yellow. Succinylated lysine residues are indicated by a red sphere at the position of the ε-amino group. For the succinylated Lys-46 residue the structure derived from the x-ray analysis is shown. a), view along the membrane normal from the external side; b), view parallel to the membrane.

conformation towards the opposite segment of the constriction site but bends towards the periplasmic space. Thereby, the succinyl-carboxylate group is located distant from the negatively charged Asp and Glu residues on the opposite segment of the constriction site. No evident attractive electrical interactions favour this conformation.

In spite of the bulky amino-succinylated residues located at the periplasmic channel mouth, the succinylation increased the single-channel conductance (see Figure 8). Moreover, the single channel conductance of the TS-porin was no longer a linear function of the bulk aqueous concentration (see Table 2). Instead a slope of about 0.5 to 0.6 was observed on a double-logarithmic scale for the conductance-versus-concentration curve, indicating that negative surface charge effects determine the properties of the TS-porin channel [29]. In addition to the succinylated Lys-46 residue in the channel eyelet, the negative charges introduced by the modification of the Lys-298, Lys-300 residues and the N-terminus are all located near the periplasmic channel mouth, resulting in substantial ionic strength-dependent surface potentials which attract cations and repel anions. Accordingly, they are expected to influence both the single-channel conductance and zero-current membrane potential.

3.1. INFLUENCE OF POINT CHARGES AT THE PORE MOUTH ON SINGLE CHANNEL CONDUCTANCE.

A quantitative description of the effect of the point charges introduced at the channel eyelet and the pore mouth on the single-channel conductance may be given on the basis of the following consideration. The surface potential, Φ, for a channel with radius, r, and a total charge, q (in As), is given by [29 30] :

$$\Phi = 2 \cdot q \exp(-r / l_D) / (4 \cdot \pi \cdot \varepsilon_0 \cdot \varepsilon \cdot r) \qquad (7)$$

where ε_0 (= 8.85 x 10^{-12} A·s/V·m) and ε (= 80) are the absolute dielectric constants of vacuum and the relative dielectric constant of water, respectively, and l_D is the so called Debeye length which controls the decay of the potential (and of the accumulated positively charged ions) in the aqueous phase:

$$l_D^2 = \varepsilon \cdot \varepsilon_0 \cdot R \cdot T / (2 \cdot F^2 \cdot c) \qquad (8)$$

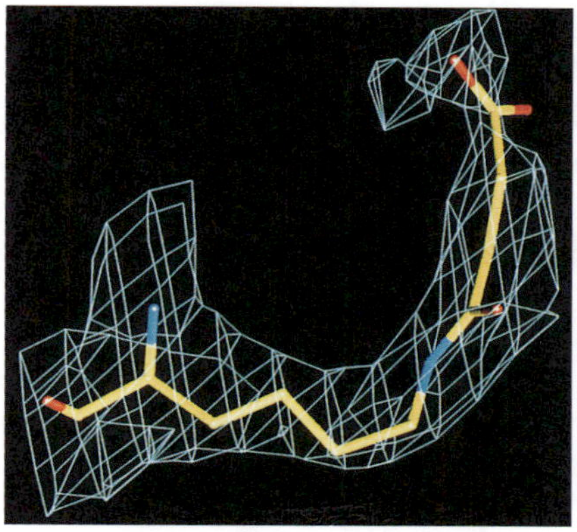

a

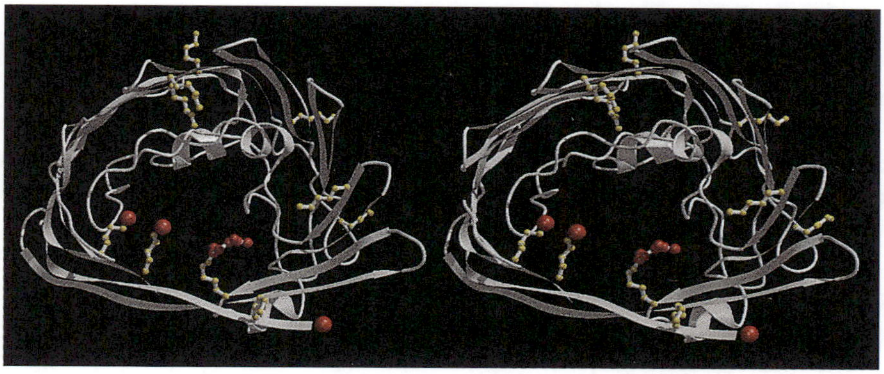

b

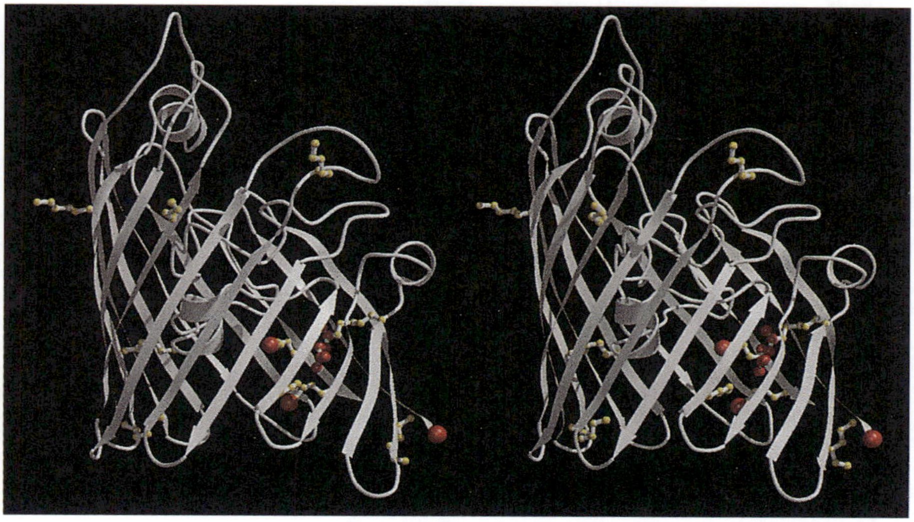

(with c = bulk aqueous salt concentration; and RT/F = 25.2 mV at 20°C). Values for l_D calculated for different ion concentrations using (8) are given in Table 5. As shown in Figure 9 particularly for TS-porin with the additional succinylation sites at the pore mouth, the single-channel conductance differs from that of native porin as long as the Debye length is larger than the radius of the channel eyelet. This contrasts with the smaller effect observed for MS-porin upon the succinylation of the inner-channel Lys-46 residue.

Table 5. Debeye lenth calculated with eqn (8) for a single elementary charge and different concentrations of a monovalent salt in water at room temperature.

c	1 mM	10 mM	0.1 M	1M
l_D	96 Å	30 Å	9.6 Å	3 Å

The surface potential around the fixed charges (7) allows the mobile counterions of the electrolyte solution to form a "cloud" there and assume a lower enthalpy locally. We assume that these ions are in equilibrium with the distant "bulk" ions (this assumption will be abandoned in the following discussion of diffusion limitation) and therefore have the same chemical potential. They thus must have a lower entropy as the bulk ions. This lower entropy is due to the higher ion concentration in the cloud of counterions. As these cloud-ions have the same chemical potential as bulk ions, they also have (approximately) the same mobility and thus the local conductivity is increased corresponding to the local higher concentration of ions [29].

The more charges are accumulated there, the higher the single-channel conductance should be in comparison to the native protein. The Boltzmann distribution relates the ratio of concentrations of monovalent cations in the bulk solution, c, with the electric potential Φ of the fixed succinyl groups at the channel mouth, c_0^+:

$$c_0^+ = c \cdot \exp(-\Phi \cdot F / (R \cdot T)) \tag{9}$$

The cation concentration c_0^+ can now be used for the calculation of the single channel conductance at bulk ion concentration c, G(c), [31]:

$$G(c) = G_0 \cdot c_0^+ / c \tag{10}$$

For the anion concentration c_0^- near the pore mouth the sign in the exponent of (9) is +, indicating that they are repelled and consequently their contribution to the current will be reduced. In our case of a very cation selective channel we neglect the contribution of the anions to the single channel conductance.

In the more general case of a nonselective or weakly selective channel, a fixed charge has a stronger attractive effect on counterions than a repulsive on ions of the same charge. This can be seen from the Boltzmann factor (9) for cations and anions or from Fig. 9 in Ch. 10 of Hille's book [32]. The anion factor differs less from 1 than the cation factor due to the course of the exponential function. Therefore we expect that in this general case upon fixing a charge to the pore mouth we still would find an increase in single channel conductance and an increase in selectivity for the counterions of the fixed charge.

A best fit of the data in Table 3 was obtained by assuming a negative elementary charge of 1.0 for TS-porin and of 0.6 for MS-porin located at the pore mouth, respectively and a radius $r = 0.5$ nm for the channel of both succinylated porins to calculate the potential with (7). This can be shown by a fit of the single channel conductance as a function of the bulk aqueous concentration c by eqn (10) which represents a combination of eqs (7) to (9) (see Fig. 9). The single channel conductance, G_0, of the modified channel but in the absence of the point charges was assumed to be 2.0 nS, instead of the 2.9 nS found for the native porin, to account for steric effects of the succinyl groups. It is noteworthy, that the number of negatively charged groups involved in the point charges is somewhat tentative since two different formalisms, the Debeye-Hückel theory and the treatment by Nelson and McQuarry [30] can be applied to the problem. Both differ in the number of charges by a factor of two, since a charge at the surface of a low dielectric membrane induces an image charge by attraction of counterions on the opposite side. On the other hand, the dielectric constant of a membrane protein is neither 2 nor 80. Even the dielectric constant of the water inside the channel is unknown and possibly smaller than 80 [33]. This means that the number of charges for TS-porin will be intermediate between 1, according to the Nelson-McQuarrie treatment, and 2, according to the Debye-Hückel treatment (see [24] for a more complete discussion of this problem).

Formally, eight negative charges were introduced in TS-porin by succinylation of four amino groups. However, only one to two negative charges are needed to explain the point charge effect on the single channel conductance. Several effects may contribute to this discrepancy: mutual charge neutralization effects, pK shifts and steric effects of the bulky succinyl groups in the channel mouth all limit the validity of the above theoretical treatment and lead obviously to a reduction of the apparent number of negatively charged groups as compared to the formally introduced ones.

The ion selectivity in the succinylated porins shifted as expected due to the replacement of up to 4 amino groups by negatively charged succinyl groups. As indicated by the increase of the P_{cation}/P_{anion} ratio, succinylation facilitates the cation relative to anion permeability, consistent with previous studies of chemically modified *E. coli* porins [34, 35] and of PhoE mutants using single-site amino acid substitutions [36].

The present study provides evidence that in the tetra-succinylated porin negative charges are placed directly at the periplasmic channel mouth of porin. The single-channel conductance is then no longer linear but shows saturation and resembles a square-root function, concomitant with an increased selectivity for cations. This effect of charges near the channel mouth has been predicted by Menestrina & Antolini [29], Dani [37] and Jordan [38] for ion-channels in general and has been experimentally verified in Gramicidin channels [31, 39] and a Ca^{++}-activated K^{+}-channel by chemical modification of surface carboxylate groups [40].

Tertiary structure-selective chemical modification such as by lysine-succinylation in this study characterised by mass spectrometric peptide mapping provides an efficient method for understanding electrical properties of channel proteins [41, 42].

3.2. PORINS AS RELATED TO OTHER CHANNEL PROTEINS:

Among channel proteins, general-diffusion porins are characterised by their large and rather non-specific single channel conductance. While sodium and potassium channel conductances at physiological conditions range from 2 to 50 pS [32], general-diffusion porin monomers at similar conditions range between 50 and 200 pS [3]. Moreover, the former channels show flux saturation [43], and have been modelled as a sequence of energy barriers connected by binding sites and the flux through them was calculated on the basis of the „reaction-rate theory“ [44, 37].

3.3. DO PORINS POSSESS ASYMMETRIC FLOW CHARACTERISTICS?

Apell's observation [39] of a dependence of the net ion flow on the orientation of the modified gramicidin channel with respect to the two aqueous phases raises the question of flow asymmetry of channels. It appeares obvious, that porin channels could be flow-optimised for certain types of solutes in their „rightside-out“ orientation, i.e. their permeability could be higher when a chemical potential gradient points from the external (e) to the periplasmic (p) side of the channel than in the opposite "inside-out" orientation. It therefore appeares to be instructive to discuss the existing theoretical models to find mechanisms which result in asymmetric flow.

3.4. WHICH THEORETICAL MODELS ARE AVAILABLE?

3.4.1. *Multiple barrier reaction rate models.*

A Multiple-barrier reaction rate theory model [44] seems to be attractive because it uses statistical mechanical concepts which are closely related with molecular interactions. The free energy barrier of a porin channel could look roughly as shown in Fig. 12. The steep flanks on both ends arise mainly from the entropy decrease associated

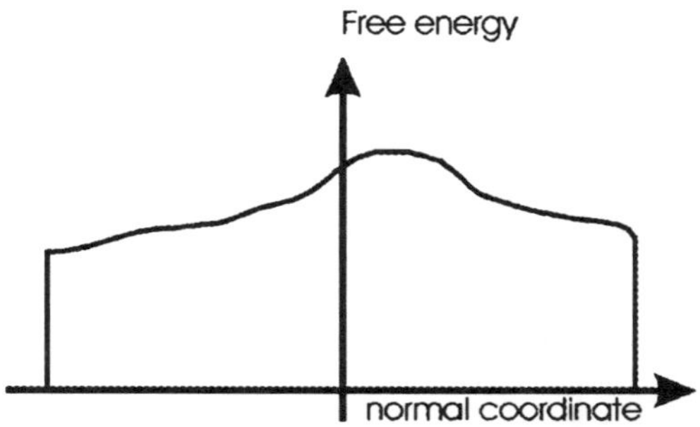

Figure 12. Hypothetical free-energy function of a solute permeating a porin channel. The potential function is assumed to depend only on the coordinate along the membrane normal. The coordinates of the steep flanks correspond to the intersections with the planes covering the channel mouths.

with the entrance of a molecule into the channel mouth. If the entrance into the channel can be treated by reaction rate theory, its rate R_{ent} will be:

$$R_{ent} = (k_{ent} \cdot c_{channel}) \cdot c_e$$

According to the activated complex theory [45, 46] k_{ent} can be written as:

$$k_{ent} = k_B T/h \cdot \exp[-\Delta G^*/RT]$$

k_B and h are Boltzmann's constant and Planck's constant, respectively, and $\Delta G^* = \Delta H^* - T\,\Delta S^*$ are the free energy, Entropy and Enthalpy differences for the state of the ion after having entered the pore mouth and before, i.e. when beeing solubilised in the e phase. The largest contribution to ΔG^* will be the entropy decrease accompanied with the funnelling of the ion into the narrow channel.

3.4.2. *The flow of solutes through a channel in the reaction-rate approach is symmetric:*

As we need to know not only the rate of channel entrance, but the flux of solutes across the whole channel, we would have to calculate reaction-rates across the free-energy profile of the type shown in Fig. 12. The linear relation of current and voltage found with nonspecific porins requires a description of the channel by many barriers of near equal height [47]. If a reaction-rate theory model was applicable, its steady-state flux would be proportional to the ΔG^* of the highest barrier in the channel [48] and therefore would not be dependent upon the orientation of the channel with respect to the chemical potential gradient as long as the contribution of the externally applied field to the barriers is neglected.

3.4.3. *The theory of electrodiffusion is more appropriate for porins than the reaction-rate theory.*

The conditions to be met for the applicability of reaction rate theory models have been discussed and are not well met in large channels such as those formed by porins where the free energy function varies smoothly as compared to the mean free path of an ion. Instead, the theory of electrodiffusion, based on the Nernst-Planck equation, should be more appropriate to describe the permeation process between the channel mouths [49].

3.4.4. *Diffusion through a channel of asymmetric geometry is symmetric*:

As a prototype of an asymmetric case, a conical channel as in Fig. 13 (a), left panel, is used in the following. We will refer to the orientation shown, with the wide mouth opening to the phase of high solute concentration c_e, as "right-side out". We will initially assume that there is no external field no membrane surface potential difference across the membrane, no electrostatic potential inside the channel and that $c_p = 0$.

Fick's law (**J** = -D **grad** c) will be used for the diffusion of a solute. Under steady-state conditions inside the channel the flow of solute into every imaginary compartment has to equal the flow out of it (**div J** = 0), i.e, $\Delta c = 0$. The Laplace equation has unique solutions for the boundary conditions which hold for the channel (see e.g. [50]): the concentration on the external and periplasmic planes spanning the channel mouths have to be c_e and 0, respectively, and on the channel walls the solute flux normal to the walls has to be zero. The latter means that the surfaces of constant concentration must meet the channel walls perpendicularly.

When we move from the wide mouth towards the opposite mouth the crossection of the conical channel will decrease. As the flow of solute through each crossection must be constant, the flux -D **grad** c must increase. The concentration c along the channel axis thus must decrease with increasing slope, as shown schematically in Fig. 13b, left panel. We now reverse the conditions i.e. apply the concentrated solute phase to the other side of the channel. The boundary condition changes as the concentration, c_e, is now facing the small mouth and the zero concentration is facing the large mouth. The following discussion is simplified if we create a more symmetrical problem by subtracting $c_e/2$ from all concentrations. We thus require the concentration c to be equal $c_e/2$ and $-c_e/2$ on the external and on the periplasmic plane, respectively. If c is the solution to Laplace's equation in the original "right-side-out" orientation, -c will be the solution to the "inside-out" situation because it shows the required boundary conditions and also because this is a solution to the Laplace equation inside the channel. As -c has reversed gradients as compared to c, the flow of solute through both cases is opposite but of equal modulus. Thus the mere diffusion model does not give asymmetric fluxes.

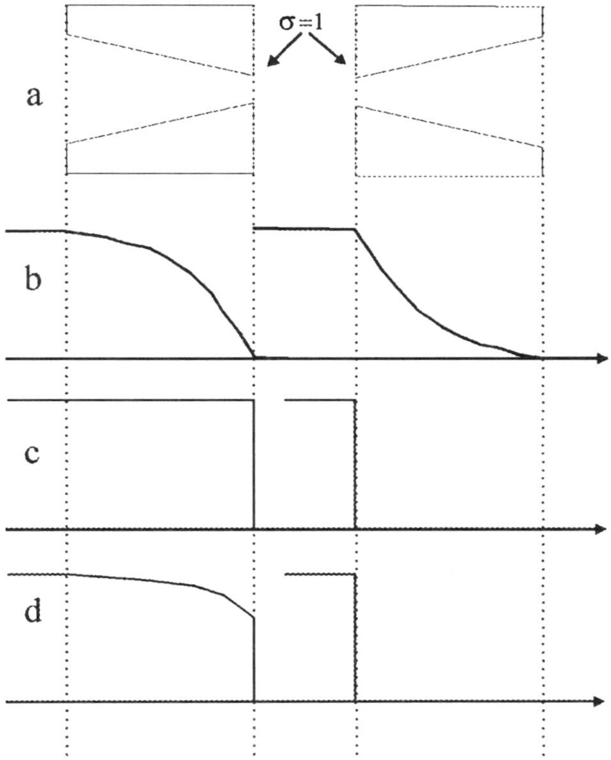

Figure 13. In the "right-side-out" orientation the "e" phase is on the left with the solute concentration c_e. The "p" phase is on the right and contains no solute molecules. (a) - (c), Hypothetical concentration profiles of solute molecules in the conical model channel in the right-side-out orientation and the inside-out-orientation. (a), without solute restriction and without osmotic flow; (b), with solute restriction and without osmotic flow, respectively; (c), with osmotic flow and with solute

As the solute in the previous model was more concentrated on one side, there will be an osmotic flow of solvent in the opposite direction [51]. A model of a conical channel involving diffusion of a solute and opposite osmotic flow has been proposed and discussed by D.G.Levitt [52]. It predicts an asymmetric flux if the flux of solute is restricted in some channel segment which is asymmetrically positioned with respect to the membrane center. The narrow mouth in the conical model given above could be such a site. The restriction for the permeation of solute is expressed by the reflection factor σ_s (see also § 10.3 in [51] for an explanation). A value of 1 blocks the permeation of the solute through the restriction site, a value of zero indicates unrestricted diffusion. The possibility of asymmetric flow becomes evident when starting with σ_s = 1 and decreasing its value slightly. The concentration profiles for σ_s near 1 with- and without osmotic flow are given in Fig. 13 (c) and (d), respectively. It is seen that the concentration gradient across the restricting narrow mouth is larger in the "inside-out" than in the "rightside-out" orientation, indicating a larger diffusive flow of solute molecules in the former case as compared to the latter, when σ_s is smaller than 1. The differences are, however, only a few percent of the flux modulus [52].

The model used above assumes the same diffusion coefficient to hold inside the channels as in the bulk aqueous phase. Interactions with the proteinacous walls can be taken into regard rudimentarily by the reflection factor. Do solutes inside the channel diffuse slower than in the bulk phases adjacent to the channel ? Are there electrostatic or other interactions (sometimes called "friction") between the permeating molecules and the channel wall? Benz *et al.* [53] have calculated the cross-section of a transmembrane aqueous cylinder just long enough to traverse a membrane and having the same conductivity as a typical general diffusion porin trimer. In the structure of porin the average channel crossection is larger than the diameter of 8,6 Å calculated for the equivalent water channel of one *Rhodobacter capsulatus* porin monomer [54]. The ions permeating the porin channel of gram-negative bacteria thus show some "friction" resulting in a smaller mobility than in the bulk solution, indicating that some interaction, i.e. some energy barrier does exist. This can either be due to an altered dynamics of water inside the channel, as demonstrated in molecular dynamics simulations [33, 55], or due to the action of an electrostatic potential on permeating charged molecules or due to transient dehydration and hydrogen bonding..

We will therefore now discuss theories which still view the permeation as a diffusion process, assuming the same diffusion constant in the channel as in bulk water, but use an electrostatic potential to model the interaction with the channel wall, i.e. electrodiffusion (see e.g. [49]).

These flux equations with the Nernst-Planck electrodiffusion theory are also symmetrical with respect to the orientation of the channel as long as the potential function of the reversed channel is just the inverse of the original channel and the contribution of the applied external field to the channel potential function can be neglected. The electrodiffusion equations thus show the same symmetry as the multiple-barrier reaction-rate results (see Läuger [48], Levitt [56] and Chen and Eisenberg [57]).

3.4.5. *Asymmetry of a reaction-rate model with fixed charges and ion binding to the channel vestibules:*

In the model of Dani [37], a two-barrier-one-site reaction rate model is used for the narrow segment of the channel between the vestibules and a Poisson-Boltzmann approach to calculate the ion concentration in the vestibules with their attached fixed negative charges. Dani's calculations show the asymmetry of flow in presence of a transmembrane electric field and identical electrolyte concentration on both sides but he does not treat the case of a channel connecting two bulk phases with electrolyte of different concentration. The current is larger when the applied field drives cations from the side with the negatively charged vestibules into the channel. As one of the vestibules is charged, the potential along the channel is asymmetric. Upon reversal of the channel it is inverted, but the contribution at the applied field is not, yielding two potentials which are not inverted with respect to each other. This causes the asymmetry.

3.4.6. *Electrodiffusion models with charged vestibules:*
Several authors have proposed theories which use the Nernst-Planck electrodiffusion equation. In an electrodiffusion model with fixed charges, mobile counterions are attracted and charges are induced.

In succinylated porin, the cloud of counterions as well as the fixed protein charges attracting them contribute to the electrostatic potential of the channel vestibules. Charges permeating the channel thus do not feel the fixed-charge potential of the protein alone, but a potential containg additional contributions by mobile clouds of counterions. The potential function thus has to be „self-consistent“, i.e. take into regard the presence of mobile and induced charges. This is the main feature of a theory described by Chen and Eisenberg based on an electrodiffusion approach [57], which has been expanded to include nonequilibrium phenomena near the vestibules [58]. Only symmetrical protein charge distributions are discussed, however. In the second article the authors elaborate on the homo-flux ratio.

3.4.7. *The homo-flux ratio is always symmetrical in absence of diffusion limitation:*
Electrodiffusion theories as well as reaction-rate theories of channels always yield the same inward to outward homo-flux ratio independent of channel orientation for the ion type I: ($J_{i,inw}/J_{i,outw}$) with or without fixed charges. This is because the flux ratio behaves analogously to the reaction rate ratio of a monomolecular reaction and, as the latter equals the Boltzmann factor with the chemical potential difference ($\mu_{i,in} - \mu_{i,ex}$) in the two phases adjacent to the channel :

$$J_{i,inw}/J_{i,outw} = \exp((\mu_{i,in} - \mu_{i,ex})/RT)$$

Using $\mu_{i,x} = \mu^0_{i,x} + \ln(c_{i,x}) + z\,F\,\varphi_{i,x}$ (x stands for "in" or "ex"), one obtains Ussings flux ratio in absence of coupling (see e.g. [32]). This result is obtained for all rate theory and electrodiffusion approaches as long as nonequilibrium phenomena like diffusion-limited access to the pore mouth [47] or phase-boundary potentials [58] are ignored. The homo-flux ratio depends only on the chemical potential of the bulk phases and is therefore independent of the channel orientation and of the free energy profile in the channel. As catalysts of membrane permeation, channels reduce the activation barrier and increase the speed of attaining the equilibrium. The individual fluxes $J_{i,inw}$, $J_{i,outw}$ and the net flux $J_i = J_{i,inw} - J_{i,outw}$ thus are influenced by all the channel parameters such as the potential function, the applied field and the presence of fixed charges near the channel mouth [58]. In particular, fixed vestibular charges can influence the potential function of the channel and consequently the fluxes in a orientation dependent manner.

3.4.8. *Summary on what has been learned about asymmetry*:
Within the framework of reaction-rate and electrodiffusion models, a flux can change upon reversal of the channel, when the barrier function or the potential function are not merely inverted upon this operation. For example the potential of a fixed charge near a channel mouth depends upon the ion concentration in the adjacent phaseand can thus change upon reversal. Moreover, the electrostatic potential inside a channel depends upon the direction of the externally applied field as it contributes to the former. A change in net flux upon reversal of the channel can thus be effected by these parameters. We therefore should have seen different current steps in single-channel reconstitution recordings of succinylated porin. As Fig. 2 shows a single maximum, we assume that the resolution in the experiment was insufficient to resolve the different conductivities for the two possible orientations. This could also have contributed to the surprisingly low number of negative charges, sufficient to explain the increased single channel conductivity in TS porin as compared to porin.

3.4.9. *Diffusion limitation*:
The homo-flux ratio can be changed under nonequilibrium conditions, i.e. under diffusion limitation [47] and with phase-boundary potentials [58].

Following the treatment by Läuger [47], the permeation of a solute through the porin channel can be conveniently divided into three steps: 1) Diffusion from the bulk solution of higher chemical potential to the channel vestibule „E“, 2) diffusion from E through the narrow channel segment to the opposite vestibule „P“, 3) diffusion from P to the bulk solution of lower chemical potential. Diffusion limitation occurs, when step 2 is so fast that the chemical potential in one or both of the vestibules differs from the adjacent bulk solution phases. The diffusion from the bulk to the vestibules can be characterised by a „capture radius“ r_0. For the simplest cylindrical pore of length l without an electrostatic potential and two identical capture radiuses Läuger has shown that diffusion limitation has to be expected for small values of l/r_0. A quantitative estimation for porin with a monovalent electrolyte and a channel capture radius of 7 Å indicates that the limit is to be expected to be $G/c \approx 20$ nS/M. This order of magnitude is reached only for TS porin (Table 2) at electrolyte concentrations lower than 30 mM. Diffusion limitation can be demonstrated by flow-saturation at increasing voltage. Our conductivity data were, however, all obtained at a relatively low voltage of 20 mV. Moreover, Läugers estimation may not apply to our situation, as according to Moore & Pearson, [59] fixed charges near the channel mouth, as in MS and TS porin, increase the diffusion limit. The radial field of the latter probably helps to funnel the diffusing ions into the vestibules and thus formally increases the capture radius. Fixed vestibular charges thus play an important role when pore mouth access is diffusion limited.

As mentioned, the data of this study cannot demonstrate that diffusion limitation has occurred in our experiments. We nevertheless wish to express the view that the phenomenon could play a role in bacterial growth, which occasionally is limited by the influx of essential small nutrient molecules through the OM, when these are available only at low concentration in the growth medium- e.g. mono-and disaccharides and nucleotides. Under these growth limitation conditions bacteria express specific porins with weak binding sites for the limiting substance. Maltoporins with a „greasy slide" chain of sugar binding sites are expressed at less than 10 μM of maltose. The internal permeability of the narrow segment of these channels for their substrate seems to be high due to the fast sliding along the binding sites [13] and diffusional access to the vestibules may well limit the flux through these specific porins due to the low concentration of the substrate molecules and because their diffusion constants are smaller than those of ions due to their larger molecular mass. The molecular structure of these porins may help to increase the diffusion limit and thus the flux of the desired substrate through the OM.

3.4.10. *Asymmetry of diffusion-limited channels:*

Diffusion-limited flux can change upon reversal of the channel as stated by Chen and Eisenberg [58] and as can be deduced from Läuger [47]. In the latter, formula (19) holds for a channel with two different capture radiuses, different concentration of electrolyte on both sides with a permeable cation, an impermeable anion and in the limit of infinite permeability of the channel, e.g. in step 2 of above. This formula indicates that the flux will be larger when the channel is oriented such that the smaller capture radius faces the phase with higher concentration of electrolyte. The P capture radius thus should be larger to allow for an effective removal of the solute from the P vestibule. In this context we should consider again the case of the sugar specific porins as they use a new concept: By setting the chemical potential inside the channel below zero (K_D ca. 200 mM in maltoporin for maltose [60]), the free-energy gradient for capturing and thus the E capture radius are both increased. In oder to remove the sugar from the channel, i.e. to get a high P capture radius, we now need a protein with a sugar affinity even higher than the channel. The periplasmic binding proteins for sugars in bacteria serve to this end with K_D values of around 3 μM [61]. Obviously such a system can lead to an increased flux of sugars into the cell only when the sugar is finally bound to a high-affinity transport system in the inner membrane as e.g. the ABC transporters [62]. It is obvious, that this system is asymmetric.

Ion selective porins as e.g porin from *Acidophorax delafildii* [26] could also serve to increase the flux of an special electrolyte through the OM with one ion type being of high molecular mass and consequent small diffusion coefficient, as e.g. an organic acid, and its small countercation. The organic acids are useful for the growth of this porin's host bacterium. Asymmetric porins and ion selective porins thus could well be a result of this evolution towards flux maximization.

4. Materials and Methods

4.1. GROWTH AND PURIFICATION OF PORIN.

Rhodobacter capsulatus 37b4 cells were grown and harvested as described [63]. Porin was purified according to [12].

4.2. SUCCINYLATION OF PORIN.

The succinylation of porin was essentially carried out as previously described [34, 17]. The reaction was started by diluting a protein stock solution (6 mg/ml) to 1 mg/ml and pipetting a 500 µl-aliquot to different amounts of succinic anhydride (0.2 and 1.4 mg, respectively), and the pH kept at 6.5 for 60 min with a pH-stat by automated addition of a 1 M NaOH solution. The succinylated protein was dialysed against 20 mM tris(hydroxymethyl)aminomethane, 300 mM LiCl, 3 mM NaN_3 containing 0.6% (w/v) n-octyltetraoxyethylene (crystallisation buffer) and finally purified by FPLC-chromatography (fast-flow Q-sepharose; Pharmacia, Freiburg, Germany).

4.3. PROTEOLYTIC DEGRADATION AND HPLC ISOLATION OF PEPTIDE FRAGMENTS.

BrCN degradation of native and succinylated porin was carried out with protein samples of 100 µg (approx. 3 nmol). Protein solutions in 10 µl crystallisation buffer (20 mM NaN_3 with 0.6% n-octyltetraoxyethylene) were mixed with 25 µl 70% formic acid and 15 µmol (5 M solution in acetonitrile) BrCN added under a continuous N_2-stream. The degradation was performed for 24 h at 20°C in subdued light, and the reaction quenched by addition of 1 ml H_2O. A 1 µl aliquot was used directly for MALDI-MS peptide mapping analysis, and the remaining peptide fragment mixture subjected to HPLC separation.

Tryptic digestion was performed with solutions of 60 µg native porin and the succinylated porin derivatives in 6 µl crystallisation buffer which was mixed with 51 µl of a 2 M urea solution in H_2O and denatured for 5 min at 95°C. The solution was cooled to room temperature, brought to pH 8 by addition of 60 µl 100 mM NH_4HCO_3, and digested for 4 h (37°C) with 3 µg trypsin (Sigma; untreated in 1 mM HCl or TPCK-treated).

HPLC separations of peptide fragments were performed with a Waters-Millipore M-590/510 solvent delivery instrument/M490 UV detection system. 100 µl aliquots of tryptic digest mixtures were separated on a 25 × 0.4 cm C8-nucleosil column (Macherey-Nagel, Duisburg, Germany), using a linear binary gradient of 0.04% aqueous TFA (A) and 0.03% TFA in acetonitrile (B), 5-95% B/60 min. A 20 × 0.4 cm C4-nucleosil column was used for the separation of BrCN peptides, at otherwise identical conditions. Peptide fractions were isolated with a Gilson-CPR fraction collector, and used directly for mass spectrometric analysis or blotted to PVDF membranes for sequence determinations. Edman sequence analyses were performed on a Knauer 910 automated sequencer.

4.4. MASS SPECTROMETRY

MALDI-MS was performed with a Bruker Biflex time-of-flight spectrometer (Bruker-Franzen, Bremen, Germany) equipped with a UV-nitrogen laser (337 nm) and a dual channel plate detector and x-mass data system for spectra acquisition and instrument control. Sample preparation was carried out with 1 μl solutions of porin and succinylated porin derivatives in crystallisation buffer (10 μg/μl) which were mixed with 19 μl of a solution (23 μg/μl) of the matrix α-cyano-4-hydroxycinnamic acid (HCCA) in acetonitril: 0.1% TFA (2:1). A 1 μl aliquot of the resulting solution spotted on a 3 mm^2 stainless steel target was brought to dryness within 3-4 min at room temperature. The crystalline sample/matrix preparation was then controlled by a microscope and inserted into the ion source. Spectra were obtained with an acceleration voltage of 10 kV, approx. 5 x 10^6 W/cm^2 laser power and 10 Hz pulse frequency (3 ns).

Spectra of proteolytic digest mixtures and isolated peptide fragments were obtained with the same solvent conditions, using 1:1-mixtures of sample:matrix solutions.

4.5. CRYSTALLISATION

Crystallisation was carried out at 17°C using the sitting-drop method. The protein concentration in the drop was 5 mg/ml in a buffer containing crystallisation buffer and 8% (w/v) of the precipitating agent polyethyleneglycol-600. The concentration of the precipitant in the reservoir solution was 35% (w/v). The best crystals were of perfect rhombohedral habit and grew in two weeks to a maximum size of 300 μm × 300 μm x 300 μm. The space group was R3.

4.6. DATA COLLECTION

X-ray diffraction data to 2.4 Å resolution were collected at room temperature on an image plate detector (STOE, Darmstadt) mounted on a rotating anode x-ray generator (Siemens). The raw data were processed with the programme XDS [64]. The 28'762 independent intensity measurements were reduced to 15'652 in the resolution range ∞

to 2.4 Å showing an overall R_{sym} of 10.6% on intensities. The overall completeness of the data set was 85.8% and 97.2% in the resolution shell ∞ to 2.7 Å. Because the R_{sym} on intensities in the outermost resolution shell (2.4 to 2.7 Å) was 37.3%, only the reflections from ∞ to 2.7 Å with an overall R of 8.8% on intensities were used. The hexagonal axes were determined as $a = b = 91.7$ Å and $c = 145.7$ Å.

4.7. REFINEMENT

The structure solution was initiated with the refined model of the native crystal form "C" (unpublished data) containing four calcium atoms, two detergent molecules and 268 water molecules. The subunit location was adjusted by 11 cycles of rigid body refinement at 2.7 Å resolution using the programme X-PLOR [65]. The resulting eulerian rotation angles (Θ_1, Θ_2, Θ_3) were (-0.02°, 0.01°, 0.19°) and the translation was (-0.01 Å, 0.09 Å, 0.00 Å).

For refinement of the succinyl-lysine, partial charges were set to zero. The corresponding 2Fobs-Fcalc-map showed for six of the ten lysine residues no electron density adjacent to their side chain. Three lysine residues (Lys-69, Lys-230, Lys-300) had some density at the ε-nitrogen but not sufficient for fitting a ε-N-(succinyl)-lysine. Only for Lys-46 contiguous density of sufficient volume was found. For the refinement all other atoms except those of succinyl-Lys-46 were fixed. After 40 cycles of conjugate gradient minimisation, 20 cycles of temperature factor refinement and an occupancy refinement for only the succinyl-Lys-46, a final free R-value [66] of 20.8% and a final working R-value of 19.9% was obtained, respectively.

The Figures were prepared with the programme O [67] on a Silicon Graphics 4D210 VGX computer.

4.8. LIPID BILAYER EXPERIMENTS

The methods used for the 'black' lipid bilayer experiments have previously been described [27]. Membranes were formed from a 1% (w/v) solution of diphytanoyl phosphatidylcholine (DiphPC, Avanti Polar Lipids, Alabaster, AL) in *n*-decane across circular holes (surface area about 0.5 to 1 mm^2) in the thin wall of a Teflon cell separating the two aqueous compartments. The temperature was kept at 25°C. All salts and buffers were of analytical grade and obtained from Merck (Darmstadt, Germany). The aqueous solutions were either unbuffered (pH ca 6), or buffered with 10 mM Hepes to pH 7. Native porin and succinylated protein derivatives were added from the stock solutions to the aqueous phase of the cis compartment (compartment to which the voltage was applied) after the membranes had turned optically black in reflected light. The current through the membranes was measured with two calomel electrodes switched in series with a voltage source and a model 427 current amplifier (Keithley, Cleveland, OH). The amplified signal was monitored with a storage oscilloscope and recorded on a strip chart recorder. For macroscopic conductance measurements the current amplifier was replaced by a Keithley electrometer (model 602). Zero-current membrane potentials were measured with the same instrument 5-10 min after the application of a salt gradient across the membrane [53].

5. Acknowledgements

This work was supported by the Deutsche Forschungsgemeinschaft (Sonderforschungsbereich 176; We962 and Pr175), Fonds der Chemischen Industrie, and by the EU-network "Peptide and Protein Structure Elucidation by Mass Spectrometry". We thank Dr D. Suckau and Dr. U. Rapp (Bruker-Franzen, Bremen) for help and assistance with the MALDI-MS instrumentation, and Dr P. Højrup (Odense University, Denmark) for help with the Edman sequence determinations. We gratefully acknowledge valuable discussions with Prof. H.-J. Apell, Prof. G. Stark, and Prof. David G. Levitt.

References

[1] Przybylski, M., Glocker, M.O., Nestel, U., Schnaible,V., Blüggel, M., Diederichs, K., Schad, M., Schmid, A., Welte,W. and Benz, R. (1996) X-Ray Crystallographic and Mass Spectrometric Structure Determination and Functional Characterisation of Succinylated Porin from *Rhodobacter capsulatus*: Implications for Ion Selectivity and Single-Channel Conductance, *Protein Sci* **5**, 1477 - 1489.

[2] Rietschel, E.T., Brade, L., Schade, U., Seydel, U., Zähringer, U., Kusumoto, S., Brade, H. (1988) Bacterial endotoxins: Properties and structure of biologically active domains, in E. Schrinner, M.H. Richmond, G. Seinbert and U. Schwarz (eds.), *Surface Structures of Microorganisms and Their Interactions with the Mammalian Host,* pp. 1-41.

[3] Benz, R. (1988) Structure and function of porins from gram-negative bacteria, *Ann. Rev. Microbiol.* **42**, 359-393.

[4] Welte, W., Nestel, U., Wacker, T., Diederichs, K. (1995) Structure and Function of the porin channel, *Kidney Int.* **48**, 930-940.

[5] Nakae, T., Ishii, J. and Ferenci, T. (1986) The Role of the Maltodextrin-binding Site in Determining the Transport Properties of the LamB Protein, *J. Biol. Chem.* **261**, 622-626.

[6] Benz, R. and Bauer, K. (1988) Permeation of hydrophilic molecules through the outer membrane of gram-negative bacteria, *Eur. J. Biochem.* **176**, 1-19.

[7] Adam, G., Läuger, P. and Stark, G. (1995) *Physikalische Chemie und Biophysik*, 3rd Ed., Springer Verlag, Berlin, Heidelberg, New York, Ch. 4.4.

[8] Schnakenberg, J. (1977) *Thermodynamic Network Analysis of Biological Systems*, Springer Verlag, Berlin, Heidelberg, New York, Ch. 2.3.

[9] Weiss, M.S., Kreusch, A., Schiltz, E., Nestel, U., Welte, W., Weckesser, J. and Schulz, G.E. (1991) The structure of porin from *Rhodobacter capsulatus* at 1.8 Å resolution, *FEBS Lett.* **280**, 379-382.

[10] Weiss, M.S., Abele, U., Weckesser, J., Welte, W., Schiltz, E. and Schulz, G.E. (1991) Molecular architecture and electrostatic properties of a bacterial porin, *Science* **254**, 1627-1630.

[11] Cowan, S.W., Schirmer, T., Rummel, G., Steiert, M., Ghosh, R., Pauptit, R.A., Jansonius, J.N. and Rosenbusch, J.P. (1992) Crystal structures explain functional properties of two *E. coli* porins, *Nature* **358**, 727-733.

[12] Kreusch, A., Neubüser, A., Schiltz, E., Weckesser, J. and Schulz, G.E. (1994) Structure of the membrane channel porin from *Rhodopseudomonas blastica* at 2.0 Å resolution, *Protein Science* **3**, 58-63.

[13] Schirmer, T., Keller, T.A., Wang, Y.-F. and Rosenbusch, J.P.(1995) Structural basis for sugar translocation through maltoporin channels at 3.1 Å resolution, *Science* **267**, 512-514.

[14] Benz, R., Woitzik, D., Flammann, H.T. and Weckesser, J. (1987) Pore forming activity of the major outer membrane protein of *Rhodobacter capsulatus* in lipid bilayer membranes, *Arch. Microbiol.* **14**, 226-230.

[15] Suckau, D., Mák, M. and Przybylski, M. (1992) Protein surface topology probing by selective chemical modification and mass spectrometric peptide mapping, *Proc. Natl. Acad. Sci. USA* **89**, 5630-5634.

[16] Nielsen, P.F., Schneider, K., Suckau, D., Landis, B. and Przybylski, M. (1990) Applications of selective chemical reactions in combination with 252-californium plasma desorption mass spectrometry in protein structure analysis, in E. Hilf (ed.), *Mass Spectrometry of Large Non-Volatile Molecules Vol 3*, World Scientif., London, pp. 194-207.

[17] Glocker, M.O., Borchers, C., Fiedler, W., Suckau, D. and Przybylski, M. (1994) Molecular characterization of surface topology in protein tertiary structures by amino-acylation and mass spectrometric peptide mapping, *Bioconj. Chem.* **5**, 583-591.

[18] Karas, M., Bahr, U. and Gießmann, U. (1991) Matrix-assisted laser desorption ionisation mass spectrometry, *Mass. Spectrom. Rev.* **10**, 335-358.

[19] Smith, R.D., Loo, J.A., Orgorzalek-Loo, R.R., Busman, M. and Udseth, H.R. (1991) Principles and practice of electrospray ionization-mass spectrometry for large polypeptides and proteins, *Mass. Spectrom. Rev.* **10**, 359-401.

[20] Przybylski, M. (1995) Mass spectrometric approaches to the characterization of tertiary and supramolecular structures of biomacromolecules, *Adv. Mass. Spectrom.* **13**, 257-285.

[21] Schiltz, E., Kreusch, A., Nestel, U. and Schulz, G.E. (1991) Primary structure of porin from *Rhodobacter capsulatus*, *Eur. J. Biochem.* **199**, 587-594.

[22] Ropele, M. and Menestrina, G. (1989) Electrical properties and molecular architecture of the channel formed by *Escherichia coli* hemolysin in planar lipid membranes, *Biochim. Biophys. Acta* **985**, 9-18.

[23] Benz, R., Schmid, A., Wagner, W. and Goebel, W. (1989) Pore formation by the *Escherichia coli* hemolysin: evidence for an association-dissociation equilibrium of the pore-forming aggregates, *Infect. Immun.* **57**, 887-895.

[24] Benz, R, Hardie, K.R and Hughes, C. (1994) Pore-formation in artificial membranes by the secreted hemolysins of *Proteus vulgaris* and *Morganella morganii*, *Eur. J. Biochem.* **220**, 339-347.

[25] Schmid, A, Benz, R. and Schink, B. (1991) Identification of two porins in *Pelobacter venetianus* fermenting high molecular mass polyethylene glycols, *J. Bacteriol.* **173**, 4909-4913.

[26] Brunen, M., Engelhardt, H., Schmid, A. and Benz, R. (1991) The major outer membrane protein of *Acidovorax delafieldii* is an anion-selective porin, *J. Bacteriol.* **173**, 4182-4187.

[27] Benz, R., Schmid, A. and Hancock, R.E.W. (1985) Ion selectivity of gram-negative bacterial porins, *J. Bacteriol.* **162**, 722-727.

[28] Weiss, M.S. and Schulz, G.E. (1992) Structure of porin refined at 1.8 Å resolution, *J. Mol. Biol.* **227**, 493-509.

[29] Menestrina, R. and Antolini, R. (1981) Ion transport through hemocyanin channels in oxidised cholesterol membranes, *Biochim. Biophys. Acta* **643**, 616-625.

[30] Nelson, A.P. and McQuarrie, D.A. (1975) The effect of discrete charges on the electrical properties of a membrane, *J. Theor. Biol.* **55**, 13-27.

[31] Apell, H.J., Bamberg, E. and Läuger, P. (1979) Effects of Surface Charge on the Conductance of the Gramicidin Channel, *Biochim. Biophys. Acta* **552**, 369-376.)

[32] Hille, B. (1992) *Ionic Channels of Excitable Membranes*, Sinauer Ass. Inc., Sunderland, Mass.,2nd Ed.

[33] Samson, M.S.P., Kerr, I.D., Breed, J., Sankaramakrishnan, R. (1995) Water in Channel-Like Cavities: Structure and Dynamics, *Biophys. J.* **70**, 693-702.

[34] Tokunaga, H., Tokunaga, M. and Nakae, T. (1981) Permeability properties of chemically modified porin trimers from *Escherichia coli* B12, *J. Biol. Chem.* **256**, 8024-8029.

[35] Benz, R., Tokunaga, H. and Nakae, T. (1984) Properties of chemically modified porin from *Escherichia coli* in lipid bilayer membranes, *Biochim. Biophys. Acta* **769**, *348-356.*

[36] Bauer, K., Struyvé, M., Bosch, D., Benz, R. and Tommassen, J. (1989) One single lysine-residue is responsible for the special interaction between polyphosphate and the outer membrane porin PhoE of *Escherichia coli*, *J. Biol. Chem.* **264**, 16393-16398.

[37] Dani, J.A. (1986) Ion-Channel Entrances Influence Permeation, *Biophys. J.* **49**, 607-618.

[38] Jordan, P. (1987) How Pore Mouth Charge Distributions Alter The Permeability of Transmembrane Ionic Channels, *Biophys. J.* **51**,297-311.

[39] Apell, H.J., Bamberg, E., Alpes, H. and Läuger, P. (1977) Formation of Ion Channels by a Negatively Charged Analog of Gramicidin A, *J. Membr. Biol.* **31**, 171-188.

[40] MacKinnon, R., Latorre, R. and Miller, C. (1989) Role of surface electrostatics in the operation of a high-conductance Ca^{++}-activated K^+ channel. *Biochemistry* **28**, 8092-8099.

[41] Przybylski, M., Borchers, C., Suckau, D., Mák, M. and Jetschke. M. (1993) Selective chemical modification and mass spectrometric peptide mapping: A new approach for the molecular characterization of surface topology and micro-environment in protein tertiary structures, in C.H. Schneider and A.N. Eberle (eds.), *Peptides 1992*, Escom Science Publ., Amsterdam, pp. 915-916.

[42] Przybylski, M. and Glocker, M.O. (1996) Electrospray mass spectrometry of non-covalent complexes of biomacromolecules - New analytical perspectives for supramolecular chemistry and molecular recognition processes, *Angew. Chem. Int Ed. Engl.* **108**, 878-899.

[43] Sakmann, B. and Trube, G. (1984) Voltage-dependent inactivation of inward-rectifying single-channel currents in the guinea-pig heart cell membrane, *J. Physiol.* **347**, 659-683.

[44] Läuger, P. (1973) Ion Transport Through Pores: A Rate-Theory Analysis, *Biochim. Biophys. Acta* **311**, 423-441.

[45] Gladstone, S., Laidler, K.J. and Eyring, H. (1941) *The Theory of rate Processes* , McGraw-Hill Book Comp., New-York and London.

[46] Gardiner, W.C. (1972) *Rates and Mechanisms of Chemical Reactions*, W.A. Benjamin Inc. Menlo Park, CA.,Ch. 6-1.

[47] Läuger, P. (1976) Diffusion-Limited Ion Flow Through Pores, *Biochim. Biophys. Acta* **455**, 493-509.

[48] Läuger, P. (1979) Transport of Noninteracting Ions through Channels, in C.F. Steven and R.W. Tsien (eds.), *Membrane Transport Processes Vol. 3*, Raven Press, New-York. pp. 17-27.

[49] Dani, J.A. and Levitt, D.G.(1990) Diffusion and Kinetic Approaches to Describe Permeation in Ionic Channels, *J. Theor. Biol.* **146**, 289-301.

[50] Morse, P.M. and Feshbach, H. (1953) *Methods of Theoretical Physics*, McGraw Hill Book Comp., New-York, Toronto, London, Ch. 6.2.

[51] Katchalsky, A. and Curran, P.F. (1975) *Nonequilibrium Thermodynamics in Biophysics*, Harvard Univ. Press, Cambridge, Mass, Ch. 10.3.

[52] Levitt, D.G. (1975) General Continuum Analysis of Transport through Pores. II. Nonuniform Pores, *Biophys. J.* **15**, 553-563.

[53] Benz, R., Janko, K., Boos, W. and Läuger, P. (1978) Formation of large ion-permeable membrane channels by the matrix porin (porin) of *Escherichia coli*, *Biochim. Biophys. Acta* **511**, 305-319.

[54] Benz, R., Woitzik, D., Flamann, H.T. and Weckesser, J. (1987) Pore forming activity of the major outer membrane protein of Rhodobacter capsulatus in lipid bilayer membranes, *Arch. Microbiol.* **148**, 226-230.

[55] Breed, J., Sankararamakrishnan, R., Kerr, I.D. and Sansom, M.S.P. (1996) Molecular Dynamics Simulations of Water within Models of Ion Channels, *Biophys. J.* **70**, 1643-1661.

[56] Levitt, G. (1991) Gneral continuum theory for multiion channel. I. Theory, *Biophys. J.* **59**, 271-277.

[57] Chen, D. and Eisenberg, R. (1993) Charges, currents and potentials in ionic channels of one conformation, *Biophys. J.* **64**, 1405-1421.

[58] Chen, D.P. and Eisenberg, R.S. (1993) Flux, Coupling and Selectivity in Ionic Channels of One Conformation, *Biophys. J.* **65**, 727-746.

[59] Moore, J.W. and Pearson, R.G. (1981) *Kinetics and Mechanism*, J. Wiley, New-York, 3rd Ed.

[60] Benz, R., Schmidt, A. and Vos-Scheperkeuter, G.H. (1987) Mechanism of Sugar Transport through the Sugar-Specific LamB Channel of *Escherichia coli* Outer Membrane, *J. Memb. Biol.* **100**, 21-29.

[61] Quiocho, F.A. (1991) Atomic structures and function of periplasmic receptors for active transport and chemotaxis, *Current Opinion in Struct. Biol.* **1**, 922-933.

[62] Boos, W. and Lucht, J.M. (1996) Periplasmic Binding Protein-Dependent ABC Transporters, in F.C. Neidhardt (ed.), *Escherichia coli and Salmonella*, ASM Press, Washington D.C., Ch. 76.

[63] Nestel, U., Wacker, T., Woitzik, D., Weckesser, J., Kreutz, W. and Welte, W. (1989) Crystallization and preliminary x-ray analysis of porin from *Rhodobacter capsulatus*, *FEBS Lett.* **242**, 405-408.
[64] Kabsch, W. (1988) Automatic indexing of rotation diffraction patterns, *J. Appl. Cryst.* **21**, 67-71
[65] Brünger, A.T. (1993) Assessment of phase accuracy by cross validation: the free r value. methods and applications, *Acta Cryst.* **D49**, 24-36.
[66] Brünger, AT. (1993) Assessment of phase accuracy by cross validation: the free r value. methods and applications, *Acta Cryst.* **D49**, 24-36.
[67] Jones, T.A., Zou, J.-Y., Cowan, S.W. and Kjeldgaard, M. (1991) Improved methods for building protein models in electron density maps and the location of errors in these models, *Acta Cryst.* **A47**, 110-119.

STRUCTURAL RELATION OF PERIDININ-CHLOROPHYLL *A*-PROTEIN (PCP) AND PROTEINS WITH GLOBIN-LIKE FOLD

A TESTABLE HYPOTHESIS FOR ITS FOLDING AND ASSEMBLY PATHWAY

KAY DIEDERICHS* and **ECKHARD HOFMANN**
Universität Konstanz, Fakultät für Biologie
Postfach 5560 (M656), D-78434 Konstanz, Germany

Abstract: Peridinin-Chlorophyll *a*-Protein (PCP) is a water-soluble, yet membrane-attached photosynthetic light-harvesting complex believed to reside in the interior of the thylakoid. We have recently reported the x-ray crystallographic structure of PCP from the dinoflagellate *Amphidinium carterae*, which crystallizes as a trimer [1]. Overall, the PCP polypeptide (312 amino acid residues) has the shape of the hull of a ship enclosing a cargo of two lipids, two chlorophyll *a* and eight carotenoid molecules. Each monomer is composed of a pseudosymmetry-related N- and C-terminal half of eight helices each adopting a peculiar topology whose relation to the globin-like fold is explored in this presentation. Based on the resemblance of folds and available data on myoglobin folding, we discuss a possible folding pathway for the PCP holoprotein which has to assemble during and after passage of the apoprotein through the thylakoid membrane.

Introduction

Higher plants and algae, as well as photosynthetic bacteria have developed light harvesting systems (LHCs), systems which work as antennae and increase the efficiency of photosynthesis by funneling collected light to the membrane-bound reaction centres. Membrane-bound LHCs have been structurally analyzed in the case of bacteria [2] and higher plants [3], and appear to be functionally similar, but structurally unrelated to water-soluble LHCs. Dinoflagellates are algal protists constituting a major portion of world's plankton. Most dinoflagellate species have a unique pigment-protein complex composed of a 32 kDa polypeptide, and two chlorophyll *a* and eight carotenoid (peridinin) molecules as chromophores. This complex, peridin-chlorophyll *a*-protein

* To whom correspondence should be addressed. email: Kay.Diederichs@uni-konstanz.de

W. Ens et al. (eds.), New Methods for the Study of Biomolecular Complexes, 277–284.

(PCP), exists in a form of half of this size in some dinoflagellate species [4]. In the long variant PCP species, the N- and the C-terminal halves of the sequence share a high degree of sequence identity (about 55%), indicating that they have evolved by gene duplication and fusion. As PCP is encoded in the nucleus, it has to be exported from the cytoplasm into the chloroplast and, as its leader sequence indicates, is exported from there into the thylakoid lumen.

PCP has no sequence or structural similarity to any other protein yet determined. Its unique presence in dinoflagellates and its apparent lack of structural relatives have been interpreted [5] such that PCP might be considered a late evolutionary enhancement of the photosynthetic machinery of dinoflagellates, unrelated to other water-soluble LHCs. Here we explore a structural similarity of PCP with proteins of the ubiquitous globin family, and demonstrate its evolutionary relationship to another group of soluble LHC, phycobiliproteins, which occurs in cyanobacteria and red algae. .

Structure of PCP

We have recently solved [1] the x-ray structure of PCP from the dinoflagellate *Amphidinium carterae*, which has a polypeptide of 312 amino acid residues, showing an internal two-fold pseudosymmetry of the N- and C-terminal domains (Fig. 1). Each domain consists of eight helices arranged in a peculiar fashion, that resembles a 'jellyroll' often encountered in the topology of β-sheets [8] and observed for the first time for helices in PCP. The juxtaposition of the N- and C-terminal domains creates the shape of a ship, whose hull is composed of the helices and whose cargo are the cofactors also showing the two-fold pseudosymmetry. In addition to the chromophores, we discovered by crystallographic means two lipid molecules (digalactosyl diacyl glycerol) which are integral structural components of the holoprotein assembly and whose existence and role in PCP is unknown. Trimers of the holoprotein complex appear to be the photosynthetically active unit, which is expected to stack onto the

Figures on following page

Figure 1: Architecture of PCP. The monomer of PCP is shown (grey), together with the chromophores (chlorophyll, green; carotenoids, red) and the lipids (blue). The N-terminal domain (left side) is related to the C-terminal domain (right) by a pseudo two-fold axis (vertical).

Figure 2: Three-dimensional (top) and topological (bottom) similarity of proteins from the globin family (left) and PCP (right). The color-ramp indicates the position along the chain from N-terminus (red) to C-terminus (blue). Ark clam hemoglobin was chosen as a typical representative of the globin family.

Figure 3: Binding of the tetrapyrrole cofactors in PCP (green), globins (red) and phycobiliproteins (purple) after superimposition of the E and F helices. The globin (ark clam) apoprotein is also shown (grey). All figures were produced with MOLSCRIPT [6] and RASTER3D [7].

Fig. 1

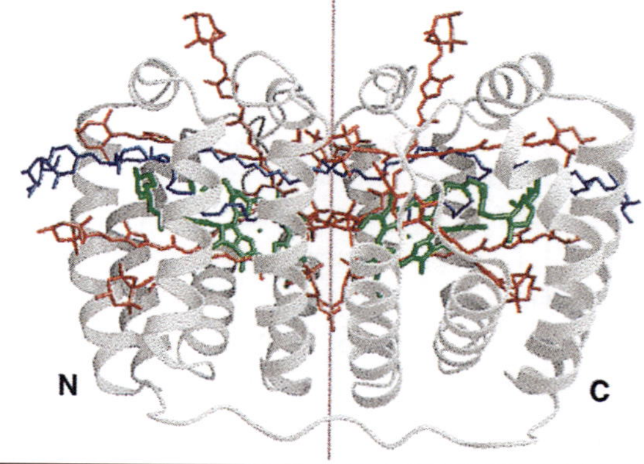

Fig. 2

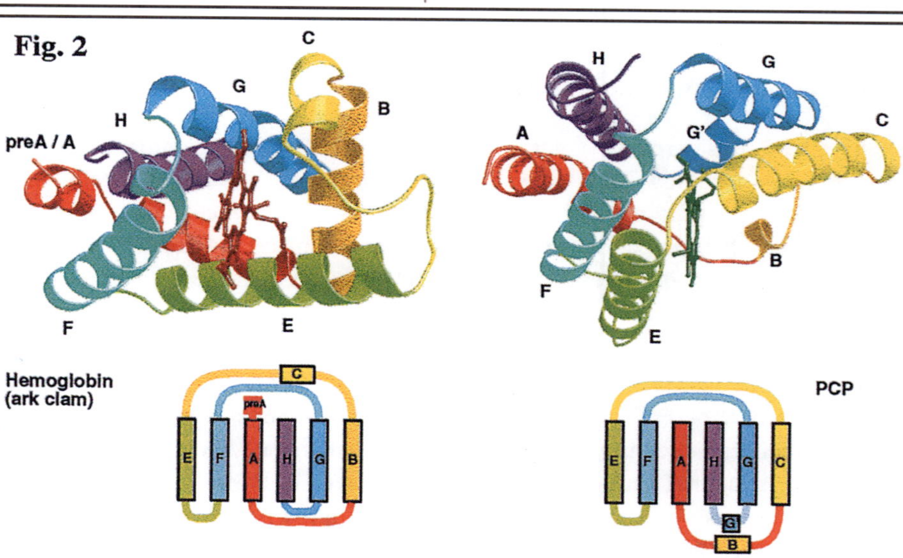

Fig. 3

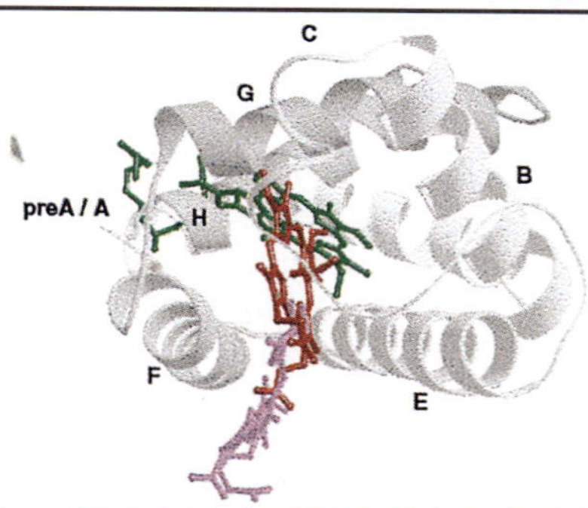

membrane of the thylakoid in order to optimize the transfer of harvested light energy to the membrane-bound components of the photosynthetic apparatus.

Is the existence of two varieties of PCP in dinoflagellates, a large and a small form, compatible with the x-ray structure of PCP from *A. carterae*? The answer is yes, as the contacts between the domains mainly involve hydrophilic interactions between the peridinin head groups, which could easily be replaced by contacts with an aqueous medium in the case of the short variants. Oligomerization provides a suitable orientation of the chromophores and allows for energy redistribution within the plane of the trimer, thus increasing the efficiency of energy transfer to the membrane-bound photosynthetic proteins. Only the C-terminal domain of the long variants is involved in trimer formation, and its sequence has most likely been optimized for this purpose. Short variant PCPs might form hexamers of the same shape as the long variant's trimers, but these hexamers would certainly be less stable. Thus, the long variant monomers appear to have functional advantages.

Comparison of PCP with the other all-α folds

Proteins can be considered as evolutionarily related if they share a significant degree of sequence similarity. However, many groups of structurally and functionally related proteins exist that lack similarity detectable with the sequence analysis programs available today [9]. In these groups, structural comparisons reveal evolutionary relationships that are hidden in the 'twilight zone' [10] (less than 25% sequence identity) of sequence alignments.

We used the SCOP [11] database (version 1.32, representing all 4432 structures in the Protein Data Bank available up to May 1996) to identify possible more distant structural relatives of PCP. To compare with the N-terminal domain of PCP (153 residues, 8 helices), we used the 71 all-α folds enumerated in SCOP as possible candidates for structural similarity with PCP. From each of these folds, representatives were taken from a non-redundant database [12], superimposed [13] (Table 1) on the N-terminal domain of PCP and then compared visually using 'O' [14]. Visual inspection was performed as the scores which are normally used for an automatic determination of topological similarity [13] were deemed inadequate for the identification of remote structural homology. Surprisingly, the topological similarity score [13] (column 5 in Table 1) matched well the visual finding that proteins of the globin-like fold (myoglobin, hemoglobin, phycobiliproteins and colicin A [15]) are structurally related to PCP (Fig. 2). In the case of the heme-binding globins, further support for the topological agreement found comes from overlapping tetrapyrrole binding sites, which are located on helices E and F. The other all-α folds with a high topological similarity score (Table 1) are much larger proteins (252 to 605 residues) with some of their helices matching those of PCP by chance. Granulocyte-colony stimulating factor, a helix-bundle protein of similar size, has only 4 helices superimposable on PCP.

Table 1: Top-scoring all-α candidates for possible similarity to PCP. The all-α entries [11] from a non-redundant set [12] representing all known protein structures up to May 1996 were compared in an improved version of SUPERIMPOSE [13] with the N-terminal domain of PCP (153 residues).

fold	proteins	PDB entries	size (residues)	topological similarity score*
methane monoxygenase hydrolase	methane monoxygenase hydrolase (methylococcus capsulatus)	1mmo	384	71.6
globin-like	myoglobin (sea hare)	2fal	145	70.1
	myoglobin (sperm whale)	1mls	154	64.7
	hemoglobin (sea cucumber)	1hlb	157	64.2
	hemoglobin (ascaris suum)	1ash	147	60.2
	hemoglobin (marine bloodworm)	2hbg	147	59.8
	hemoglobin (innkeeper worm)	1ith	141	59.5
	hemoglobin (yellow lupine)	2gdm	153	58.7
	hemoglobin (ark clam)	3sdh	145	58.6
4-helical cytokine	granulocyte-colony stimulating factor (bovine)	1bgc	158	64.1
aldehyde ferredoxin oxidoreductase	aldehyde ferredoxin oxidoreductase (pyrococcus furiosis)	1aor	605	61.5
cytochrome P450	cytochrome P450 (saccarapolyspora erythraea)	1oxa	403	59.2
chorismate mutase II	chorismate mutase II (yeast)	1csm	252	59.2
glycosyltransferase of the superhelical fold	CelD cellulase (clostridium)	1clc	541	58.5

* for definition, see reference 13.

It is worth discussing that the globin-fold has often been described as 3 helices (A,E,F) sandwiching at right angles on a layer of 3 other helices (B,G,H). This '3-on-3' description was found adequate for demonstrating structural relatedness of colicin A and globins [15], as well as that of phycobiliproteins and globins [16]. However, an early review of protein folding patterns [8] gave an alternative descriptions of the globin fold, the 'greek key' (so called according to its resemblance with decorative patterns on greek vases). We note that the greek key description is more informative, as it focuses on the connections between the helices inasmuch as on their packing, whereas the '3-on-3' term merely refers to the arrangement of helices in space, neglecting possibly differing topologies caused by different connections. In our first description of PCP [1], we identified its fold as a 'jellyroll', which is a special case of a greek key fold. However, a 3-on-3 helical representation of PCP would, given its resemblance with the hull of a ship with helices A, F, E on one side and C, G, H on the other, therefore be possible as well. On the other hand, the angle between these 'sandwich layers' is closer to 0° than to 90° which is why we favour the greek key description of the PCP/globin fold. In Fig. 2, we therefore compare the folds of globins (in this case hemoglobin of ark clam) and PCP. As can be seen, both structures can be represented by nearly identical greek key topology diagrams (Fig. 2).

Other proteins exhibiting a globin-like fold and binding a related cofactor are the phycobiliproteins. While these LHCs are functionally closely related to PCP, they do not structurally agree better with PCP's N-terminal domain than do globins (r.m.s.d.=4.8Å for phycocyanin [17] *vs.* approx. 4.3Å for globins, superimposing about 100 residues). However, the fact that a very similar function is displayed by structurally related proteins makes a very strong point towards their phylogenetic relationship. In further support of this view, phycobiliproteins have developed an α- and a β-subunit of around 162 and 172 amino acids, respectively, which are quite similar in structure (r.m.s.d.= 1.3Å for phycocyanin), matching the development of long variant forms of PCP with closely related N- and C-terminal domains (r.m.s.d.= 0.9Å [1]).

Figure 3 shows a superimposition of the three different tetrapyrrole cofactors of PCP, myoglobin and phycocyanin. We find that the binding position is between the E and F helices in all three cases. However, the binding mode varies from buried in the case of PCP, to exposed in the case of the phycobiliproteins. Obviously, the key to the observation of structural deviations between PCP and phycobiliproteins lies in the different chromophores: whereas PCP provides a wide open pocket for a large number of hydrophobic pigments, the phycobiliproteins bind their extended open-chain tetrapyrroles on the outside of a compact apoprotein, with the globins occupying an intermediate position.

Based on these observations, we propose that PCP and the phycobiliproteins derive from a globin-like ancestor, and that the observed structural differences are largely due to the adaptation to different cofactors during divergent evolutionary pathways.

How does the cargo get on board?

Little is known about the folding pathway of most proteins. The general understanding is that folding from an extended state initiates at one or more places of the sequence in so-called 'folding nuclei', leading to a partial condensation of secondary structure elements which finally, driven by minimization of the free energy, collapse to the native state in a cooperative fashion. However, in the case of the globins, new methods (stopped-flow circular dichroism, hydrogen-exchange pulse labeling and temperature-jump fluorescence) for the study of folding intermediates in the 15-ns to 5-ms time scale have been developed and applied [18,19], resulting in a more detailed description of the folding pathway than available for any other family of proteins.

The emerging picture is that of a 'molten globule' state as an early protein-folding intermediate, with the A,G, and H helices forming the first native-like contacts. Using the topology diagrams (Fig. 2) for comparison, this means that condensation starts at the center of the greek key for proteins of the globin fold. In analogy to the globins, we therefore expect PCP to fold its A, G and H helices first. As the only substantial openings of the PCP monomer, those harbouring the lipid and the conjugated systems of the carotenoids (Fig. 1), are lined by the H and G helices, we postulate that these cofactors must be present during the first folding steps. It appears likely that the remaining steps during folding resemble those of the globins: addition of the B helix (C helix in PCP), and then the remaining F and E helices, thus completing the greek key. In the case of PCP, these steps are probably facilitated and guided by the presence of and the interaction with the cargo. As a result, the polypeptide chain wraps around the cofactors. We note that this condensation can happen simultaneously in the N- and C-terminal domains as the last helix of the N-terminal domain is connected by a flexible linker of 13 residues (the 'keel', see Fig. 1) to the first helix of the C-terminal domain. Thus, the packing of the N- against the C- terminal carotenoids can be optimized by their interaction, and is not constrained by a rigid scaffold of the polypeptide.

Future prospects

Electrospray ionization-mass spectrometry has been shown [20] to be capable of observing intact protein molecules as well as identifying noncovalently attached cofactors. In collaboration with the group of M. Przybylski at the University of Konstanz, we are currently investigating in parallel the unfolding of myoglobin and PCP under native conditions and under mild proteolysis. Unfolding experiments can yield accurate information about the last steps of the folding process which are under consideration here, and we expect them to be compatible with our folding hypothesis outlined above. By detecting unfolding intermediates of PCP and its constituents, we especially hope to clarify the role of the lipids, which are integral components of the holoprotein with completely unknown function.

Acknowledgements: We thank Dr. M. Glocker for discussions.

Abbreviations:

PCP	-	Peridinin Chlorophyll *a* Protein
LHC	-	Light Harvesting Complex
r.m.s.d.	-	root mean square deviation

References

1 E. Hofmann, P.M. Wrench, F.P. Sharples, R.G. Hiller, W. Welte, and K. Diederichs (1996) Structural basis of light harvesting by carotenoids: peridinin-chlorophyll-protein from *Amphidinium carterae*. *Science* **272**, 1788-1791.

2 G. McDermott, S.M. Prince, A.A. Freer, A.M. Hawthornthwaite-Lawless, M.Z. Papiz, R.J. Cogdell, and N.W. Isaacs (1995) Crystal structure of an integral membrane light-harvesting complex from photosynthetic bacteria. *Nature* **374**, 517-521.

3 W. Kühlbrandt, D.N. Wang, and Y. Fujiyoshi (1994) Atomic model of plant light-harvesting complex by electron crystallography. *Nature* **367**, 614-621.

4 N.S. Govind, S.J. Roman, R. Iglesias-Prieto, R.K. Trench, E.L. Triplett, and B.B. Prézelin (1990) An analysis of the light-harvesting peridinin-chlorophyll a-proteins from dinoflagellates by immunoblotting techniques. *Proc. R. Soc. Lond.* **B240**, 187-195.

5 Larkum, T. (1996) How dinoflagellates make light work with peridinin. *Trends in Plant Science* **8**, 247-248.

6 P.J. Kraulis (1991) MOLSCRIPT: a program to produce both detailed and schematic plots of protein structures. *J. Appl. Cryst.* **24**, 946-950.

7 E.A. Merritt and M.E.P. Murphy (1994) RASTER3D version 2.0 - a program for photorealistic molecular graphics. *A. Crystallogr.* **D50**, 869-873.

8 J. Richardson (1981) The anatomy and taxonomy of protein structures. *Adv. Prot. Chem.* **34**, 167-339.

9 L. Holm and C. Sander (1996) Mapping the protein universe. *Science* **273**, 595-602.

10 U. Hobohm and C. Sander (1995) A sequence property approach to searching protein databases. *J. Mol. Biol.* **251**, 390-399.

11 A.G. Murzin, S.E. Brenner, T. Hubbard, and C. Chothia (1995) Scop: a structural classification of proteins database for the investigation of sequences and structures. *J.Mol.Biol.* **247**, 536-540.

12 U. Hobohm, M. Scharf, R. Schneider, and C. Sander (1992) Selection of a representative set of structures from the Brookhaven Protein Data Bank. *Protein Science* **1**, 409-417.

13 K. Diederichs (1995) Structural superposition of proteins with unknown alignment and detection of topological similarity using a six-dimensional search algorithm. *Proteins: Structure, Function and Genetics* **23**, 187-195.

14 T.A. Jones, J.Y. Zou, S.W. Cowan, and M. Kjeldgaard (1991) Improved methods for building protein models in electron density maps and the location of errors in these models. *Acta Crystallogr.* **A47**, 110-119.

15 L. Holm and C. Sander (1993) Structural alignment of globins, phycocyanins and colicin A. *FEBS Letters* **315**, 301-306.

16 T. Schirmer, W. Bode, R. Huber, W. Sidler, and H.J. Zuber (1985) X-ray crystallographic structure of the light-harvesting biliprotein C-phycocyanin from the thermophilic cyanobacterium *Mastigocladus laminosus* and its resemblance to globin structures. *J. Mol. Biol.* **184**, 257-277.

17 M. Duerring, G.B. Schmidt and R. Huber (1991) Isolation, crystallization, crystal structure analysis and refinement of constitutive C-phycocyanin from the chromatically adapting cyanobacterium *Fremyella diplosiphon* at 1.66Å resolution. *J. Mol. Biol.* **217**, 577-592.

18 P.A. Jennings and P.E. Wright (1993) Formation of a molten globule intermediate early in the kinetic folding pathway of apomyoglobin. *Science* **262**, 892-896.

19 R.M. Ballew, J. Sabelko, and M. Grueble (1996) Direct observation of fast protein folding: the initial collapse of apomyoglobin. *Proc. Natl. Acad. Sci. USA* **93**, 5759-5764.

20 several articles in this volume

THE DEVELOPMENT OF NMR METHODS TO STUDY PROTEIN STRUCTURE AND DYNAMICS

LEWIS E. KAY
Departments of Medical Genetics, Biochemistry and Chemistry
University of Toronto
Toronto, Ontario, Canada, M5S 1A8

Abstract

An understanding of the role played by a protein in cellular function requires a detailed picture of its three-dimensional structure as well as an appreciation of how the structure varies as a function of time due to molecular dynamics. Over the past several years multi-dimensional, multi-nuclear solution NMR spectroscopy has become a powerful technology for obtaining both structural and dynamical information on proteins and protein-ligand systems. However, until recently the methods were limited to the study of molecules having molecular weights on the order of 25 kDa or less. Recent developments making use of fractional or complete deuteration have increased the scope of structural studies by NMR and have also improved studies of sidechain protein dynamics.

Introduction

In the past decade NMR spectroscopy has emerged as a powerful technique for the determination of the three-dimensional structures of proteins in solution. In the early 1980s, studies by R. R. Ernst [1] and K. Wüthrich [2] demonstrated that it is possible to obtain atomic resolution structures of small proteins, with molecular weights less than approximately 10 kDa (~ 100 amino acids). The work of Ernst provided the framework for the extension of the NMR technique from one to two and three frequency dimensions. Recent developments in NMR spectroscopy have had a significant impact on solution structural studies of proteins [3,4]. The improvements in the technology have been several fold and include: (i) the increase in the dimensionality of experiments from two to three and four, providing improved resolution for complex spectra; (ii) the uniform incorporation of ^{15}N, ^{13}C and ^{2}H labels

W. Ens et al. (eds.), New Methods for the Study of Biomolecular Complexes, 285–293.

into the biological system of interest coupled with the development of sophisticated NMR pulse schemes to transfer magnetization between scalar (through bond) and dipolar (through space) coupled spins; and (iii) significantly improved radio frequency (RF) electronics and increased magnetic field strengths as well as the development of commercially available hardware such as pulsed field gradients. Using this technology a large number of structures of proteins or protein complexes in the molecular weight range of ~15-20 kDa have been reported in the past several years [5,6]. Recently backbone assignments for two 269 residue proteins [3] and a symmetric dimer of molecular weight 38 kDa (Copie and Torchia, personal communication) have been reported.

These new methods utilizing uniform ^{15}N and ^{13}C labeling of proteins are called triple resonance (^{1}H, ^{15}N, ^{13}C) techniques, since the naturally-present ^{1}H resonances of the molecule are recorded along with those of incorporated ^{15}N and ^{13}C nuclei. Triple resonance methods require uniform labeling of protein samples with both ^{15}N and ^{13}C and make use of the large couplings that exist between the ^{15}N and ^{13}C nuclei and between these nuclei and their directly attached protons for efficient magnetization transfer. In this approach, assignment of backbone NH, $^{13}C^{\alpha}$, $^{13}C^{\beta}$, $^{13}C'$ (carbonyl), $^{1}H^{\alpha}$ and $^{1}H^{\beta}$ chemical shifts is accomplished using 3D experiments which correlate nuclei three at a time. Because of their excellent resolution and sensitivity and the redundancy of information, backbone assignment is straightforward. Sidechain assignments utilize the 3D HCCH-TOCSY as well as experiments correlating sidechain ^{1}H and ^{13}C shifts with backbone ^{15}N and NH shifts [3]. Structural information is obtained from nuclear Overhauser effect (NOE) experiments which provide distance correlations between protons within 5 Å of each other. Various dihedral angle restraints can be derived from experiments which measure homo- or hetero-nuclear coupling constants, as reviewed by Bax et al. [7].

Development of ^{15}N, ^{13}C, ^{2}H NMR Spectroscopy

As discussed above, ^{1}H, ^{13}C, ^{15}N triple resonance 3D and 4D spectroscopy has increased the size limits of protein structures that can be determined by the NMR technique to approximately 25 kDa. There are two reasons for this limit using the triple resonance approach. First, as the molecular weight increases, the number of cross peaks in spectra also increases. In the case of the triple resonance experiments developed for backbone assignment the number of peaks increases in a linear fashion with molecular weight. However, considering the excellent resolution afforded in such spectra, this does not represent a serious problem. The number of cross peaks in NOE-type spectra increases rapidly with size and poses a more serious problem for studying proteins of increasing size. However, the second factor, the rapid decay of the NMR signal that occurs during the multitude of transfer steps in a complex NMR experiment, has the most significant effect on the size of molecules that can be studied by NMR. A typical multi-dimensional NMR experiment can be schematized as follows:

$$\mathbf{A} \rightarrow \mathbf{B} \rightarrow \mathbf{C} \rightarrow \mathbf{D} \rightarrow \mathbf{E} \rightarrow \ldots \rightarrow \mathbf{Z}$$

where the transfer of magnetization proceeds from **A** to **Z** via **B**, **C**, **D**, etc. The amount of time required to transfer magnetization along each link in the chain, from say **B** to **C**, is a function of the strength of the coupling between the participating links. The transfer can vary from ~4 ms if **B** = ^{1}H and **C** = ^{13}C to 20-30 ms if **B** = ^{15}N and **C** = ^{13}C, for example. During this transfer, the signal decays via relaxation processes whose efficiency, for macromolecules, increases linearly with molecular weight. The decay time of the signal varies depending on the type of nucleus (*ie*, whether **A** = ^{1}H, ^{13}C, ^{15}N etc); for ^{13}C nuclei coupled to protons the decay time can be as short as ~ 15 ms for proteins in the 20-30 kDa range. If the decay rate is the same order of magnitude as the transfer rate, a significant attenuation of the signal can be expected. In principle, there are three approaches to increasing the amount of signal observed at the end of the transfer. The first is to increase the inherent sensitivity of the experiment. This approach has been discussed in detail in the literature [8]. The second is to increase the strength of the coupling between the spins involved in the transfer. Unfortunately, these couplings are fixed by spin type and can not be manipulated. The third approach is to decrease the efficiency of relaxation loss that occurs during the transfer. This can be manipulated experimentally, through the substitution of deuterium for carbon bound protons in the protein [9,10].

The relaxation of nuclei is caused by fluctuating magnetic fields that are the result of the overall molecular tumbling in solution as well as internal dynamics. For the case of heteronuclei such as ^{15}N or ^{13}C directly coupled to ^{1}H spins, the major source of relaxation is due to fluctuating dipolar fields caused by the ^{1}H spins. The sizes of these dipolar fields are proportional to the gyromagnetic ratio of the spins which give rise to the fields, in this case ^{1}H spins. By substituting ^{13}C bound protons with deuterons, the size of the dipolar fields which cause relaxation of the attached ^{13}C nucleus is reduced by a factor of ~6.5, since the gyromagnetic ratio of deuterons (^{2}H) is ~6.5 times smaller than that of protons. This results in a substantial decrease in the relaxation rates of the ^{13}C spins. In addition, the proton nuclei remaining in the molecule have much slower decay rates as well, since many of the relaxation pathways which would normally involve adjacent protons are significantly attenuated by the incorporation of deuterium.

The idea of using ^{2}H labeling of proteins as a means of spectral editing dates back to the late 1960s with the pioneering experiments of Crespi *et al.* [11] and Markley and coworkers [12]. However, it was really the elegant work of LeMaster in the late 1980s [13] that demonstrated the utility of random fractional deuteration of proteins to aid in the sequential assignment of what was then considered to be a large protein for NMR studies, *E. Coli* thioredoxin (108 residues). To obtain a balance between reduced line widths and reduced sensitivity, Lemaster prepared a 75% deuterium labeled sample of thioredoxin. He noted that the resolution obtained is significantly better than observed for fully protonated samples and that the sensitivity of the amide-amide region of the NOE spectrum was substantially better. Moreover, the intensities of NOE cross peaks connecting amide protons and carbon bound protons were comparable to intensities of the corresponding cross peaks in unlabeled protein. Finally, cross peak intensities of aliphatic-aliphatic NOEs were only reduced by a factor of three.

We have recently developed a suite of triple resonance experiments for the backbone assignment of ^{15}N, ^{13}C, ^{2}H labeled proteins with high sensitivity and significantly improved resolution [10]. The methods have been applied to study the 37 kDa ternary complex of the trp-repressor, corepressor and trp-operator DNA and more recently the backbone resonances of a 64 kDa trp-repressor complex have been nearly completely assigned [14] (>95%). Studies of the 64 kDa trp-repressor complex were only possible when performed on a >90% deuterated ^{15}N, ^{13}C labeled sample, stressing the requirement for high levels of deuteration for the study of large molecular weight proteins and protein-ligand complexes. While such a perdeuteration scheme is of benefit for the assignment of backbone NH, ^{15}N and ^{13}C chemical shifts the depletion of protons results in molecules with significant levels of protonation only at exchangeable NH sites. Structural studies by NMR depend to a large extent on the generation of inter-proton distance restraints, established via NOE-based experiments. The absence of large numbers of protons, poses therefore, a very serious problem for obtaining three dimensional structures of proteins using current NMR approaches. With these limitations in mind we have currently developed a biosynthetic approach in which overexpression of proteins in D_2O and with protonated, ^{13}C-pyruvate as the sole carbon source results in molecules which are highly deuterated at the majority of positions, with the exception of methyl groups of Ala, Val, Leu and Ile ($\gamma 2$) [15]. Molecules labeled in this way can be assigned in a straightforward manner using recently developed ^{15}N, ^{13}C, ^{2}H NMR experiments. Most important, is that it is possible to obtain NH-NH, NH-methyl and methyl-methyl NOEs in such systems in order to determine overall folds of the protein. Experimental results on the ^{15}N, ^{13}C, ^{2}H, $^{1}H_3C$-labeled C-terminal SH2 domain from phospholipase Cγ1 and calculations performed on a number of proteins ranging in molecular weight from 15-40 kDa indicate that it will be possible to generate global folds of the majority of proteins in this fashion.

Understanding Protein Dynamics

The identification of Src homology 2 (SH2) domains in many proteins involved in signal transduction has led to a rapid increase in our understanding of the molecular basis of oncogenesis. SH2 domains are regions of approximately 100 residues that bind to specific phosphotyrosine (pTyr) containing sequences. Recently the three dimensional structures of a number of SH2 domains have been solved by nuclear magnetic resonance (NMR) and X-ray crystallographic methods [16]. These structural studies have revealed a similar overall topology for the SH2 domains consisting of a large central b-sheet and an associated b-sheet, flanked by two a-helices. The phosphotyrosine binding site involves a network of charge-charge and hydrogen bonding interactions between residues of the SH2 domain, including arginine residues and the phosphate oxygens and aromatic electrons of the pTyr ring from the peptide. In a collaboration between the laboratories of Julie Forman-Kay and L.E.K. at the University of Toronto, the three dimensional (3D) solution structure of the C-terminal SH2 domain of phospholipase C-γ1 (PLCC) in complex with a phosphopeptide derived from the platelet-derived growth factor (PDGF) receptor Tyr-1021 site (pY1021) has been solved by heteronuclear NMR methods [17]. The topology of this SH2-pTyr peptide complex is similar to that reported for the SH2 domains from Src [18] and Lck [19]. However, the binding site for residues C-terminal to the pTyr is an extended groove that contacts the peptide at residues extending from the pTyr to positions six residues C-terminal to the pTyr. A similar extended binding site has been observed for the amino-terminal SH2 domain of the Syp tyrosine phosphatase (NSyp) in complex with a number of high affinity peptides [20]. The PLCC and NSyp SH2 peptide binding sites are different from the sites observed in structures of Src and Lck complexed with a peptide containing the sequence pTyr-Glu-Glu-Ile. For the SH2 domains of Src and Lck the mode of peptide binding has been described as a 'two pronged plug' interaction, with the pTyr inserting into a large pocket and the Ile into a separate and smaller pocket.

Despite the importance of these static 3D structures, it must be recognized that the picture obtained is not complete. Protein molecules are not static in solution and, indeed, the key to a protein's functionality may lie in its dynamic properties. Recently, Shoelson and coworkers (personal communication) have carried out binding studies where the parent high affinity peptides that bind to a particular SH2 domain were either (i) truncated one residue at a time or (ii) substituted with an alternative amino acid. In both cases binding affinities were measured and compared with the parent peptide. In the case of the Syp SH2 domain it was found that hydrophobic residues at the +5 position (i.e., the position five residues C-terminal to the pTyr) and the +3 position are required for high affinity binding. This is expected on the basis of the extended hydrophobic binding groove of the NSyp SH2 domain which contacts residues on the peptide up to the +5 position. However, a tripeptide centered on the pTyr was found to bind to the PLCC SH2 domain with only a fifteen fold reduction in affinity (K_d ~ 15 μM) relative to the parent peptide (1 μM). Moreover, truncation of residues at the +2 through +6 positions had significantly smaller effects on binding for the PLCC SH2 domain relative to the NSyp SH2 domain. The results for the PLCC SH2 binding are surprising given the fact that both the PLCC SH2 and the NSyp SH2 domains have similar extended binding sites.

In an effort to understand why the extensive contacts between the PLCC SH2 domain and peptide residues at the +2 through +6 positions do not confer significant binding energy or specificity we have undertaken a study of the dynamics of methyl sidechains in both the free and complexed forms of the PLCC and NSyp SH2 domains. In principle, NMR spectroscopy is a powerful technique for probing sidechain dynamics of proteins and a number of studies of methyl dynamics at specific ^{13}C labeled sites in proteins have appeared in the literature [21]. In practice, however, a number of difficulties with such techniques have emerged. First, only a small subset of the sidechains can be examined in this way. Second, ^{13}C spin relaxation methods are hampered by the effects of cross correlation between ^{1}H - ^{13}C dipoles in methylene and methyl groups which can result in systematic errors in extracted motional parameters. To circumvent these problems we have developed a new strategy for studying picosecond-nanosecond sidechain dynamics based on the fractional incorporation of deuterium into uniformly ^{15}N, ^{13}C labeled proteins [22]. Because the deuterium lines in a protein are extremely broad, the experiments record the relaxation properties of the deuterons indirectly, through measurement of a series of high resolution, constant time ^{13}C, ^{1}H correlation maps where the intensity of the correlations relate to the relevant ^{2}H relaxation property, T_1 or $T_{1\rho}$. In principle, dynamics information about any labeled site in the molecule can be obtained. To date, we have restricted our analysis to methyl groups, due to the excellent resolution and sensitivity in this region of the correlation spectrum. The method allows dynamic information to be extracted from all methyl positions in the molecule simultaneously, in a manner which is free from the effects of cross correlation [23].

Application of these experiments to the PLCC SH2 domain establishes that certain of the residues of the SH2 domain which line the binding site contacting the +1 through +6 positions of the peptide are highly flexible [24]. It is very likely that the high degree of flexibility of the binding surface contributes to the relaxed specificity of the domain for target phosphotyrosine containing sequences. In addition the combination of significant amplitude motions and the steep distance dependence of the van der Waals potential may well result in a substantial decrease in the interaction energy which would otherwise manifest in a static site. In contrast, preliminary dynamics studies of the NSyp SH2 domain indicate that the hydrophobic binding interface is more rigid in this system than in the PLCC SH2:peptide complex; this increased rigidity may explain why hydrophobic interactions in the NSyp SH2 domain are more stabilizing than in the PLCC SH2 domain.

NMR relaxation parameters can be interpreted in terms of (i) an order parameter, S, which is related to the amplitude of the motion, and (ii) correlation times describing the time scale of the dynamics [25]. We have recently developed a relation between changes in order parameters and changes in conformational entropy, thereby providing a link between motional parameters and thermodynamics [26]. Experimentally, global thermodynamic values which characterize a net change in the system associated with a transition between two states, have been measured from optical methods such as circular dichroism and fluorescence or by calorimetric approaches such as scanning calorimetry and titration calorimetry [27]. On the theoretical side, molecular dynamics and Monte Carlo simulations allow estimation of thermodynamic parameters in some detail and offer the possibility of localizing changes in such values to particular sites in the molecule [28]. However, extraction of meaningful values from such simulations does require the use of accurate force fields, simulations of sufficient length and care that the system under study is in equilibrium. The NMR methods developed enable the determination of changes in entropy arising from changes in nano-second to pico-second timescale dynamics on a per residue basis.

Summary

A full understanding of protein function requires both structural and dynamics studies. At present there are only two methods that are able to provide a molecular picture at atomic resolution: X-ray crystallography and NMR spectroscopy. Methodology for extending structural studies by NMR has been developed which involves labeling with ^{15}N, ^{13}C and ^{2}H. The methods promise to extend the molecular weight limitations currently imposed on structural studies by NMR. In addition to structural information NMR can also provide insight into molecular dynamics. It is clear that the static structures of a number of related SH2 domains do not explain their different peptide binding properties; an understanding of the dynamics at the binding interface may be important in this regard.

Acknowledgments

This research was supported by grants from the Natural Sciences and Engineering Research Council of Canada, the Medical Research Council of Canada and the National Cancer Institute of Canada. The author is extremely grateful to his colleagues, Ranjith Muhandiram, Kevin Gardner, Mike Rosen, Cheryl Arrowsmith, Julie Forman-Kay and Daiwen Yang for important contributions to the research described herein.

References

1. Ernst, R. R., Bodenhausen, G. & Wokaun. (1987) *Principles of Magnetic Resonance in One and Two Dimensions*, Clarendon Press:bb Oxford.
2. Wuthrich, K. (1986) *NMR of Proteins and Nucleic Acids*, Wiley, New York.
3. Bax, A. (1994) Multidimensional Nuclear Magnetic Resonance Methods for Protein Studies, *Curr. Opin. Struct. Biol.* **4**, 738-744.
4. Kay, L. E. (1995) Field Gradient Techniques in NMR Spectroscopy, *Curr. Opin. Struct. Biol.* **5**, 674-681.
5. Ikura, M., Clore, G. M., Gronenborn, A. M., Zhu, G., Klee, C. B. & Bax, A. (1992) Solution Structure of a Calmodulin-Target Peptide Complex by Multidimensional NMR, *Science* **256**, 632-638.
6. Zhang, H., Zao, D., Revington, M., Lee, W., Jia, X., Arrowsmith, C. H. & Jardetzky, O. (1994) The Solution Structure of the trp Repressor-Operator DNA Complex. *J. Mol. Biol.* **229**, 735-746.
7. Bax, A., Vuister, G. W., Grzesiek, S., Delaglio, F., Wang, A. C., Tschudin, R. & Zhu, G. (1994) Measurement of Homo- and Heteronuclear J Couplings from Quantitative J Correlation. *Meth. Enzym.* **239**, 79-92.
8. Kay, L. E. (1995) Pulsed Field Gradient Multi-Dimensional NMR Methods for the Study of Protein Structure and Dynamics in Solution. *Prog. Biophys. molec. Biol.* **63**, 277-299.
9. Grzesiek, S., Anglister, J, Ren, H. and Bax, A. (1993) ^{13}C Line Narrowing by ^{2}H Decoupling in $^{2}H/^{13}C/^{15}N$ Enriched Proteins. Application to Triple Resonance 4D J Correlation of Sequential Amides. *J. Am. Chem. Soc.* **115**, 4369-4370.
10. Yamazaki, T., Lee, W., Arrowsmith, C. H., Muhandiram, D. R. & Kay, L. E. (1994) A Suite of Triple Resonance NMR Experiments for the Backbone Assignment of ^{15}N, ^{13}C, ^{2}H Labeled Proteins with High Sensitivity, *J. Am. Chem. Soc.* **116**, 11655-11666.
11. Crespi, H. L., Rosenberg, R. M. and Katz, J. J. (1968) Proton Magnetic Resonance of Proteins Fully Deuterated Except for ^{1}H-Leucine Side Chains *Science*, **161**, 795-796.
12. Markley, J. L., Putter, I. and Jardetzky, O. (1968) High Resolution Nuclear Magnetic Resonance Spectra of Selectively Deuterated Staphylococcal Nuclease, *Science* **161**, 1249-1251.
13. LeMaster, D. M. and Richards, F. M. (1988) NMR Sequential Assignment of *Escherichia coli* Thioredoxin Utilizing Random Fractional Deuteration, *Biochemistry* **27**, 142-150.
14. Shan, X., Gardner, K. H., Muhandiram, D. R., Rao, N. S., Arrowsmith, C. H. and Kay, L. E. (1996) Assignment of ^{15}N, $^{13}C\alpha$,$^{13}C\alpha$ and HN Resonances in an ^{15}N, ^{13}C, ^{2}H Labeled 64 kDa Trp Repressor-Operator Complex Using Triple Resonance NMR Spectroscopy and ^{2}H Decoupling (1996), *J. Am. Chem. Soc.* **28**, 6570-6579.
15. Rosen, M. K., Gardner, K. H., Willis, R. C., Parris, W. E., Pawson, T. and Kay, L. E. (1996) Selective Methyl Group Protonation of Perdeuterated Proteins. *J. Mol. Biol.* In Press.
16. Yu, H. and Schreiber, S. L. (1994) Signaling an interest, *Nature struct. Biol.* **1**, 417-420.
17. Pascal, S. M., Singer, A. U., Gish, G., Yamazaki, T., Shoelson, S. E., Pawson, T., Kay, L. E. & Forman-Kay, J. D. (1994) Nuclear Magnetic Resonance Structure of an SH2 Domain of Phospholipase C-γ1 Complexed with a High Affinity Binding Peptide, *Cell* **77**, 461-472.
18. Waksman, G., Kominos, D., Robertson, S. C., Pant, N., Baltimore, D., Birge, R. B., Cowburn,D., Hanafusa, H., Mayer, B. J., Overduin, M., Resh, M. D., Rios, C. B., Silverman, L and Kuriyan, J. (1992) Crystal Structure of the Phosphotyrosine Recognition Domain SH2 of v-src Complexed with Tyrosine Phosphorylated Peptides. *Nature* **358**, 646-653.
19. Eck, M. J., Atwell, S. K., Shoelson, S. E. & Harrison, S. C. (1994) Structure of the Regulatory Domains of the Src-family Tyrosine Kinase Lck, *Nature* **368**, 764-769.

20. Lee, C. H., Kominos, D., Jacques, S., Margolis, B., Schlessinger, J., Shoelson, S. E. & Kuriyan, J. (1994) Crystal Structures of Peptide Complexes of the Amino-Terminal SH2 Domain of the Syp Tyrosine Phosphatase, *Structure* **2**, 423-438.
21. Nicholson, L. K., Kay, L. E., Baldisseri, D. M., Arango, J., Young, P. E., Bax, A. & Torchia, D. A. (1992) Dynamics of Methyl Groups in Proteins as Studied by Proton-Detected ^{13}C NMR Spectroscopy. Application to the Leucine Residues of Staphylococcal Nuclease. *Biochemistry* **31**, 5253-5263.
22. Muhandiram, D. R., Yamazaki, T., Sykes, B. D. & Kay, L. E. (1995) Measurement of ^{2}H T_1 and $T_{1\rho}$ Relaxation Times in Uniformly ^{13}C-Labeled and Fractionally ^{2}H-Labeled Proteins in Solution. *J. Am. Chem. Soc.* **117**, 11536-11544.
23. Yang, D. & Kay, L. E. (1996) The Effects of Cross Correlation and Cross Relaxation on the Measurement of Deuterium T_1 and $T_{1\rho}$ Relaxation Times in $^{13}CH_2D$ Spin Systems. *J. Magn. Reson. Series B* **110**, 213-218.
24. Kay, L. E., Muhandiram, D. R., Farrow, N. A., Aubin, Y. & Forman-Kay, J. D. (1996) Correlation between Dynamics and High Affinity Binding in an SH2 Domain Interaction, *Biochemistry* **35**, 361-368.
25. Lipari, G. & Szabo, A.(1982) Model Free Approach to the Interpretation of Nuclear Magnetic Resonance Relaxation in Macromolecules. *J. Am. Chem. Soc.* **104**, 4546-4570.
26. Yang, D. and Kay, L. E. (1996) Contributions to Conformational Entropy Arising from Bond Vector Fluctuations Measured from NMR-Derived Order Parameters: Application to Protein Folding. *J. Mol. Biol.* In Press.
27. Plum, G. E. & Breslauer, K. J. (1995) Calorimetry of Proteins and Nucleic Acids. *Curr. Opin. Struct. Biol.* **5**, 682-690.
28. Brooks, C. L., Karplus, M. & Pettitt, B. M. (1988) Proteins, *A Theoretical Perspective of Dynamics, Structure, and Thermodynamics*, John Wiley & Sons, New York.

BIACORE TECHNOLOGY - A MARRIAGE PARTNER TO MASS SPECTROMETRY ?

ROBERT KARLSSON
Pharmacia Biosensor AB,
S751 82 Uppsala, Sweden

1. Introduction

A number of optical detection techniques [1,2,3] permit direct observation of binding events at sensor surfaces. Combined with immobilization chemistry and efficient sample handling these techniques can be applied to concentration analysis, epitope mapping and affinity and kinetic analysis [4]. The basic procedure for all types of interaction analysis is simple. The sensor surface is sequentially contacted with buffers, reagents and samples. Binding events at the sensor surface are immediately displayed on a computer screen. The surface can often be regenerated so that many injections of sample can be performed on one surface. Detection is direct and no labeling of reagents is required. The response is proportional to the mass of bound analyte [5].

a. Traditional assay b. BIAcore assay

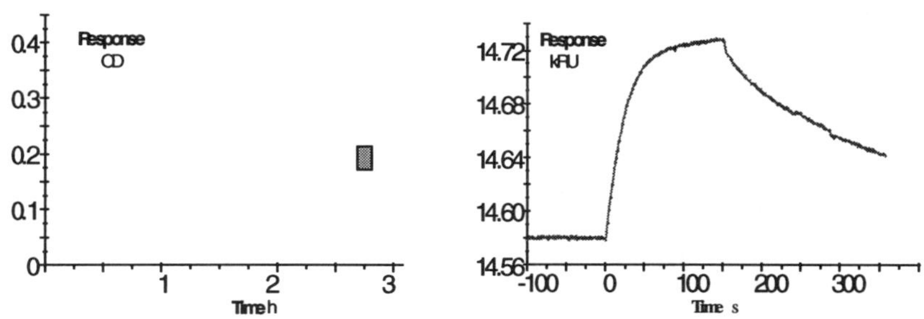

Figure 1. In traditional assays a label is often used for detection. The assay is an end point assay. Only one data point and no kinetic information is obtained. BIAcore assays are more informative, since real time binding and dissociation curves are obtained, and easier to perform, since labels are not required.

W. Ens et al. (eds.), New Methods for the Study of Biomolecular Complexes, 295–303.

The use of optical biosensors for interaction analysis (fig. 1) has made it possible to investigate a large number of protein-protein, protein-peptide and protein-DNA systems [6,7]. Binding curves are often easy to interpret, but in cases when biosensor technology is used to study complex interactions, or for screening purposes, additional information regarding the nature of the bound complex or bound analyte is often required. In this paper I will describe BIAcore, a biosensor system based on surface plasmon resonance (SPR) detection, and briefly discuss a few applications where the combination of BIAcore with mass spectrometry (MS) may provide more information than the Biosensor technology itself.

2. BIAcore technology

BIAcore system combines SPR detection with a surface and a liquid delivery technology (fig. 2a).

2.1. SURFACES AND REAGENTS

The surface in the SPR sensor consists of a thin gold film on a glass support (fig. 3). Surfaces with different chemistries are available. The HPA (hydrophobic adsorption) surface is obtained by attaching a thiolalkane layer on top of the gold surface. This surface is suitable for use with lipid films where one interaction partner is embedded in a lipid environment [8].

On other sensor surfaces a carboxylated dextran matrix is covalently attached to a modified thiolalkane layer. The dextran matrix extends approximately 100 nm out from the thiolalkane layer and provides a hydrophilic environment in which molecular interactions take place. The dextran polymer is carboxylated to facilitate covalent coupling of ligands to the sensor surface. During immobilization the carboxyl groups are converted to N-hydroxy-succinimide esters. These esters react with amine groups on proteins and direct immobilization of proteins is therefore possible [9]. The ester groups can also be derivatized prior to protein immobilization to allow for subsequent thiol or aldehyde coupling. The binding capacity of the dextran surface is very high and exceeds 30 ng/mm^2.

In many applications it is an advantage to use capture surfaces or reagents in order to bind the ligand in an oriented manner, so that immobilization does not interfere with the binding site. The SA chip and the NTA chip are examples of pre-made capture surfaces with streptavidin or NTA, a derivative of nitrilo-tri-acetic acid, immobilized to the dextran matrix. These surfaces are intended for studies where one interaction partner has been linked to a biotin (often DNA) or to a hexahistidine tag. In other cases capture antibodies like anti mouse Fc (a-MFc) antibody and anti glutathione S-transferase (a-GST) antibody are immobilized to facilitate the study of MAb-antigen interactions and interactions involving GST-fusion partners.

Which of the two interaction partners that is immobilized is often governed by the purpose of the experiment or by how easy the surface is to regenerate, the size of the protein or the amount of sample available for analysis. Since detection is mass sensitive it is often an advantage to immobilize the smaller partner.

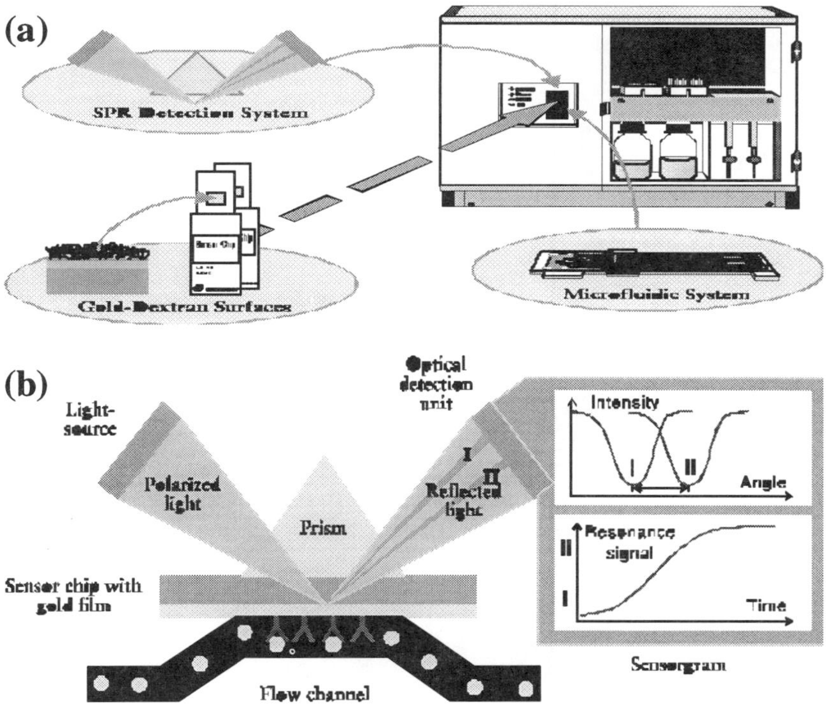

Figure 2. a. BIAcore technology combines surface plasmon resonance detection, sensor surface and liquid handling technology into one analytical system. *b.* Light is coupled to the sensor surface via a prism. The position of the reflectance minimum is monitored continuously by a photodiode array. When a molecule, introduced via the flow cell, binds at the sensor surface, the position of the reflectance minimum will change. The binding curve is obtained by plotting the position of the minimum versus time.

2.2 SAMPLE AND LIQUID DELIVERY SYSTEMS

The liquid delivery system consists of two syringe pumps, an autoinjector and an Integrated microFluidic System an (IFC) (fig. 2a). One pump maintains a continuous flow over the sensor surface and the other operates in combination with the autoinjector. The autoinjector can be interfaced with a number of vials including Eppendorf tubes and Elisa plates. The IFC contains tubing and ports for a pump, injection, sample recovery and waste. Four 60 nl flow cells are formed by grooves in the IFC and by the surface itself. In each flow cell approximately one mm^2 of the surface becomes exposed to liquid. Integrated tubing and valves makes the IFC into a switchboard. The flow can be directed either to one flow cell (FC) at a time or in series to a combination of flow cells. Once the sample has passed through the flow cell(s) it can be directed either to waste or to the recovery station. Working flow rates are from 1 to 100 µl/min and typical sample volumes range from 5 to 250 µl although larger volumes can be injected.

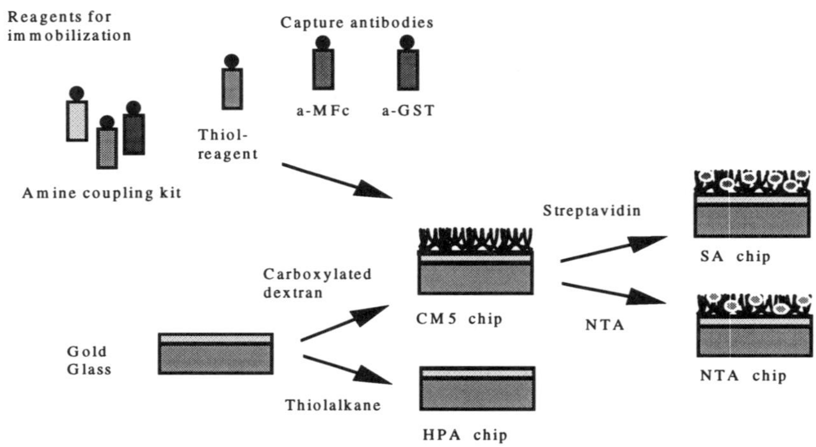

Figure 3. Schematic drawing illustrating how reagents and surfaces are combined. Starting from the gold surface the HPA chip is obtained by adding a thiolalkane layer. This surface is hydrophobic. Hydrophilic surfaces are obtained by coupling a dextran polymer to a modified thiolalkane layer. An amine coupling kit and a thiol reagent provide the basis for covalent immobilization of ligands to the surface. Capture antibody reagents or ready made surfaces with pre immobilized streptavidin and NTA facilitate analysis.

2.3. SURFACE PLASMON RESONANCE DETECTION

Surface plasmon resonance is an optical phenomenon [9]. It occurs in connection with total internal reflection of light at a metal film - liquid interface. Normally, light traveling through an optically denser medium, (in fig. 2b the glass prism) is totally reflected back, when reaching an interface to an optically less dense medium (in fig. 2b the buffer) provided that the angle of incidence is larger than a critical angle. Although the light is totally reflected, a component of the incident light momentum called the evanescent wave penetrates a distance of the order of one wavelength into the buffer. If the interface is coated with a thin metal film (in fig. 2b the gold film) this evanescent wave will interact with oscillating free electrons in the metal film surface. When surface plasmon resonance occurs, light energy is lost to the metal film and the reflected light intensity is thus decreased. The resonance phenomena occurs for light incident at a sharply defined angle. The position of the minimum of reflected light intensity is sensitive to changes in refractive index in the buffer close to the surface. A change in buffer composition or binding of an analyte to the gold film will both result in a change in refractive index. Thus when the position of the light intensity minimum is plotted versus time the graph will contain components reflecting buffer changes and binding events. By using a reference surface where no ligand is immobilized, and no binding occurs, the buffer effect can be monitored separately and therefore subtracted away. The binding event can then be visualized free from added buffer effects.

2.4. USING BIACORE

The BIAcore system is very flexible and an overview of suitable experimental conditions for typical applications is presented in fig. 4.

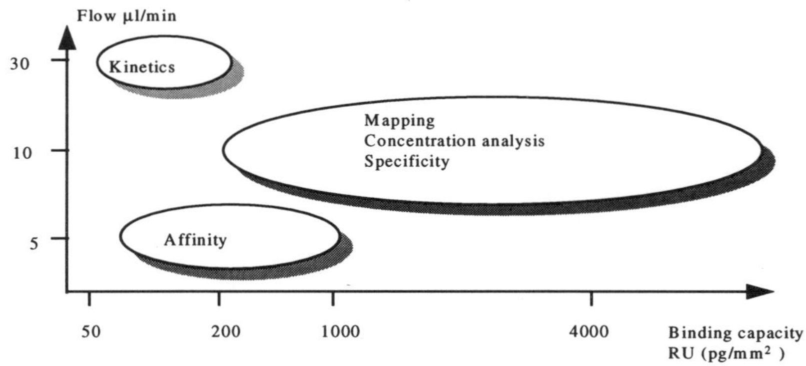

Figure 4. The flow rate and the binding capacity are important experimental parameters. The figure illustrates how these parameters can be selected to match different applications.

2.4.1 Kinetic analysis

For kinetic analysis experimental conditions are chosen so that the balance between how fast the analyte is consumed by the reaction and how efficient the analyte can be transported to the surface is optimized. If the association rate constant is very high, binding will be rapid, and the concentration of analyte close to the surface, C_S, may drop below the injected concentration C_0. Under such circumstances the binding rate will reflect the diffusion of analyte molecules to the surface and not the kinetic properties of the interaction partners. Low levels of immobilized ligand and high flow rates are therefore recommended [11]. With fewer binding agents immobilized, sample depletion becomes less severe, since fewer analyte molecules bind, and by increasing the flow rate the extension of the depleted layer is reduced so that diffusion occurs over shorter distances.

2.4.2 Affinity analysis.

In affinity analysis an accurate determination of the steady state response is important. When the dissociation rate constant is high steady state levels are reached within a few seconds and binding curves resemble a square wave. This type of response is similar to that obtained when a buffer with refractive index different from the running buffer is injected. It may therefore be difficult to establish if binding has occurred. This dilemma is resolved [12] when the sample is injected over a series of surfaces with different amounts of ligand immobilized (fig. 5a). The magnitude of a binding event will vary with the level of immobilized ligand whereas a buffer effect will not.

On the other extreme i.e. when the dissociation rate constant is low it will take a long time to reach steady state. Experimentally both situations can be addressed with low flow rates. In the first case the flow rate is not important and in the second case a low flow rate ensures a long injection time (twelve hours continuous injection is possible at a flow rate of one µl/min). Since mass transfer don't effect the steady state level it is not necessary to keep immobilization levels as low as in kinetic analysis.

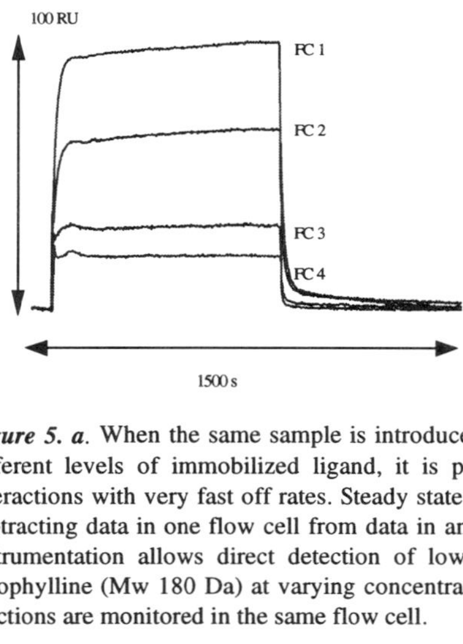

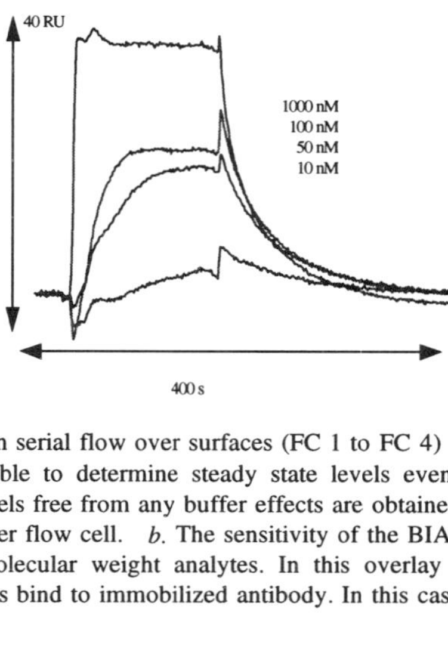

Figure 5. a. When the same sample is introduced in serial flow over surfaces (FC 1 to FC 4) with different levels of immobilized ligand, it is possible to determine steady state levels even for interactions with very fast off rates. Steady state levels free from any buffer effects are obtained by subtracting data in one flow cell from data in another flow cell. *b.* The sensitivity of the BIAcore instrumentation allows direct detection of low molecular weight analytes. In this overlay plot theophylline (Mw 180 Da) at varying concentrations bind to immobilized antibody. In this case all reactions are monitored in the same flow cell.

2.4.3 Other applications.

In applications like drug screening, mapping studies or when the function of large molecular complexes is investigated, the quantitative aspects of binding is less critical and a wider range of experimental conditions can be used. When the analyte is of low molecular weight as in fig. 5b the response levels will always be small and it is necessary to use high levels of immobilized ligand.

For concentration analysis it is an advantage to perform the analysis under mass transfer limiting conditions. This makes the analysis robust since the response will mainly depend on the diffusion properties of the analyte. Again this indicates the use of low moderate flow rates and high levels of immobilized ligand.

3. BIAcore and Mass Spectrometry

3.1. A NEED FOR COMPLEMENTARY INFORMATION

Four typical BIAcore applications are illustrated in fig. 6. In drug screening (fig. 6a) [13] the injected sample is normally a mixture of different components. When binding has been observed the next step will be to identify the bound component. In gene assembly (fig. 6b) [14] it is interesting to know whether the reaction goes to completion and how homogeneous the final product is. When large complexes are formed and the effect of phosphorylation is studied (fig. 6c) [15] a further characterization of complexes formed and released is desirable. In kinetic analysis experimental data may indicate analyte heterogeneity and additional evidence for this would be valuable. In all these cases mass spectrometry could give complementary information and add value to the BIAcore analysis.

3.2. COMBINING BIA AND MS

To interface BIA and MS two strategies are evident. One is to use the recovery function of the BIAcore and to elute the bound sample prior to MS analysis. The BIAcore surface can then be used repeatedly. This is an advantage in high throughput assays, for instance in drug screening and in applications where MS is used to confirm the specificity of a reaction. The other approach is to remove the surface from the BIAcore system and to use it as substrate for MS. In this context MALDI-MS may be used to characterize complex molecular aggregates.

That both strategies are feasible is illustrated elsewhere in this volume [16] and in [17]. The usefulness of a BIAcore-MS combination still has to be proven since very few results have been published. As the methodology develops, the possibilities and the limitations will become clearer. The interface between the technologies is really not developed. BIAcore and MS technologies may have to adapt to each other, so that the combined technologies can be used to advantage in the study of molecular interactions and molecular aggregates.

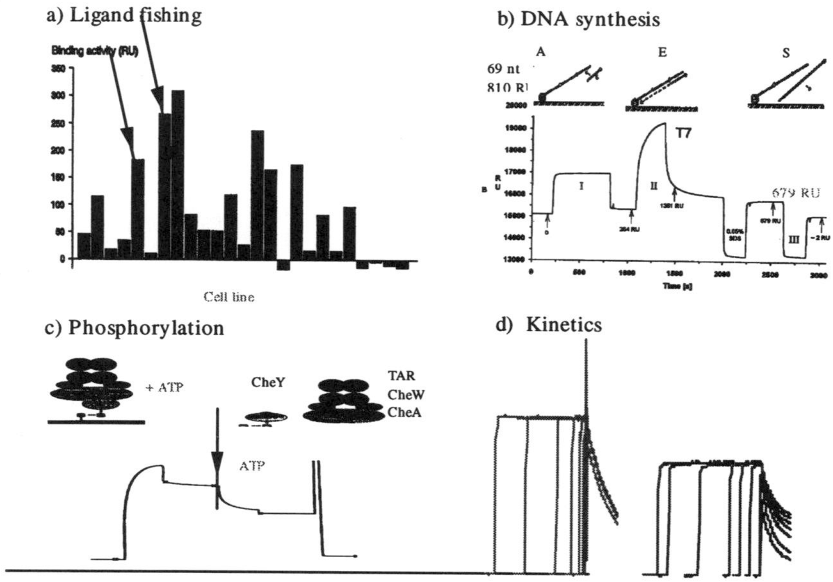

Figure 6. Applications where mass spectrometry data could provide additional information. ***a.*** Response values recorded when culture media from different cell lines were injected over a surface with an immobilized orphan receptor. ***b.*** DNA synthesis on a chip. 69 nucleotides are first captured on the surface. A complimentary primer is added and annealed (A). DNA is elongated by injection of T7 polymerase and reagents (E). After a washing step, DNA strands are separated by the injection of dilute hydrochloric acid (S). By comparing the signal for capture of the first DNA-strand (810 RU) with the drop in signal level during strand separation (679 RU) it is not possible to determine the completeness of the reaction or the homogeneity of the reaction product. It is only possible to conclude that the reaction has taken place and must be almost complete. ***c.*** A complex of at least four proteins (CheY, CheA, CheW and TAR) have been assembled on the chip. When ATP is added partial dissociation is observed. The degree of CheY phosphorylation and the composition of eluted and remaining aggregates is unknown. ***d.*** Kinetic of analysis of two antibody - antigen interactions. The right panel data suggests that the analyte may be heterogeneous whereas the left panel data, where all dissociation curves coincide indicates that both antibody and antigen are homogeneous.

References

1. Jönsson, U. and Malmqvist, M., (1992), Real-time biospecific interaction analysis. The integration of surface plasmon resonance detection , general biospecific interface chemistry and microfluidics into one analytical system. *Advances in Biosensors*, **2**, 291-336.
2. Cush, R., Cronin, J.M., Stewart, W.J., Maule, C.H., Molloy, J. and Goddard, N.J. (1993), The resonant mirror: a novel optical biosensor for direct sensing of biomolecular interactions. Part I: Principle of operation and associated instrumentation. *Biosens. Bioelectron.*, **8**, 347-354.
3. Tiefenthaler, K., (1993), Grating couplers as label-free biochemical waveguide sensors. *Biosens. Bioelectron.*, **8**, xxxv-xxxvii.
4. Fägerstam, L., Frostell, Å., Karlsson, R., Kullman, M., Larsson, A., Malmqvist, M. and Butt, H., (1990), Detection of antigen-antibody interactions by surface plasmon resonance. Application to epitope mapping. *J. Mol. Recognit.*, **3**, 208-214.

5. Stenberg, E., Persson, P., Roos, H. and Urbaniczky, C., (1991), Quantitative determination of surface concentration of protein with surface plasmon resonance using radiolabeled proteins. *J. Colloid Interface Sci.,* **143**, 513-526.
6. Szabo, A, S., Stolz, L. and Granzow, R., (1995),), Surface plasmon resonance and its use in biomolecular interaction analysis (BIA). ***Curr. Opin. Struct. Biol.*** **5**, 699-705
7. Fisher, R.J. and Fivash, M., (1994), Surface plasmon resonance based methods for measuring the kinetics and binding affinities of biomolecular interactions. *Curr. Opin. Biotech*, **5**, 389-395.
8. Kuziemko, G.M., Stroh, M. and Stevens, R.C., (1996), Cholera toxin binding affinity and specificity for gangliosides determined by surface plasmon resonance. *Biochem.* **35**, 6375-6384.
9. Löfås, S. and Johnsson, B., (1990), A novel hydrogel matrix on gold surfaces in surface plasmon resonance sensors for fast and efficient covalent immobilization of ligands. *J. Chem. Soc. Chem. Commun.*, **21**, 1526-1528.
10. Raether, H., (1988), Surface plasmons on smooth and rough surfaces and on gratings, Springer, Berlin.
11. Karlsson, R., Roos, H., Fägerstam, L. and Persson, B., (1994), Kinetic and concentration analysis using BIA technology. Methods: A companion to methods in enzymology, 6, 99-110.
12. Karlsson, R. and Ståhlberg, R. Surface plasmon resonance detection and multi-spot sensing for direct monitoring of interactions involving low molecular weight analytes and for determination of low affinities. (1995), Anal. Biochem., 228, 274-280.
13. Bartley, T.D., Hunt, R.W., Welcher, A.A., Boyle, W.J., Parker, V.P., Lindberg, R.A., Lu, H.S., Colombero, A.M., Elliott, R.L., Guthrie, B.A., Holst, P.L., Skrine, J.D., Toso, R.J., Zhang, M., Fernandez, E., Trall, G., Varnum, B., Yarden, Y., Hunter, T. and Fox, G.M., (1994), B61 is a ligand for the ECK receptor protein-tyrosine kinase. Nature, 368, 558-560.
14. Nilsson, P., Persson, B., Uhlén, M. and Nygren, P-Å., (1994), Real-time monitoring of DNA-manipulations using biosensor technology. Anal. Biochem., 224, 400-408.
15. Schuster, S., Swanson, R., Alex, A., Bourret, R. and Simon, M., (1993), Assembly and function of a quaternary signal transduction complex monitored by surface plasmon resonance. Nature, 365, 343-347.
16. Sönksen, C.P., Jansson, Ö., Stenberg, E. and Roepstorff, P., (1996), Interaction studies combining plasmon resonance in BIAcore instrument with MALDI- and nano ESI-MS. This volume.
17. Krone, J.R., Nelson, R., Dogruel, D., Williams, P. and Granzow, R. (1996), Interfacing mass spectrometric immunoassays with BIA. BIAJournal, 3, 16-17.

BIOSPECIFIC INTERACTIONS MEASURED BY CAPILLARY ELECTROPHORESIS

N.H.H. HEEGAARD
Department of Autoimmunology
Bldg. 81, Rm. 581
Statens Serum Institut
Artillerivej 5
DK-2300 Copenhagen S
Denmark

1. Introduction

A characterization of how biomolecules interact combined with a knowledge of their concentration, distribution, environment, and regulation of synthesis is required to understand how biological systems work. The characterization of molecular interactions involves demonstration of binding, identification of binding sites and estimation of quantitative parameters such as binding strength, rate constants, and binding stoichiometry. Equilibrium binding constants are informative regarding the physiological importance of a given acceptor-ligand interaction and various methods (TABLE 1) have been devised for the measurement of this parameter. All methods listed may allow for determination of binding constants but some are more widely applicable than others. The ideal method should be applicable to a wide range of interacting biomolecules under physiological conditions using low amounts of native material at high precision and speed. Of the methods listed in TABLE 1 only mass spectrometry (MS) and capillary electrophoresis (CE) are in principle applicable for the study of native and not necessarily purified biomolecules using minute amounts of material. However, the analytical conditions in the two methods are quite different and none of the methods can be said to be generally applicable even though there are now many examples of the uses of MS (this book) and CE (TABLE 2) for the measurement of various biospecific interactions.

As one of the more recent methods for the binding studies CE offers distinct advantages and unique potential problems. This review will illustrate the applicability of CE for qualitative and quantitative binding studies and highlight advantages and problems.

W. Ens et al. (eds.), New Methods for the Study of Biomolecular Complexes, 305–318.

TABLE 1. Comparison of frequently used methods for binding studies

METHOD	BASIS	LABEL/ IMMOBIL.	CONSUMP	TIME	PURIF.
Equilibrium dialysis	Size	+/-	Moderate-high	Days	Yes
Electro-phoresis	Charge/ mass	-/-	Low	Min (CE)	No
Calorimetry	Transit. temp.	-/-	Moderate	Min	Yes
Size excl. chromatogr. & Filtration	Size	+/-	Moderate-High	Hrs	Yes
Spectroscopy	Quench	-/-	Low	Min	Yes
Surface plasmon resonance	Mass	-/+	Low	Hrs	Yes
Mass spectrometry	Mass	-/-	Low	Min	No
Solid-phase immobilization	Retention	(+)/+	Moderate	Hrs	Yes

TABLE 2. Examples of classes of biomolecular interactions studied by means of CE

INTERACTION	REF.
Antigen-antibody	[1-6]
Carbohydrate-lectin	[7,8]
Protein-metal ions	[9,10]
Peptide-heparin	[11-13]
Protein-ligand	[14]
Peptide-ligand	[15-16]
DNA-DNA	[17]
DNA-intercalators	[18]
DNA-protein	[19]
Albumin-drugs	[20,21]

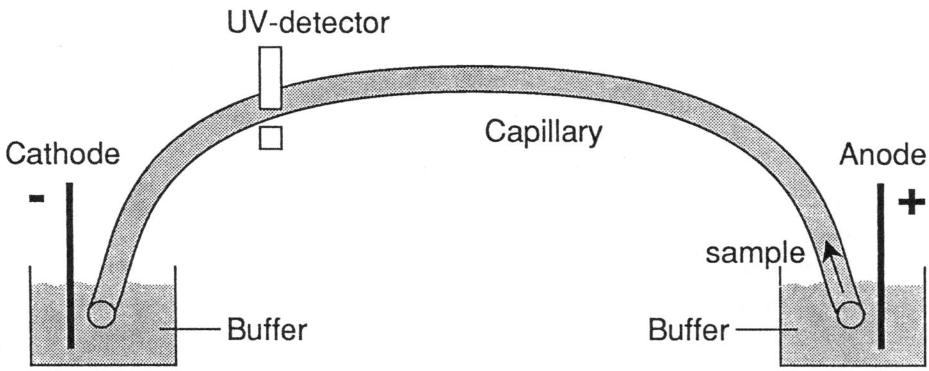

Figure 1. Principle of capillary zone electrophoresis.

2. Principle of CE

The three fundamental features of CE (the use of narrow tubes, high field strengths, and on-line detection) were devised already in the thirties by Tiselius [22] and made practical by the advent of modern capillary technology in the eighties [23,24]. The basic set-up is as shown in *Figure 1.* Capillaries may be filled with gels or entangled polymer material but very often only with buffer (free solution CE) as in all examples given in this paper. The fused silica capillaries are typically of 50 μm internal diameter, *i.e.* containing 1 μl buffer/50 cm. The electrodes are most often placed with the anode at the sample injection end (contrary to the usual situation in conventional gel electrophoresis) and detection is accomplished through a window near the end of the capillary. Usually UV-absorbance or laser-induced fluorescence detection is used. Samples are introduced in one end of the capillary by electrokinetic injections or as in these studies by controlled pressure injections resulting in typical cases in the consumption of 1-10 nl sample/injection. Separations take place using voltages in the range of 100-500 V/cm. These high field strengths are possible because of the narrow dimensions of the capillary that allow for an efficient heat dissipation and make it possible to perform high resolution separations of a very wide range of different molecules.

At neutral pH two forces, electrophoresis and electroosmosis, the latter originating from the fixed negative charges of the ionized silanol groups on the inner glass surface, drive molecules through the capillaries so that even negatively charged species normally migrate toward the cathode. Data output in CE is in the form of electropherograms corresponding to the chromatograms of chromatography systems except that the peak appearance times are a result of differences in migration velocities and not in retention times. For more details readers are referred to several recent excellent reviews and textbooks on the theory and applicability of CE [25-30].

3. Qualitative Binding Studies Using CE

The basic concept behind all uses of electrophoresis for binding studies (affinity electrophoresis) is that a complexed molecule differs from an unbound molecule in electrophoretic behavior. Electrophoresis is then used as a means to separate bound from free molecules. In this section the use of CE for qualitative binding studies *i.e.* for the demonstration and structural characterization of binding will be briefly illustrated.

3.1. SCREENING FOR BINDING ACTIVITY

Screening for activity of individual components in a mixture of molecules may be done by CE by analysis of the mixture incubated with the putative ligand followed by electrophoresis and inspection of the separation pattern for changes of individual components with respect to position, area, or shape of peaks. Almost any change to a peak in the pattern would indicate interactions. However, one may overlook weak interactions using this approach (*cf.* below). With this in mind CE is a quite powerful method for screening for binding activity of individual compounds in complex mixtures [6,12-15] especially because it is a fast and simple technique and only requires a small amount of material for testing binding of a large number of candidate ligands. If an individual peak in a complex mixture has been identified as having binding activity it may be isolated by preparative CE and characterized further *e.g.* by MS [13]. In *Figure 2* is shown an example of a mixture of two crude synthetic peptides (each consisting of two main components) screened for heparin binding activty by analyzing them in the absence (A) and presence (B) of heparin with which the sample had preincubated. By complexation with the heavily sulfated heparin molecules it is clear that some of the peptides from the fraction representing the disulfide-linked peptide dimer (major peaks in this preparation are labelled "2" in the figure) are intercating with heparin. In the analysis there are simply no peaks left from this peptide preparation after heparin incubation (*Figure 2B*). This kind of analysis could also be performed using small phosphorylated monosaccharides as putative ligands [31].

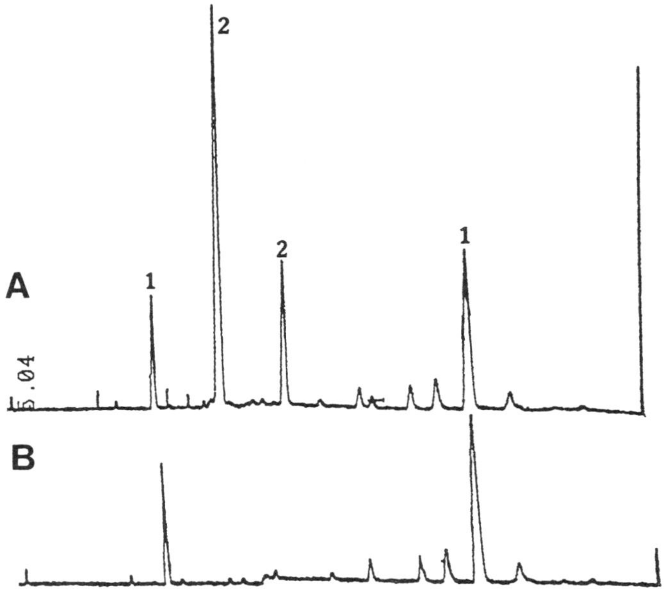

Figure 2. CE-based screening for heparin binding activity of a mixture of crude synthetic peptides: 1, Glu-Lys-Pro-Leu-Gln-Asn-Phe-Thr-Leu-Ser-Phe-Arg-NH_2, 2, disulfide-linked dimer of Glu-Lys-Pro-Leu-Gln-Asn-Phe-Thr-Leu-Cys-Phe-Arg-NH_2 both consisting of the intended product and major impurities. Peptides were dissolved at 1 mg/ml in water and mixed 1:1 and then incubated 1:1 with buffer (A) or heparin (0.5 mg/ml in pH 2.5 buffer) (B) before being injected using an 8 s vacuum injection. Electrophoresis performed at 20 kV in 20 mM sodium citrate, pH 2.5 with detection at 210 nm.

3.2. BINDING SITE SPECIFICITY

As an example of the validation of the specificity of an anti-phosphotyrosine monoclonal antibody (Ab), the analysis of *Figure 3* is shown [6]. Here, the ligand is added to the electrophoresis buffer and the ensuing electropherogram is as before inspected for changes in peak positions, shape or area. The experiment shows that only upon addition of phosphotyrosine to the buffer but not on addition of an equimolar concentration of phosphoserine the Ab peak was affected. The figure illustrates how stronger interactions may be revealed in affinity CE. A severely broadened peak is observed when the specific ligand is included into the electrophoresis buffer (*Figure 3B*). It must be stressed that a meaningful interpretation of experiments such as these require that the two putative ligands that are compared for binding both are expected to induce migration changes upon binding. In the case of a protein interacting with a small molecule this will normally only be the case if the ligand is charged because the mass increase alone will not be sufficient to alter the mass-to-charge ratio which is the most important parameter determinating migration in CE separations of proteins [32,33]. In the present example both putative ligands are phosphorylated so complexation of any of these with the antibody would be expected to change its migration. The addition of a possible ligand to the electrophoresis buffer will normally be the first approach to be chosen when examining for binding because it will also reveal weaker interactions. The usefulnes of this has been illustrated in studies pinpointing active ligand-binding peptides in complex mixtures such as those generated by proteolytic digests of proteins [12,13,34]

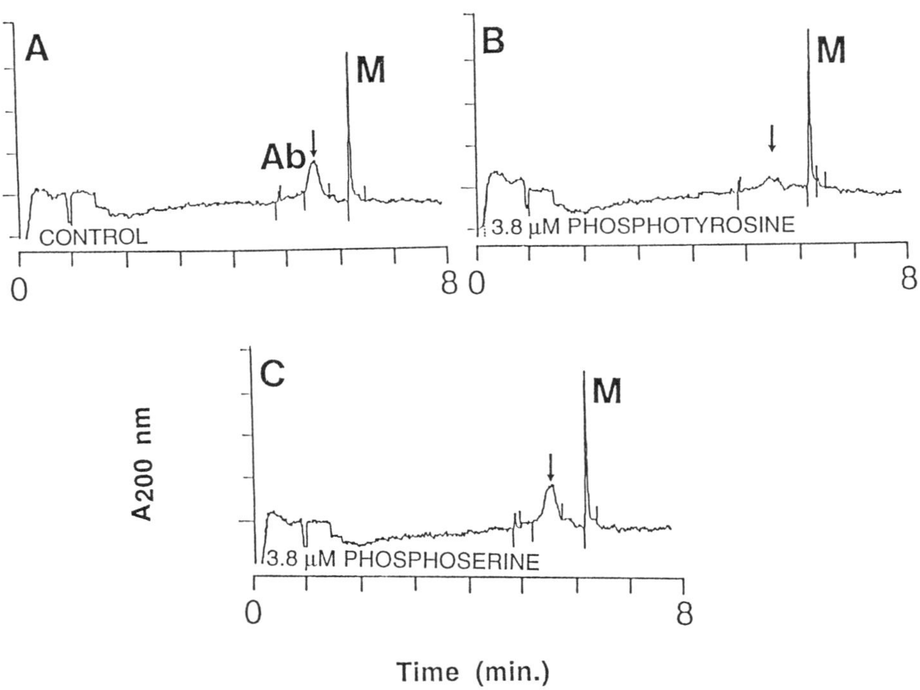

Figure 3. Analysis of the binding site specificity of a monoclonal anti- phosphotyrosine antibody by CE. The sample was a monoclonal anti-phosphotyrosine (UBI, clone 4G10) antibody (Ab), 20 μl (approximately 7 μM) + 1μl Asp-Ala-Glu-Phe-Arg-NH_2 synthetic peptide (1 mg/ml) (added as an internal marker, M). 5 s pressure injections into a pH 8.8 isotonic borate buffer containing additions of either phosphotyrosine (B) or phosphoserine (C) as indicated. 25 kV separation, detection at 200 nm.

4. Quantitative Binding Studies Using CE

Conventional electrophoresis has been used for quantitative binding studies before the advent of capillary electrophoresis and such methods are still widely used *e.g.* gel shift assays of DNA-binding proteins where migration changes caused by the mass increase that follows from stable complex formation in preequilibrated samples are measured at different reactant ratios [35,36]. Another type of migration shift electrophoresis which is carried out in loose agarose gels has also been used quite frequently to characterize low-to-intermediate affinity interactions *e.g.* between DNA and peptides and between lectins and glycoproteins [37-39]. In these lower affinity interactions the binding kinetics is typically sufficiently fast to estimate binding constants from migration shifts of the analyte induced by different amounts of ligand present during electrophoresis [40-42]. Thus, stable complexes and less tight interactions are amenable to two different experimental set-ups.

With CE the analysis and measurements of peak areas or migration times are feasible for a wider range of molecules with much greater precision, and - importantly - in a wider range of buffers so that it may be possible to closely mimic physiological conditions. However, as in conventional affinity electrophoresis the rate constants for the interaction still determine the approach that must be used to generate data from which binding constants are estimated. Thus, for interactions where the half-time of a complex is much smaller than the time of the electrophoresis experiment, migration shift experiments must be done to calculate binding constants because pre-equilibrated samples will dissociate in the electrophoresis buffer immediately after being injected. For tighter binding the complexes have a longer lifetime and do not dissociate appreciably during electrophoresis. An equilibrium between unbound and bound molecules during the electrophoresis is therefore not going to be realized and pre-equilibration of samples followed by measurement of free (or bound) molecules forms the basis of binding assays. In the following sections examples of both situations are given. It must be noted, however that the characteristics of the kinetics of other interactions may preclude quantitative measurements by either electrophoretic approach.

4.1. LONG-LIVED COMPLEXES

A number of peptides derived from human amyloid P component were synthesized and examined in CE for heparin-binding activity [11,34]. One example of an analysis is shown in *Figure 4*. A mixture of two synthetic peptides, one being a C-terminal carboxylate the other a C-terminal carboxamide but otherwise identical, were incubated with heparin and then analyzed by CE at different times after the start of the incubation. The experiment illustrates the fast and efficient separation of two peptides that differ by one charge but are otherwise identical. Over time there is a decrease in the peak corresponding to peptide A so this peptide appears to form stable complexes with heparin while the other more negatively charged peptide B does not appear to be affected by heparin during the incubation time. With CE the time courses of interactions are conveniently followed because one may perform many repeated injections of *e.g.* 5 nl from an incubation mixture of 10-20 µl. Thus the time necessary to attain an equilibrium is easily determined as the time point where no further changes take place. No new peak representing the molecular complex appears. This may be due to a dissociation that is too slow to contribute to the area of the free peptide peak but fast enough to distribute dissociated peptide over *e.g.* a couple of minutes so that the broad peak disappears in the baseline. It could also be because the ligand (heparin) is a heterogeneous mixture of differently sulfated glycosaminoglycan chains of varying lengths and dissaccharide composition [43] resulting in peptide-heparin complexes with an array of different mobilities.

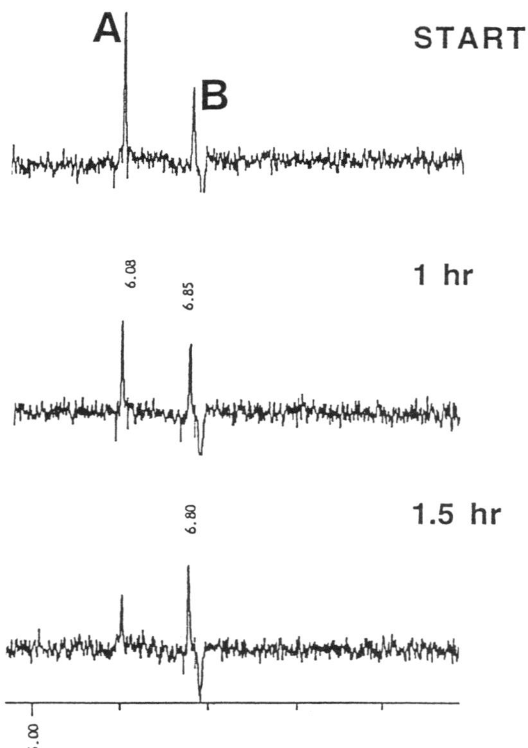

Figure 4. Binding of heparin by a synthetic peptide demonstrated by CE. A mixture of Gly-Tyr-Val-Ile-Ile-Lys-Pro-Leu-Val-Trp-Val as the carboxamide (A) or the carboxylate (B) form (0.04 mg/ml) and heparin (0.02 mg/ml) in 0.1 M sodium phosphate, pH 7.3 was analyzed at different times after start of incubation. 5 s pressure injected aliquots were analyzed at 20 kV. Detection was at 200 nm.

4.1.1. *Data Handling*

In experiments such as the one shown in *Figure 4* a correlation between amount of heparin in the preincubation mixture and the area of free peptide that is electrophoretically separated from bound species existed. In these cases the correlation between concentration of ligand and amount of free and bound acceptor molecules at fixed acceptor concentrations after establishment of equilibrium is derived from the peak areas in the CE analyses. CE simply functions as a means of simultaneous separation and quantitation of free and bound species. Traditional binding curves of bound acceptor *vs.* concentration of ligand or data plotted according to the Scatchard equation [44] may then be analyzed and used to estimate binding constants [11]. In many cases it will not be possible to measure precisely the peak area corresponding to bound material, but as long as the peak representing free material is measurable the method can be used. The main problem with this approach has to do with the detector sensitivity because UV detection normally only offers a concentration sensitivity for peptides down to μM [11,32]. For strong interactions it will therefore not be possible to detect analytes below saturating concentrations. The concentration sensitivity may be increased 100-1000 times using laser-induced fluorescence detection [32], but this will in most cases entail labelling of the acceptor molecule or exploiting native fluorescence. Another potential problem occurs when analyzing less tight binding (smaller $1/k_{off}$). Preformed complexes of this nature may as previously mentioned dissociate upon injection and electrophoresis in a buffer devoid of ligand and no evidence of binding will be found. In these situations the approach described below must be employed.

4.2. SHORT-LIVED COMPLEXES

Migration shift analyses are suitable for interactions that have so fast on- and off-rates that during electrophoresis molecules partition between the unbound and the bound state, and all spend the same amount of time in complexes depending on the ligand concentration. In cases where the velocity of the complex is different from that of the free molecules the described interaction behavior will result in shifts in the migration time of interacting acceptor molecules when they are electrophoresed in a buffer where ligand has been added while the shape of their peaks will be unaltered (*Figure 5*). Distorted peak shapes (asymmetric and broadened) are indicative of an uneven distribution of time spent in complexes and in unbound form between molecules. This happens with rate constants that are slower than in the ideal case mentioned above where the half-life of complexed molecules are shorter than the time of the electrophoresis experiment. The evolution of peak broadening and asymmetry as a function of ligand concentration is unique for a given set of rate constants [45] and the experimental data may thus in selected cases - based on computer simulations of the observed peak shape changes - be used to estimate the rate constants [46].

A migration shift experiment typically of low affinity interactions is shown in *Figure 5* where a peptide fragment isolated from the C-terminus of human amyloid P component [12] was subjected to CE in the presence of increasing concentrations of heparin. Interactions with the negatively charged ligand lead to longer and longer migration times of the peptide which consequently can be observed to cross the deflection of the baseline corresponding to the movement of neutral molecules (water).

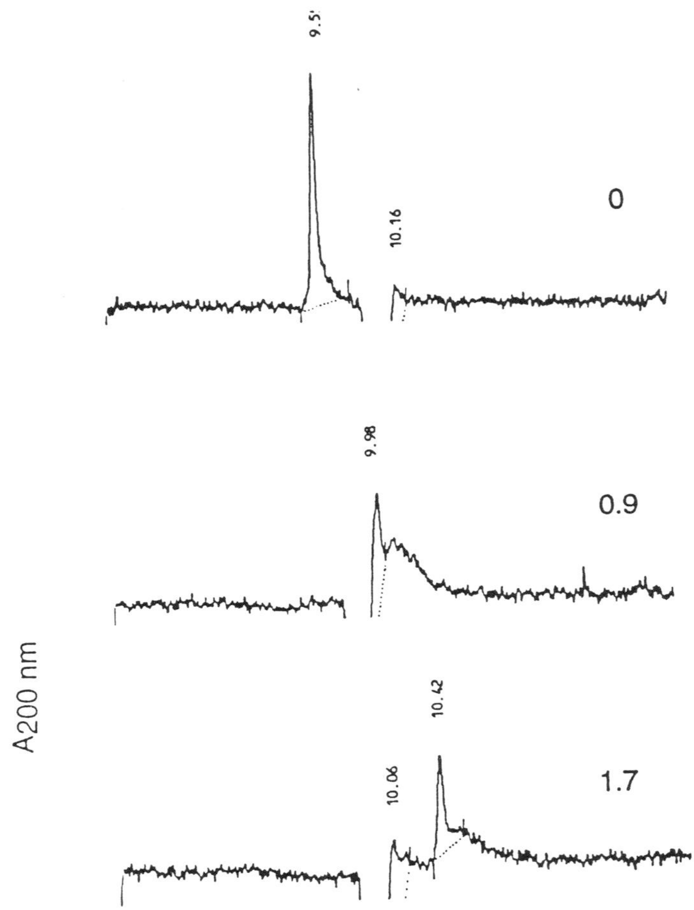

Figure 5. Demonstration of heparin binding by affinity CE. A fragment from *Staphylococcus V8* proteinase-treated human amyloid P component [12] was isolated and analyzed by CE. Analysis at 5 kV of 8 s pressure injected aliquots using 0.1 M sodium phosphate, pH 7.5 containing additions of heparin (mg/ml) as stated.

4.2.1. *Data Handling*

Migration shifts of non-distorted peaks are amenable to analysis on the basis of equations derived from conventional gel affinity electrophoresis. Migration shifts may be expressed as a function of the concentration of ligand by including the dissociation constant, K_D for the interaction in the equation [40-42,47]. The equation may be used with the migration data from CE experiments and the assumptions and requirements [45,48] underlying the theory remain the same, *i.e.*, a sufficiently short half-life of complexes, constant velocity of free and complexed molecules throughout the electrophoresis, no influence of electrical fields on the binding interactions, a much higher ligand concentration than the concentration of acceptor, only one binding site or if more, homogeneous and independent sites, and finally no participation by either ligand or acceptor in unwanted interactions such as with the capillary wall, with buffer components or within the molecules themselves (resulting in aggregation and precipitation). In migration shift experiments it is very important to include a marker molecule in the sample that does not interact with ligand and thus makes it possible to compensate for migration changes that are caused by small variations in buffer ionic composition and temperature and not by specific ligand interactions. Also, since the electrophoretic mobility μ is proportional to the inverse of the migration time t the equation [40,41,47]:

$$\Delta\mu = \Delta\mu_{max}c/(K_D + c)$$

becomes:

$$\Delta(1/t) = \Delta(1/t)_{max}c/(K_D + c)$$

The best way [49] to arrange this equation appears to be according to the Eadie-Woolf-Hofstee equation [50,51]:

$$\Delta(1/t) = \Delta(1/t)_{max} - K_D[\Delta(1/t)/c]$$

where c is the concentration of ligand and the $\Delta(1/t)$ are the differences between the inverse migration time (corrected for the marker migration time) in reference experiments with no ligand added and in experiments performed in the presence of the concentration c of ligand [7]. Thus, when experimental data are plotted with $\Delta(1/t)$ as a function of $\Delta(1/t)/c$ one gets the K_D as the slope of the best fitted straight line [3]. The K_D values obtained in this way have been verified by independent methods [3]. Stoichiometries for both low and higher affinity binding may be determined by titration experiments as demonstrated by Chu and co-workers [5]. One major practical advantage of the approach is that the concentration of acceptor molecules does not need to be known, and that the binding constant determination can be performed also on complex mixtures *e.g.* of peptides because the electrophoresis separates the mixture and visualizes interactions of individual components in one step (see *e.g.* [16]).

5. Conclusions

CE has proven its value for binding constant determination in a number of specific applications both using the pre-equilibration approach and the migration shift approach. The method, however, still can not be said to be generally applicable as conditions for solubility, reproducible analyses, and suitability for one or the other approaches must be worked out in each case. In addition there may be problems of identifiyng components of samples, and limitations in detector sensitivity may preclude the study of strong interactions. Also, there will be certain ranges of rate constants that will not be compatible with any of the CE methods for affinity studies. On the other hand there is no doubt that CE is unique in its ability to yield binding constants for interactions of very small amounts of unlabelled components in complex mixtures with specific ligands under conditions of physiological pH and ionic strength. CE is complementary to MS in binding studies in the sense that CE often will be best suited for the study of intermediate-to-low affinity interactions while measurement after the ionization process in MS in most cases will require quite strong interactions.

6. References

1. Nielsen, R. G., Rickard, E. C., Santa, P. F., Sharknas, D. A., and Sittaampalam, G. S. (1991) Separation of antibody-antigen complexes by capillary zone electrophoresis, isoelectric focusing and high-performance size-exclusion chromatography, *J. Chromatogr.* **539**, 177-185.
2. Schultz, N. M., Huang, L., and Kennedy, R. T. (1995) Capillary electrophoresis-based immunoassay to determine insulin content and insulin seretion from single islets of Langerhans, *Anal. Chem.* **67**, 924-929.
3. Heegaard, N. H. H., Olsen, D. T., and Larsen, K.-L. P. (1996) Immuno-capillary electrophoresis for the characterization of a monoclonal antibody against DNA, *J. Chromatogr.* (in press)
4. Mammen, M., Gomez, F. A., and Whitesides, G. M. (1995) Determination of the binding of ligands containing the *N*-2,4-Dinitrophenyl group to bivalent monoclonal rat anti-DNP antibody using affinity capillary electrophoresis, *Anal. Chem.* **67**, 3526-3535.
5. Chu, Y.-H., Lees, W. J., Stassinopoulus, A., and Walsh, C. T. (1994) Using affinity capillary electrophoresis to determine binding stoichiometries of protein-ligand interactions, *Biochemistry* **33**, 10616-10621.
6. Heegaard, N. H. H. (1994) Determination of antigen-antibody affinity by immuno-capillary electrophoresis, *J. Chromatogr.* **680**, 405-412.
7. Shimura, K. and Kasai, K. (1995) Determination of the affinity constants of Concanavalin A for monosaccharides by fluorescence affinity probe capillary electrophoresis, *Anal. Biochem.* **227**, 186-194.
8. Honda, S., Taga, A., Suzuki, K., Suzuki, S., and Kakehi, K. (1992) Determination of the association constant of monovalent mode protein-sugar interaction by capillary zone electrophoresis, *J. Chromatogr.* **597**, 377-382.
9. Heegaard, N. H. H. and Robey, F. A. (1993) A capillary electrophoresis-based assay for the binding of Ca^{2+} and phosphorylcholine to human C-reactive protein, *J. Immunol. Methods* **166**, 103-110.
10. Kajiwara, H. (1991) Application of high-performance capillary electrophoresis to the analysis of aonformation and interaction of metal-binding proteins, *J. Chromatogr.* **559**, 345-356.
11. Heegaard, N. H. H. and Robey, F. A. (1992) Use of capillary zone electrophoresis to evaluate the binding of anionic carbohydrates to synthetic peptides derived from serum amyloid P component, *Anal. Chem.* **64**, 2479-2482.

12. Heegaard, N. H. H., Mortensen, H. D., and Roepstorff, P. (1995) Demonstration of a heparin-binding site in serum amyloid P component using affinity capillary electrophoresis as an adjunct technique, *J. Chromatogr.* **717**, 83-90.
13. Heegaard, N. H. H. and Roepstorff, P. (1995) Preparative capillary electrophoresis and mass spectrometry for the identification of a putative heparin-binding site in amyloid P component, *J. Cap. Elec.* **2**, 219-223.
14. Chu, Y.-H., Avila, L. Z., Biebuyck, H. A., and Whitesides, G. M. (1992) Use of affinity capillary electrophoresis to measure binding constants of ligands to proteins, *J. Med. Chem.* **35**, 2915-2917.
15. Chu, Y.-H., Avila, L. Z., Biebuyck, H. A., and Whitesides, G. M. (1993) Using affinity capillary electrophoresis to identify the peptide in a peptide library that binds most tightly to vancomycin, *J. Org. Chem.* **58**, 648-652.
16. Chu, Y.-H. and Whitesides, G. M. (1992) Affinity capillary electrophoresis can simultaneously measure binding constants of multiple peptides to vancomycin, *J. Org. Chem.* **57**, 3524-3525.
17. Baba, Y., Tsuhako, M., Sawa, T., Akashi, M., and Yashima, E. (1992) Specific base recognition of oligodeoxynucleotides by capillary affinity gel electrophoresis using polyacrylamide-poly(9-vinyladenine) conjugated gel, *Anal. Chem.* **64**, 1920-1925.
18. Guttman, A. and Cooke, N. (1991) Capillary gel affinity electrophoresis of DNA fragments, *Anal. Chem.* **63**, 2038-2042.
19. Xian, J., Harrington, M. G., and Davidson, E. H. (1996) DNA-protein binding assays from a single sea urchin egg: A high-sensitivity capillary electrophoresis method, *Proc. Natl. Acad. Sci. USA* **93**, 86-90.
20. Ljungberg, H. and Nilsson, S. (1995) Protein-based capillary affinity gel electrophoresis for chiral separation of β-adrenergic blockers, *J. Liq. Chromatogr.* **18**, 3685-3698.
21. Barker, G. E., Russo, P., and Hartwick, R. A. (1992) Chiral separation of leucovorin with bovine serum albumin using affinity capillary electrophoresis, *Anal. Chem.* **64**, 3024.
22. Tiselius, A. (1937) A new apparatus for electrophoretic analysis of colloidal mixtures, *Trans. Faraday Soc.* **33**, 524-531.
23. Mikkers, F. E. P., Everaerts, F. M., and Verheggen, T. P. E. M. (1979) High performance zone electrophoresis, *J. Chromatogr.* **169**, 11-20.
24. Jorgenson, J. W. and Lukacs, K. D. (1981) Zone electrophoresis in open-tubular glass capillaries, *Anal. Chem.* **53**, 1298-1302.
25. Landers, J. P., Oda, R. P., Spelsberg, T. C., Nolan, J. A., and Ulfelder, K. J. (1993) Capillary electrophoresis: A powerful microanalytical technique for biologically active molecules, *Biotechniques* **14**, 98-111.
26. Grossman, P. D.(1992) Factors affecting the performance of capillary electrophoresis separations: Joule heating, electroosmosis, and zone dispersion, in P.D. Grossman and J.C. Colburn (eds.), *Capillary Electrophoresis*, Academic Press, Inc., San Diego, CA, pp. 3-43.
27. Xu, Y. (1995) Capillary electrophoresis, *Anal. Chem.* **67**, 463R-473R.
28. Kuhr, W. G. and Monnig, C. A. (1992) Capillary electrophoresis, *Anal. Chem.* **64**, 389R-407R.
29. Guzman, N.A. (ed.) (1993) *Capillary Electrophoresis Technology*, Marcel Dekker, Inc., New York, NY.
30. Grossman, P.D. and Colburn, J.C. (1992) *Capillary electrophoresis: Theory and practice*, Academic Press, Inc., San Diego, CA.
31. Heegaard, N. H. H. and Robey, F. A. (1994) The emerging role of capillary electrophoresis as a tool for the study of biomolecular noncovalent interactions, *American Lab.* **26**, T28-X28.
32. Gordon, M. J., Huang, X., Pentoney, S. L.,Jr., and Zare, R. N. (1988) Capillary electrophoresis, *Science* **242**, 224-228.
33. Cohen, A. S., Paulus, A., and Karger, B. L. (1987) High-performance capillary electrophoresis using open tubes and gels, *Chromatographia* **24**, 15-24.
34. Heegaard, N. H. H., Heegaard, P. M. H., Roepstorff, P., and Robey, F. A. (In press, 1996) Ligand binding sites in human serum amyloid P component, *Eur. J. Biochem.*
35. Stevens, S. Y., Swanson, P. C., and Glick, G. D. (1994) Application of the gel shift assay to study the affinity and specificity of anti-DNA antibodies, *J. Immunol. Methods* **177**, 185-190.
36. Fried, M. G. (1989) Measurement of protein-DNA interaction parameters by electrophoresis mobility shift assay, *Electrophoresis* **10**, 366-376.
37. Nedved, M. L. and Moe, G. R.(1995) in Techniques in Protein Chemistry VI (pp. 393-400, Academic Press, Inc., San Diego, CA.

38. Takeo, K. and Kabat, E. A. (1978) Binding constants of dextran and isomaltose oligosaccharides to dextran-specific myeloma proteins determined by affinity electrophoresis, *J. Immunol.* **121**, 2305-2310.
39. Bøg-Hansen, T. C., Bjerrum, O. J., and Ramlau, J. (1975) Detection of biospecific interaction during the first dimension electrophoresis in crossed immunoelectrophoresis, *Scand. J. Immunol.* **4, suppl.2**, 141-147.
40. Bøg-Hansen, T. C. and Takeo, K. (1980) Determination of dissociation constants by affinity electrophoresis: Complexes between human serum proteins and concanavalin A, *Electrophoresis* **1**, 67-71.
41. Horejsí, V. (1979) Some theoretical aspects of affinity electrophoresis, *J. Chromatogr.* **178**, 1-13.
42. Takeo, K. and Nakamura, S. (1972) Dissociation constants of glucan phosphorylases of rabbit tissues studied by polyacrylamide gel disc electrophoresis, *Arch. Biochem. Biophys.* **153**, 1-7.
43. Hardingham, T. E. and Fosang, A. J. (1992) Proteoglycans: many forms and many functions, *FASEB J.* **6**, 861-870.
44. Scatchard, G. (1949) The attraction of proteins for small molecules and ions, *Ann. N. Y. Acad. Sci.* **51**, 660-672.
45. Matousek, V. and Horejsí, V. (1982) Affinity electrophoresis: A theoretical study of the efffects of the kinetics of protein-ligand complex formation and dissociation reactions, *J. Chromatogr.* **245**, 271-290.
46. Avila, L. Z., Chu, Y.-H., Blossey, E. C., and Whitesides, G. M. (1993) Use of affinity capillary electrophoresis to determine kinetic and equilibrium constants for binding of arylsulfonamides to bovine carbonic anhydrase, *J. Med. Chem.* **36**, 126-133.
47. Takeo, K. (1987) Affinity electrophoresis, in A. Chrambach, M.J. Dunn, and B.J. Radola (eds.) *Advances in electrophoresis, vol. 1*, VCH Verlag, Weinheim, pp. 229-279.
48. Heegaard, N. H. H. (1994) Characterization of biomolecules by electrophoretic analysis of reversible interactions, *Appl. Theor. Electrophoresis* **4**, 43-63.
49. Dowd, J. E. and Riggs, D. S. (1965) A comparison of estimates of Michaelis-Menten kinetic constants from various linear transformations, *J. Biol. Chem.* **240**, 863-869.
50. Dixon, M. and Webb, E. C. (1979) Enzymes, p. 60, Longman Group Ltd., London.
51. Cantor, C. R. and Schimmel, P. R. (1980) Biophysical Chemistry. Part III: The behavior of biological macromolecules, p. 897, W.H.Freeman and Co., New York.

APPLICATION OF NEUTRON SCATTERING TECHNIQUES TO PROBLEMS OF BIOLOGICAL RELEVANCE

BRIAN M. POWELL
Atomic Energy of Canada Ltd.
Chalk River Laboratories
Chalk River
Ontario
Canada K0J 1J0

ABSTRACT. A brief introduction is given to neutrons and their properties, particularly as they can be applied to biological problems. The technique of neutron scattering is outlined. Three examples of the application of neutron scattering methods to biological problems are given to illustrate the information that can be derived. Two examples, (i) a determination of the nature of intermembrane forces and (ii) the solution structure of calmodulin, illustrate the application of elastic scattering techniques. The third example illustrates the use of inelastic neutron scattering methods in an investigation of the dynamics of myoglobin.

1. Introduction

Neutron scattering techniques were first applied to biological systems in the late 1960's. Structural neutron biology began with investigations of the membrane myelin (Parsons and Akers, 1969), of haemoglobin solutions (Schmatz, 1969) and of single crystal myoglobin (Schoenborn, 1969). Studies to explore the dynamics of biological systems began almost simultaneously. Spectra from simple polypeptides in different conformational states were measured by Gupta et al., (1968) and Boutin and Yip, (1968). In poly-L-glutamic acid, water modes were investigated by taking difference spectra from wet and dry samples (Whittemore, 1968). The first conference dedicated to neutron structural biology was held in 1975 (Schoenborn, 1976) and a major review of the field appeared in the same year (Jacrot, 1976). Developments in understanding the dynamics of biological systems were slower in coming and the first reviews of these studies did not appear until the mid 1980's (Middendorf, 1984). The most recent review of the subject is concerned with protein dynamics (Martel, 1992).

W. Ens et al. (eds.), New Methods for the Study of Biomolecular Complexes, 319–341.

In the last five years there has been a significant increase in the applications of neutron scattering methods to biology, both for structural and dynamical studies. The technique is now becoming accepted as one of the standard experimental tools for advancing our understanding of function in biological systems. However, since most attendees of this workshop are probably unfamiliar with neutron scattering methods, the present paper will begin with an introduction to neutrons and neutron scattering, particularly with reference to biological systems. Three specific examples will then be discussed to illustrate the information that can be derived by application of neutron scattering techniques.

2. Neutrons

2.1. NEUTRON PROPERTIES

Neutrons are a nuclear particle, produced in copious quantities by the nuclear fission reactions in reactors and by certain high energy accelerators. Thermal neutrons can provide information about biological systems, or indeed, any condensed matter systems, because they possess a unique collection of properties.

(i) Neutrons are a neutral particle, consequently they are a weak probe of matter. This makes them suitable for investigations of bulk matter and simplifies the interpretation of data. However, it also means that the intensity of neutron scattering is generally weak.

(ii) Neutrons are scattered by nuclei. The interaction is characterized by a single quantity, the neutron scattering length (usually designated as b). This is different for every isotope in the periodic table and is known for every isotope. In general the neutron scattering length is real and positive, but for a few isotopes it is real and negative and for a few others it is complex. Values for the neutron scattering lengths have been compiled by Sears (1992).

(iii) The neutrons produced in a reactor or accelerator have a distribution of wavelengths, typically from 0.3 Å to 20 Å. This range covers interatomic spacings in many condensed systems.

(iv) The energy of a neutron is proportional to the square of its wavevector, i.e. inversely proportional to the square of its wavelength. Consequently, neutrons have an energy range from 900 meV to 0.2 meV. This spans the energy of many of the fundamental excitations in condensed matter systems. It is this simultaneous matching of both the length and energy scales that gives neutrons their particular power.

(v) Although not discussed in this review, it should be noted that neutrons possess a magnetic moment and can be polarized. Consequently, neutrons are a powerful probe of magnetic properties and some biological systems are intrinsically magnetic.

Table 1 shows the neutron scattering cross-sections for the first few elements of the periodic table. The cross-section, $\sigma = 4\pi b^2$. The coherent cross-section, σ_{coh}, is that appropriate for collective motions. Its value varies from isotope to isotope, but has no systematic variation. For neutrons, hydrogen scatters just as effectively as iron or lead or uranium. The incoherent cross-section, σ_{inc}, is that appropriate for single particle motions. Its value is generally less than that of the corresponding coherent cross-section. The third column shows the absorption cross-section σ_a. It is generally small, but for a few specific isotopes has a large value. For example it can be seen that natural boron has a large absorption cross-section for neutrons, but this comes entirely from the ^{10}B isotope.

TABLE 1 The neutron cross-sections for some of the light elements. Units are barns (= 10^{-24} cm^2).

Element	σ_{coh}	σ_{inc}	σ_a
H	1.7568 (−)	80.26	0.3326
D	5.592	2.05	0.0005
B	3.54	1.70	767.0
C	5.550	0.00	0.0035
N	11.01	0.50	1.90
O	4.232	0.00	0.0002

The incoherent cross-section for hydrogen is large compared with all other cross-sections except the absorption ones. This means that for any material containing a significant number of hydrogens, e.g. all biological materials, the neutron scattering will be predominantly from the hydrogens. Consequently, in neutron measurements on biological systems, it will be the hydrogen motions that are probed preferentially.

The cross-section is proportional to the square of the neutron scattering length. The negative sign beside the hydrogen coherent cross-section in Table 1 signifies that its coherent scattering length, b, is negative. But the deuterium scattering length is positive. Consequently, a change from hydrogen to deuterium results in a huge change in the neutron scattering. This change leads to the concept of contrast.

For biological materials that cannot, in general, be crystallized and must be investigated in solution, a fundamental quantity describing the neutron scattering is the scattering length density. It is defined by

$$\rho(\mathbf{r}) = \frac{1}{V}\int_{\upsilon} b_i(\mathbf{r})d^3\mathbf{r} \qquad (1)$$

where b_i is the scattering length of atom i and V is the volume of the sample.

Figure 1 shows the scattering length density as a function of the D_2O content of the solvent, for several common molecules found in biological material. It is assumed that all exchangeable protons are exchanged in proportion to the D_2O content of the solvent. It can be seen that the line for water crosses those for many biological constituents. At any given D_2O content the difference between the scattering length densities of water and of a specific molecule is called the *contrast* for that molecule. The point at which the two lines intersect is called the *contrast match point* and at this D_2O concentration the molecule will become invisible to neutrons. This feature of contrast matching is extensively utilized in measurements of neutron scattering from solutions.

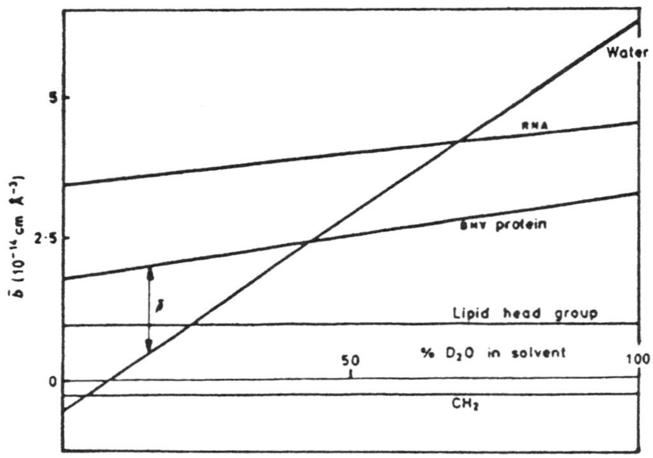

Figure 1 Scattering length density $(\bar{b})$ for some common biological molecules. $\bar{\rho}$ is the *contrast* between a molecule and the solvent. The intersection with the water line is the *contrast match point.* (Jacrot, 1976).

2.2. NEUTRON SCATTERING

Let us now discuss the nature of a neutron scattering experiment. In Figure 2(a) we show a conceptual experimental configuration in real space. Neutrons with wave vector $\mathbf{k}_0$ are incident on the sample S. We shall not describe here how neutrons with a specified wave vector may be selected from the distribution produced in a reactor or accelerator. Instead, we refer the reader to one of the comprehensive reviews of neutron scattering methods (Squires, 1978; Lovesey, 1984; Sköld and Price, 1986). The neutrons are scattered from the sample and we choose to measure those scattered through an angle ϕ, the scattering angle. These neutrons have wave vector $\mathbf{k}'$ and are scattered into solid angle $d\Omega$ and detected by the detector D. The neutron experiment will thus consist of measuring the intensity of the scattered neutrons as a function of the scattering angle ϕ.

Since the neutron is a scattering probe, it is usually easier to think of the experiment in terms of reciprocal space rather than real space. The reciprocal space representation of the experiment is shown in Figure 2(b). It defines the wave vector transfer $\mathbf{Q}$, (or the momentum transfer $\hbar\,\mathbf{Q}$), usually known as the scattering vector in X-ray diffraction. The scattering triangle shown in Figure 2(b) has $k' \neq k_0$. Since the energy of the neutron is proportional to the square of its wave vector, this means that $E' \neq E_0$ i.e. the neutron has exchanged energy with the sample. The experimental configuration shown in Figure 2(a) is thus oversimplified. Rather than the scattered neutrons entering the detector directly, as is shown, an energy analysis of the scattered beam should be made before the neutrons are detected. The experiment will thus measure both the change in energy of the neutron and its change in momentum, $\hbar \mathbf{Q}$.

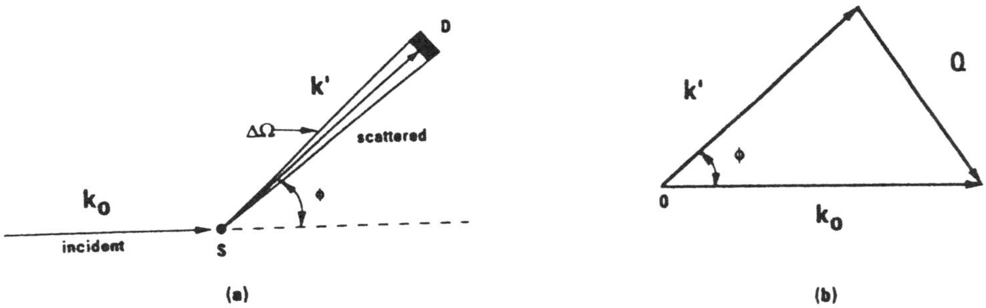

Figure 2 Schematic diagram of a neutron scattering experiment, (a) in real space, (b) in reciprocal space.

The neutron scattering event satisfies the two conservation conditions:

$$\hbar \mathbf{Q} = \hbar(\mathbf{k}_0 - \mathbf{k}') \tag{2}$$

$$\hbar\omega = E_0 - E' = \frac{\hbar^2}{2m_n}\left(k_0^2 - k'^2\right). \tag{3}$$

where m_n is the mass of the neutron.

Equation 2 represents the conservation of momentum. It specifies the momentum change of the neutron and this change in momentum is given to the sample in the scattering event. Equation 3 represents the conservation of energy. It specifies the energy change of the neutron, $\hbar\omega$, and this change in energy is given to the sample in the scattering event. Note that the energy change can be of either sign, i.e. the neutron may give energy to the sample or may take energy from the sample.

The two quantities $\mathbf{Q}$ and ω are the dynamical variables describing the excitation spectrum of the sample. The intensity of the neutron scattering is characterized by the double differential cross-section, the number of neutrons scattered into solid angle $d\Omega$ with energy lying between E' and $E' + dE'$. It is given by,

$$\frac{d^2\sigma}{d\Omega dE'} = C\frac{k'}{k_0} S(\mathbf{Q},\omega) \tag{4}$$

where C is a constant containing various instrumental parameters and $S(\mathbf{Q},\omega)$, the dynamical structure factor, contains all the information about the sample. Thus, apart from the trivial ratio of wave vectors, the measured neutron intensity gives information directly about the quantity that describes the properties of interest in the sample. For further details of the theory and practice of neutron scattering, the reader is referred to one of the major reviews of the field (Squires, 1978; Lovesey, 1984; Sköld and Price, 1986).

To scientists familiar only with the scale of standard laboratory equipment, neutron instruments will appear as very large devices. Figure 3 is a photograph of the DUALSPEC spectrometers at Chalk River Laboratories. They are two spectrometers stacked one on top of the other but designed to operate independently. Their size can be judged from the scientists posed beside the instrument. Despite their size, all angles specifying the positional settings of the instruments can be set to a precision of 0.002°. The instruments run continuously under computer control, collecting and storing data, changing the orientation and the environment of the sample, as specified in the control sequence chosen by the researcher.

Figure 3 The DUALSPEC spectrometers at Chalk River Laboratories. The upper instrument (on the right) is a high-resolution powder diffractometer. The lower instrument (on the left) is a polarized beam triple-axis spectrometer.

3. Bilayer Membranes

3.1. MODEL SYSTEMS

A membrane is a selectively permeable barrier that serves to enclose and protect the cell, but it also permits passage of the chemicals necessary for the healthy functioning of the cell. The bilayer membrane is a molecular sheet made up of a double layer of lipid molecules. Figure 4 is an artist's conception of such a membrane. Presently there is great interest in the properties of these membranes, the structures they can assume and their dynamical fluctuations. Neutron diffraction is an effective probe for such investigations.

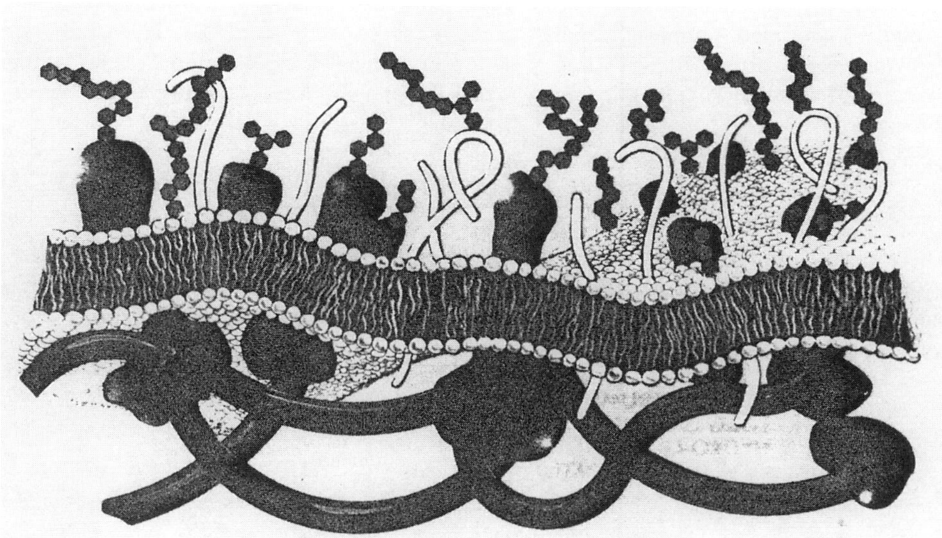

Figure 4 Schematic illustration of a biological membrane. The rippled sheet formed by the double layer of lipid molecules is the bilayer membrane. It is traversed by various membrane proteins. (Bloom, 1992).

A point of particular interest is the nature of the intermembrane interactions, specifically, the character of the membrane-membrane repulsive force at small intermembrane separations. This is of importance in cell-cell interactions and cell fusion. There are two views of this repulsive force. In the first, the polar head groups of the molecules comprising the layer are thought to disturb the water structure near the lipid, perhaps by a charge transfer mechanism. The result is a repulsive force known as the hydration force. In the second view the membrane is recognized as a dynamical entity with membrane fluctuations and molecular diffusion constantly occurring. The result of these entropic fluctuations is the repulsive force. The two views can be distinguished by the dependence of the applied pressure (which is determined from the relative humidity), as a function of membrane separation. For the hydration force the dependence will be exponential, for the entropic force the dependence will be a power law. Consequently, a choice between the two views can be made by measuring this functional dependence. Katsaras and Jeffrey (1996) have recently carried out neutron diffraction measurements for the model membrane dipalmitoylphosphotidylcholine (DPPC) to resolve this question. The samples were oriented multi-bilayer membranes of DPPC and fully hydrogenated specimens were utilized as well as specimens with the three methyl groups deuterated.

3.2. NEUTRON MEASUREMENTS

The neutron diffraction measurements were made with the T3 spectrometer at the NRU reactor, Chalk River. Figure 5 compares the data for the hydrogenated and partially deuterated samples. The dramatic changes as a result of deuteration are evident. Eight orders of scattering were observed, allowing a determination of the variation of the scattering length density across the bilayer. The scattering length density is expressed as

$$\rho(x) = C + \frac{2}{d_n} \sum_{n=1}^{N} F_n \cos\left(\frac{2\pi n x}{d_n}\right) \tag{5}$$

where x is measured from the centre of the bilayer, d_n is the Bragg repeat distance and F_n is the structure factor of reflection n and N is the number of orders observed. C is a constant (= F_0/d) which depends on the unobservable structure factor of the zeroth order peak. It sets the absolute scale for $\rho(x)$.

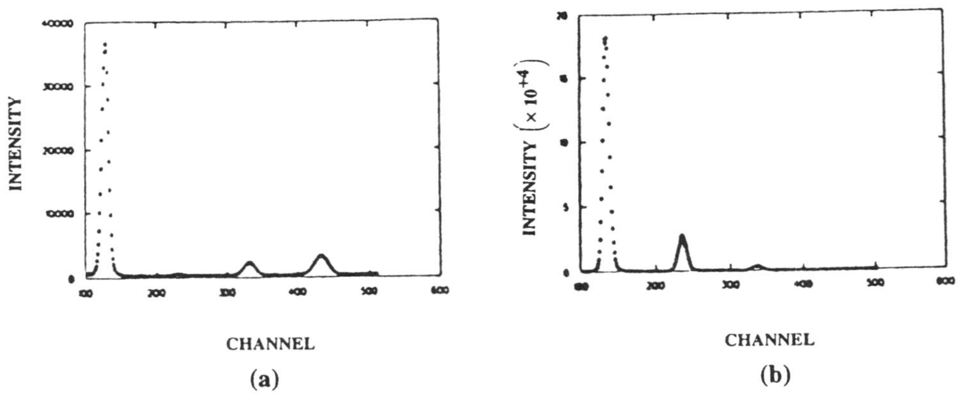

Figure 5 Neutron diffraction data from oriented, multi-bilayers of DPPC, (a) fully hydrogenated (b) d_9-DPPC, all methyl groups deuterated (Katsaras and Jeffrey, 1996).

Since the F_n are proportional to the square roots of the experimental intensities, the signs of the F_n are not determined. This is the phase problem of conventional crystallography. It is conveniently solved in the present case by measuring the dependence of the intensities on the D_2O concentration in the solvent. Since the system is centrosymmetric, there are simple rules that permit the determination of the appropriate signs for the F_n from this data (Worcester and Franks, 1976). Inserting values into the right hand side of equation 5 will then determine the scattering density $\rho(x)$. However, in the present case, there is experimental data for both deuterated and hydrogenated samples and consequently, the *difference* in scattering densities from these two data sets can be derived. Since only the methyl groups were deuterated, this difference will show only the methyl groups. The result is shown in Figure 6. The oscillations of scattering length density about zero show that the subtraction is not quite exact, but the position and the shape of the methyl group distribution about the centre of symmetry is seen to be accurately determined. The intermembrane separation can thus be derived.

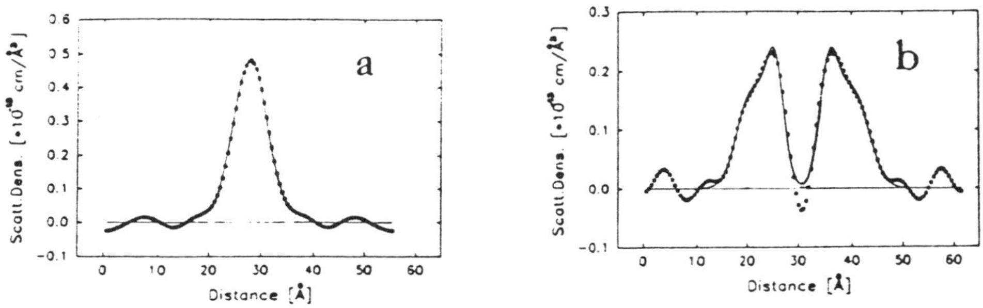

Figure 6 The profile of the scattering length density for the methyl groups across a bilayer of DPPC derived from the difference density profiles (a) bilayers at 0° relative humidity (spacing is 55.7 Å) (b) bilayers at 100° relative humidity (spacing is 61.3 Å). (Katsaras and Jeffrey, 1996)

Figure 7 shows a plot of the logarithm of the pressure versus the intermethyl separation. It can be seen that the plot is a straight line over four decades of pressure. This strongly supports the view that the intermembrane repulsive force is the hydration force. The separation of other constituents of the DPPC molecule can also be determined, e.g. the phosphate separation and the ester separation. However, since neither of these chemical groups can be deuterated, the separations cannot be determined by the difference method. Instead, the complete scattering length density profile must be computed from equation 5 and the separations then derived by measuring the appropriate peak positions on either side of zero. These experimental separations are thus less accurate than those determined for the methyl groups. But the same plots for phosphate and ester separations also show a straight line dependence. This supports the view that the repulsive force has the character of the hydration force.

This application of neutron scattering is the one-dimensional analog of a single crystal structure determination. It utilizes a highly oriented specimen, the data is in the form of "Bragg" intensities and single crystal methods of analysis are employed. But if the sample is non-crystalline, e.g. a solution, then different techniques must be used.

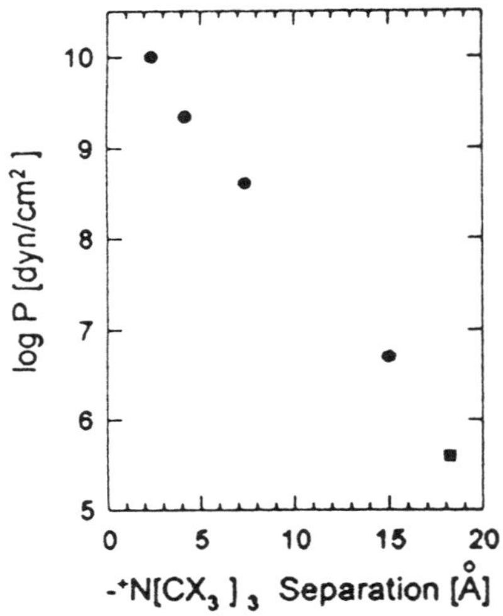

Figure 7 Plot of log P versus intermethyl separation for DPPC bilayers. (Katsaras and Jeffrey, 1996).

4. Solution Structures

The structure of biomolecules in solution is another field in which neutron diffraction can provide data. The technique is very different from that discussed in section 3 above. It measures the neutrons scattered at very small angles, i.e. the angle, ϕ, in Figure 2 is small, typically $\lesssim 2°$. This neutron technique is known as Small Angle Neutron Scattering (SANS).

4.1. FORMALISM

The intensity of the scattered neutrons, $I(\mathbf{Q})$, is given by,

$$I(\mathbf{Q}) = \left| \int_{V_p} \left(b_p(\mathbf{r}) - b_s \right) e^{i\mathbf{Q}\cdot\mathbf{r}} d^3\mathbf{r} \right|^2 \quad (6)$$

where $b_p(\mathbf{r})$ is the scattering length density of particle p and b_s that of the matrix (solution). $\left(b_p(\mathbf{r}) - b_s\right)$ is the contrast for particle p and V_p is the volume of the particle. $\mathbf{Q}$ is the wave vector transfer vector defined in equation 2, its magnitude, $Q = 4\pi \sin\left(\phi/2\right)/\lambda$, where λ is the neutron wavelength.

Biomolecules in solution will be randomly oriented and the neutron scattering from them will be spherically averaged. The spherical average of equation 6 can be expressed as;

$$I(\mathbf{Q}) = 4\pi \int_0^{d_{max}} P(R)\frac{\sin(QR)}{QR}dR \qquad (7)$$

where $P(R)$, the pair distribution function, is a distribution of distances $|\mathbf{r}-\mathbf{r}'|$ between all pairs of nuclei, weighted by the product of their respective contrast factors. d_{max} is the maximum dimension of the particle.

Taking the inverse Fourier transform of equation 7 gives,

$$P(R) = \frac{1}{2\pi^2}\int_0^{\infty} I(Q)(QR)\sin(QR)dQ \qquad (8)$$

$P(R)$ is a complicated function, but it contains much structural information about the particle shape. The intensity of neutrons scattered through zero degrees (i.e. Q = 0) is given by,

$$I(0) = \left[\left(\bar{b}_p - b_s\right)V_p\right]^2 \qquad (9)$$

where $\left(\bar{b}_p - b_s\right)$ is the mean contrast factor. But we also have,

$$I(0) = \int_0^{d_{max}} P(R)dR \qquad (10)$$

$I(0)$ cannot be measured directly, but must be determined by extrapolation to Q = 0. Another commonly used parameter is the radius of gyration, R_g, defined by,

$$R_g^2 = \frac{\int_0^{d_{max}} P(R)R^2 dR}{\int_0^{d_{max}} P(R)dR} \ . \qquad (11)$$

4.2. Ca^{2+}-CALMODULIN STRUCTURE

The protein calmodulin mediates the regulatory effect of calcium ions on many biochemical processes. It is believed that when Ca^{2+} binds to calmodulin, conformational changes are induced in the protein structure that enables it to bind to, and hence activate, a target enzyme. But little is known about the interactions of calmodulin with its target proteins. The structure of Ca^{2+}-calmodulin crystallized at low pH is known from X-ray diffraction (Babu et al, 1985,1988). It has two globular lobes connected by an extended seven-turn α-helix. Each lobe contains two Ca^{2+} binding sites. Heidorn et al. (1989) used SANS measurements to determine the conformation of Ca^{2+}-calmodulin bound to a target enzyme. The enzyme chosen for the study was myosin light chain kinase (MLCK), or more accurately, a synthetic peptide corresponding to the sequence of residues of the binding domain of MLCK. This complex is believed to be a good model with which to study calmodulin-enzyme interactions.

The SANS data was collected on the Low-Q diffractometer at the Los Alamos Neutron Scattering Centre. The calmodulin was deuterated and the MLCK was proteated. Data is shown in Figure 8 as log (intensity) versus Q^2 (this is known as a Guinier plot) for several concentrations of D_2O in the solvent. Since calmodulin was deuterated, the contrast was greatest (and hence the intensity highest) when the solvent was pure H_2O and progressively decreases as the fraction of D_2O increases. At 37% D_2O, the MLCK is contrast matched and the observed scattering arises only from the Ca^{2+}-calmodulin. The lines represent least-squares fits of a model P(R) to the scattering data.

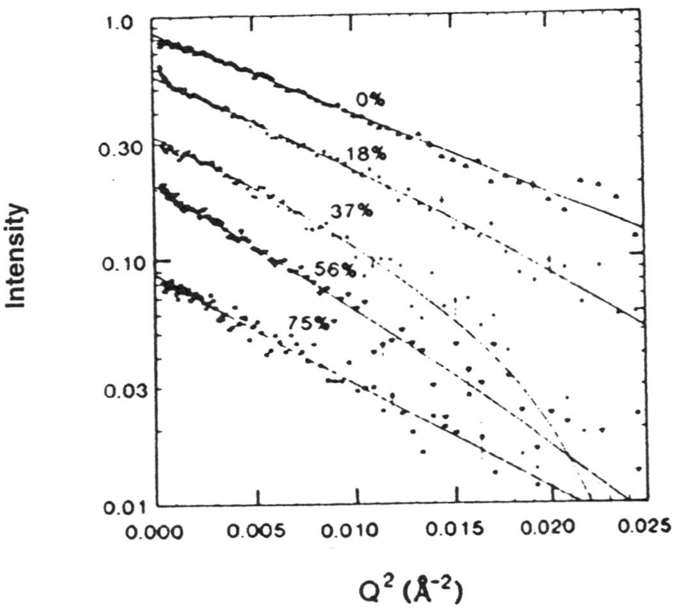

Figure 8 A Guinier plot (log I versus Q^2) of the SANS data from the Ca^{2+}-calmoldulin — MLCK complex for various contrasts. The lines are the fit of a model $P(R)$, (equation 8). 37% D_2O in the solvent is the contrast match point for proteated MLCK. (Heidorn et al., 1989).

Pair distribution functions determined by the fits are shown in Figure 9. The curve in Figure 9(a) shows P(R) for deuterated Ca^{2+}-calmodulin in pure H_2O and Figure 9(b) the corresponding curve for deuterated Ca^{2+}-calmodulin complexed with proteated MLCK in 37% D_2O. The curves indicate a dramatic conformational change in the complex. There is a large (~30%) reduction in d_{max} , the maximum dimension of the complex and the shape of the complex also changes dramatically. The uncomplexed calmodulin has a double-peaked $P(R)$ (Figure 9(a)) that indicates a more extended structure containing two distinct lobes. The complexed calmodulin has a very symmetric $P(R)$ (Figure 9(b)) that indicates a more compact, spherical-like structure. At 37% D_2O, the contrast match point, the MLCK is invisible to the neutrons. Thus, this $P(R)$ corresponds to Ca^{2+}-calmodulin alone.

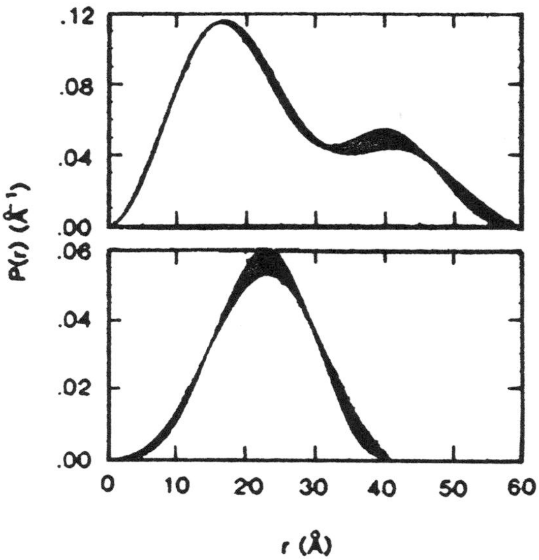

Figure 9 Pair distribution functions, $P(R)$, determined from the data of Figure 8. The band represents one standard deviation and indicates the error in $P(R)$. (a) deuterated Ca^{2+}-calmodulin in H_2O (b) deuterated Ca^{2+}-calmodulin — proteated MLCK in 37% D_2O. (Heidorn et al. 1989).

In Figure 10 the square root of the zero angle scattering is plotted as a function of solvent scattering density. As expected, the dependence is linear. The intercept on the contrast axis gives the mean scattering length density $\bar{b}_p$, leading to a value for the volume of the complex of 2.2×10^4 Å^3. This volume corresponds to the unhydrated volume of the complex. It agrees well with the value estimated from the partial specific volume and molecular weight.

The radius of gyration is calculated from the pair distribution function by means of equation 11. It has been shown that R_g^2 can be separated into a term independent of variations in scattering length density, R_v^2 and additional terms that are not (Ibel and Stuhrmann, 1975). The former is a measure of the overall shape of the complex and the coefficients of the latter are related to the internal density fluctuations of the particle.

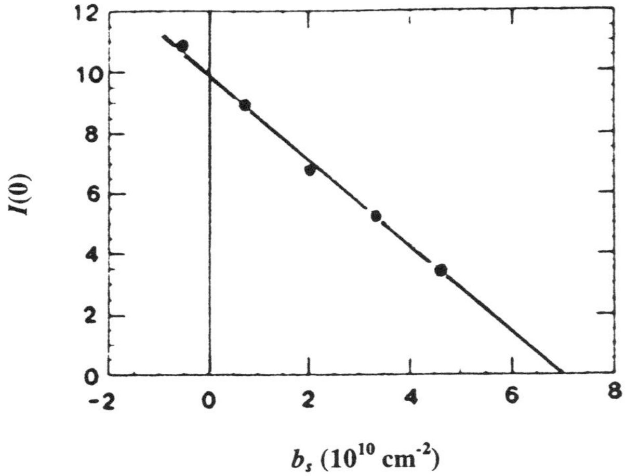

Figure 10 The square root of the forward scattering intensity, $I(0)$, plotted against the solvent scattering density. (Heidorn et al. 1989).

$$R_g^2 = R_v^2 + \frac{\alpha}{b_c} - \frac{\beta}{b_c^2} \tag{12}$$

where $b_c = \overline{b}_p - b_s$. The coefficient α, is the first moment of density fluctuations about the mean and β is a measure of the separation of the centres of the scattering densities. Figure 11 shows R_g^2 plotted against $1/b_c$ the inverse of the mean contrast.

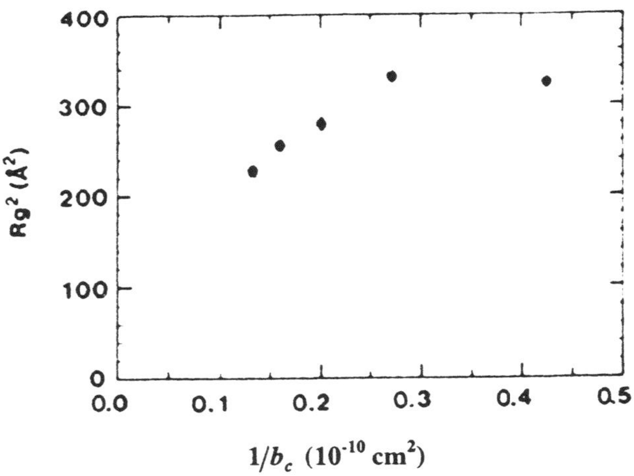

Figure 11 The square of the radius of gyration plotted against the inverse of the mean contrast factor. (Heidorn, 1989).

In principle, it is possible to fit the expansion of equation 12 to this data and determine values for R_v^2, α and β. However, the errors in the parameters derived from this data are too large to give meaningful values. To determine the parameters more quantitatively requires data at negative contrast. This would require proteated calmodulin and deuterated MLCK. However, the analysis did show that α was positive and that β was nonzero. The positive value for α shows that the component with the larger scattering length density (Ca^{2+}-calmodulin) lies towards the outside of the complex, while the nonzero value for β shows that the scattering centres of the two components are not coincident.

The SANS measurements did not succeed in determining unambiguously the solution structure of calmodulin complexed with MLCK. However, the results are consistent with the view that MLCK binds near calmodulin's interconnecting helix and the two globular lobes of calmodulin form additional contacts by closing around the peptide. The SANS measurements thus yielded an unexpected view of the interaction of Ca^{2+}-calmodulin and the binding domain of one of its target enzymes.

5. Inelastic Neutron Scattering Measurements

The two examples discussed in sections 3 and 4 utilize elastic neutron scattering methods, i.e. no energy analysis is made of the scattered neutrons. Thus, they provide information about the structural properties of the systems, but can say nothing of their dynamical fluctuations. However, for many biosystems, it is believed that their biological function is intimately related to the dynamical motions of the constituents. Consequently, there is an increasing effort to measure and to understand quantitatively the dynamics of biomolecules.

There is a large body of experimental data on the dynamics of "small" molecules, e.g. benzene, and Figure 12 shows a typical inelastic scan for such a molecule. It shows a very intense and sharp (resolution width) elastic peak. In the wings of this is a quasi-elastic scattering distribution. This often has a Lorentzian line shape and arises from the molecular diffusive motions. These motions are truly inelastic ($\omega \neq 0$), but by convention, it is only scattering with ω outside the quasi-elastic regime that is referred to as inelastic. Inter-molecular modes are those in which the molecule moves as a rigid body, motions involving translations and librations. Their energy is determined by the intermolecular forces (usually weak). The intra-molecular modes, usually at higher energy, are those involving stretching, bending and torsions of the molecular bonds. Their energy is determined by the internal forces within the molecule.

For a biomolecule, the distinction between inter- and intra-molecular motions is probably not a useful concept. Further, since such molecules are large, very low energy motions are expected to exist. Consequently, an energy separation between diffusive motions (quasi-elastic scattering) and low energy inter/intra molecular motions, may not exist. Intuitively, we may expect that the approach that has proved so useful for small molecules, the analysis of molecular motions into several simpler component motions, will be much less plausible for biomolecules. Despite the potential difficulties of interpretation, inelastic neutron scattering measurements increasingly are being made on biomolecules.

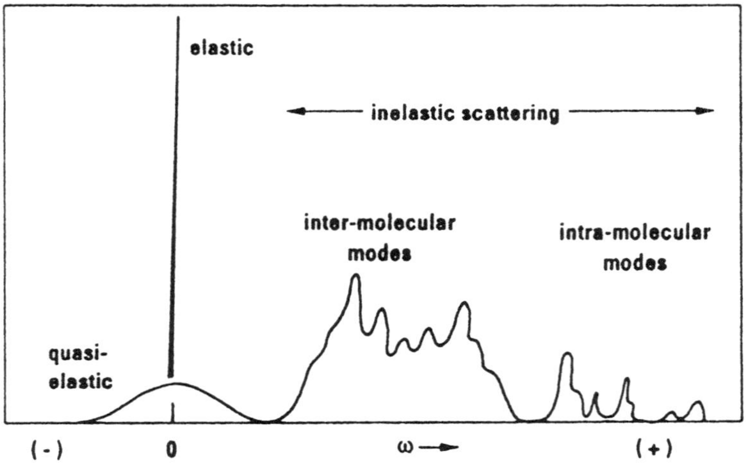

Figure 12 A typical inelastic neutron scattering scan for a "small" molecule, indicating the molecular motions and their relative frequency ranges.

5.1. DYNAMICS OF MYOGLOBIN

A particularly detailed study was made by Cusack and Doster (1990) of the temperature dependence of the low energy dynamics of myoglobin, deuterated by hydration in D_2O. The measurements were made on spectrometers at the Institute Laue-Langevin, Grenoble. In Figure 13 is shown the dynamical structure factor, $S(Q, \omega)$, for D_2O hydrated myoglobin at temperatures between 100 K and 350 K. The intensity is dominated by the incoherent scattering from hydrogen (see Table 1) and has been put on an absolute scale by comparison with a vanadium standard. The spectra at low temperatures (100 K and 180 K) show no broadening of the elastic line and have a well-defined, although broad, peak at 25 cm^{-1}. As the temperature increases, this peak becomes increasingly indistinct as quasi-elastic scattering develops.

The spectra of Figure 13 are very different from the idealized one shown in Figure 12. The most striking feature is the peak at 25 cm^{-1}. The contribution of a mode to $S(Q, \omega)$ is weighted by the amplitude of the atomic motions in that mode. The occurrence of a peak thus indicates there are many motions in myoglobin that occur with this energy. The peak is broader than resolution (defined by the width of the elastic peak) and so does not correspond to a single mode. But its presence shows that the dominant motions in myoglobin have this energy, at least at low temperatures.

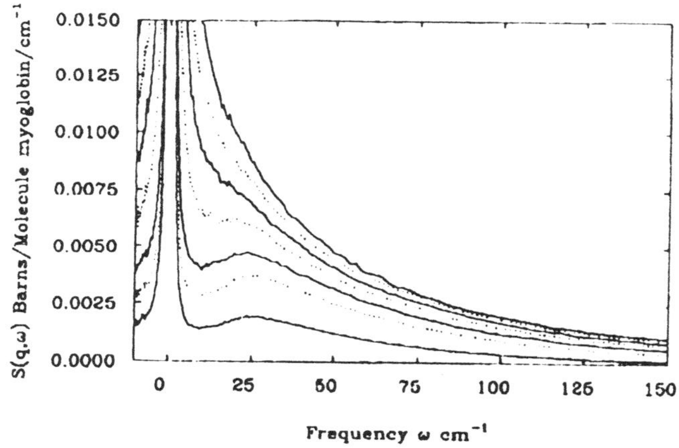

Figure 13 The temperature dependence of the dynamical structure factor $S(Q,\omega)$ for D_2O-hydrated myoglobin. In order of increasing intensity the temperatures are 100, 180, 220, 270, 300, 320 and 350 K. (Cusack and Doster, 1990).

The specimen used by Cusak and Doster was D_2O hydrated sperm whale myoglobin. However, a similar feature was observed in dry equine myoglobin by Martel and Lin (1989). In their inelastic neutron spectra at low temperature (77 K), they found a broad peak with a frequency of 29 cm^{-1}, but at 298 K this peak had become a shoulder on the side of the elastic peak. The similarity in the behaviour of the peak in dry and hydrated myoglobin strongly suggests that the feature is intrinsic to the dynamical properties of myoglobin and is not some artefact of the experimental conditions.

The interpretation of data such as that shown in Figure 13 is far from straightforward. There is an elastic response that, at low temperatures is distinct. At higher temperatures however there is no obvious distinction between the elastic response and the quasi elastic scattering, or between the latter and the scattering at higher frequencies that is expected to arise from vibrational modes. Formally, the dynamical structure factor can be separated into components as follows,

$$S(Q,\omega,T) = e^{-Q^2 U_{\text{vib}}(T)}\left\{S_0(Q,T)\delta(\omega) + S_{\text{qel}}(Q,\omega,T) + S_{\text{vib}}(Q,\omega,T)\right\} \qquad (13)$$

where S_0 is the elastic incoherent structure factor, S_{qel} is the quasi-elastic response and S_{vib} arises from vibrational modes. The exponential term is the Debye-Waller factor and U_{vib} is related to the mean square atomic displacements in the vibrational modes. It can be shown that the one-phonon vibrational structure factor can be written as,

$$S_{\text{vib}}(Q,\omega,T) = \frac{\hbar Q^2}{6\omega\left(e^{\hbar\omega/k_B T} - 1\right)} G(\omega) \qquad (14)$$

where $G(\omega)$ is the amplitude-weighted frequency distribution of the vibrational modes and $\left[e^{\hbar\omega/k_B T} - 1\right]$ is the Bose factor.

5.2. ELASTIC RESPONSE

If we assume that myoglobin behaves as a harmonic solid, then the incoherent elastic intensity, $S(Q,\omega \simeq 0) \propto e^{-Q^2 U_{vib}}$ and a plot of log $S(Q,\omega \simeq 0)$ versus Q^2 will be a straight line with gradient; U_{vib}. This analysis was made by Doster et al (1989) using a fixed window technique. The result is shown in Figure 14. At low temperatures the expected straight line is observed, but at higher temperatures the harmonic approximation is no longer valid. This is interpreted as the result of a dynamical transition occuring at ~180 K (no structural transition is observed) in which new, non-vibrational degrees of freedom are excited.

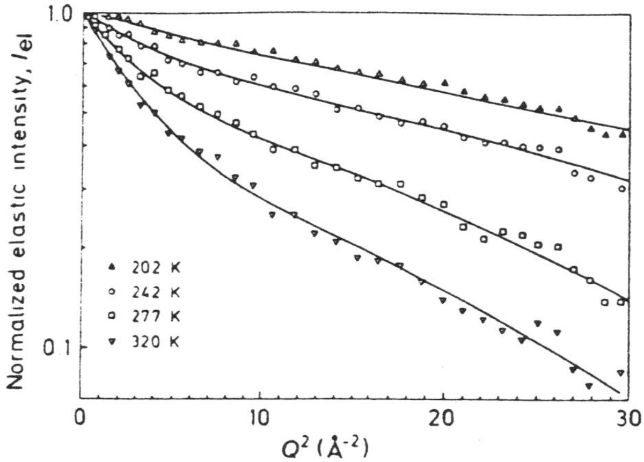

Figure 14 The temperature dependence of the elastic intensity , $S(Q,\omega \simeq 0)$ from myoglobin plotted against Q^2. The lines are a fit to the data with the expression of equation 15. (Doster et al., 1989).

The simplest model that might account for these observations involves the jumping of hydrogens between distinct sites of different energy. If the sites are separated by a distance, d, and free energy, ΔG, then the elastic intensity is given by,

$$S(Q,0) = e^{-Q^2(\Delta x^2)}\left\{1 - 2p_1p_2\left[1 - \frac{\sin(Qd)}{Qd}\right]\right\} \quad (15)$$

where Δx^2 is a mean square atomic displacement. It is equal to U_{vib} at low temperatures. p_1, p_2 specify the probability of finding a hydrogen in the ground and excited state respectively.

$$\frac{p_2}{p_1} \propto e^{(-\Delta G/RT)} \quad (16)$$

The solid lines in Figure 14 are the least-squares fits to the data with the expressions of equations 15 and 16.

The value of the jump distance, d, is found to be 1.5 Å and a van't Hoff plot of p_2/p_1 gives an energy asymmetry of 12 kJ mol^{-1}. This suggests the existence of torsional degrees of freedom.

5.3. VIBRATIONAL MOTIONS

The analysis of the "elastic" structure factor described above shows that non-vibrational motions exist in myoglobin, at least at high temperature. But we know that harmonic, vibrational motions must also exist. For "small" molecules these would have energies in the inelastic scattering range, as illustrated in Figure 12. However, as seen from Figure 13, for myoglobin the separation into quasi-elastic and inelastic scattering is not distinct. Consequently, analysis of the data must be made in a less direct manner.

We assume that the vibrational frequency distribution, $G(\omega)$, is temperature independent and that the temperature dependence observed for the vibrational scattering is due entirely to the Bose and Debye-Waller factors of equations 13 and 14. The temperature dependence of the Bose factor is explicit, while that of the Debye-waller factor is linear for harmonic vibrational motions. If the ratio $S(Q,\omega,T)/S(Q,\omega,T_0)$ is calculated, where T_0 is a reference temperature for which the quasi-elastic scattering is negligible (e.g. T_0 = 180 K), then the ratio should be linear as a function of ω within the energy range for which harmonic vibrational scattering occurs.

Cusack and Doster (1990) calculated this ratio and found the linear relationship was satisfied within the energy range ~100 cm^{-1} to ~32 cm^{-1}. Below this energy excess scattering is observed whose intensity increases with temperature. This is interpreted as the quasi-elastic scattering.

The frequency distribution function can be determined by fitting the expressions of equations 13 and 14 to the data of Figure 13. Cusack and Doster carried out this fitting and their result is shown in Figure 15, as a log/log plot. The dashed line has a slope of 2 and is expected for a Debye solid, i.e. one in which $G(\omega) \propto \omega^2$. At 180 K it can be seen that for low frequencies (~20 cm^{-1}) the measured distribution is Debye-like. There is an enhancement of modes at ~25 cm^{-1}, as expected from the presence of the broad peak in $S(Q, \omega)$ at this frequency and a deficit of modes at higher frequencies. As the temperature increases, myoglobin becomes less and less Debye-like at low frequencies, while at high frequencies $G(\omega)$ begins to deviate increasingly from that of the quasiharmonic solid at 180 K. The increasing number of modes at low frequencies arises because the quasi-elastic scattering has been analyzed in terms of the vibrational distribution function, although scattering below ~32 cm^{-1} does not appear to be part of the vibrational scattering. In a conventional solid, an "excess" of modes in $G(\omega)$ at low frequency is the signature for the existence of a glassy or amorphous phase. Thus, it is tempting to try and interpret this "non-vibrational" scattering in terms of excitations appropriate for an amorphous system.

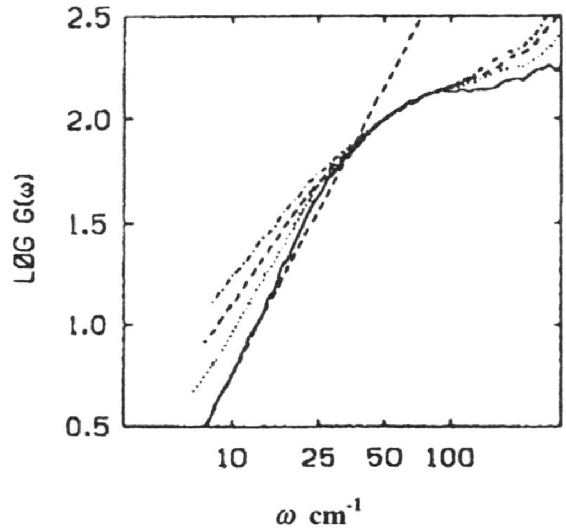

Figure 15 The temperature dependence of the vibrational distribution function, $G(\omega)$ shown as a log/log plot. In order of increase in $G(\omega)$ the temperatures are 180, 250, 300 and 350 K. The dashed line shows the distribution function of a Debye solid. (Cusack and Doster, 1990).

5.4. QUASI-ELASTIC SCATTERING

The scattering remaining after subtraction of the elastic response and of the vibrational scattering is interpreted as quasi-elastic scattering. It was derived by Cusack and Doster (1990) and analysed by Doster et al. (1989). The data, combined from two spectrometers, is shown on a log/log plot in Figure 16. It shows two distinct spectral components, a slow α-process and a faster β-process, with different shapes and temperature dpendences. The temperature dependence of the intensity of the β-component is consistent with local jumps between two states separated in energy by ~12 kJ mol^{-1}. This motion may be the torsional motion of the hydrogens postulated from the temperature dependence of the elastic response. The lineshape is intermediate between Gaussian and Lorentzian and the linewidth, although not well defined, appears to be temperature independent with a value of ~2 meV. This corresponds to an estimated correlation time of ~0.35 psec.

The α-spectral component is more complex. It is not Lorentzian in shape and so is not derived from a simple diffusion process. Doster et al. (1989) fitted the lineshape to a Cole-Davidson function of the form,

$$S(Q,\omega) = A(Q)\,\mathrm{Im}\left\{(1 + i\omega\tau_0)^{-b}\right\}/\omega \tag{17}$$

where Im signifies the imaginary part and τ_0 is a relaxation time. At large ω this expression shows a power law dependence with an exponent $-(1 + b)$. A Lorentzian lineshape would have $b = 1$, but the value obtained by fitting to the data of Figure 16 is $b = 0.25$. This value is temperature independent for temperatures greater than 250 K.

This temperature independence virtually excludes an interpretation of the component in terms of a distribution of correlation times and the effective correlation time is of order several psecs.

The shape of the total spectrum is qualitatively consistent with the results derived from mode-coupling theories of the liquid-glass transition in a hard core liquid. In this picture the α-component arises from a collective process resulting from non-linear coupling between density fluctuations. The β-component arises from local motions of atoms within the cage formed by their neighbours. The inelastic neutron scattering from myoglobin thus gives a picture of the dynamical excitations of the molecule. There is vibrational motion of the constituent atoms combined with some form of slow collective motion within the molecule, but superimposed on this collective motion there are also local, non-vibrational motions. A complete molecular dynamics simulation calculation for myoglobin is essential if this picture of the motions is to be understood in detail. Such calculations are in progress.

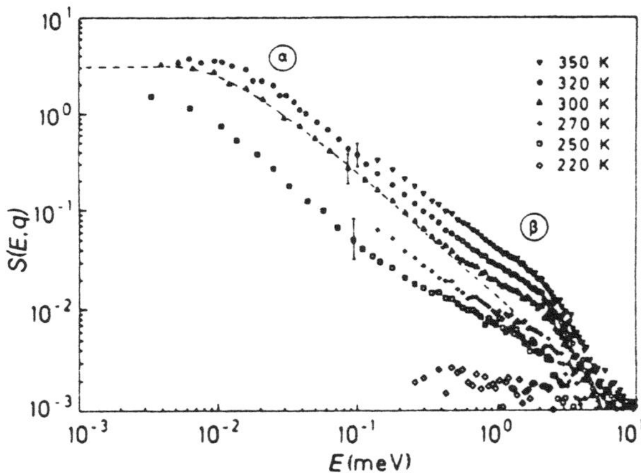

Figure 16 The temperature dependence of the quasi-elastic scattering from myoglobin, shown as a log/log plot. The dashed line shows a fit to the data with the expression of equation 17, convoluted with the experimental resolution function. (Doster et al., 1989).

6. Conclusion

In this brief review I have given a very superficial overview of the properties of neutrons and of neutron scattering, particularly as the technique may be applied to biological problems. I have discussed three different examples of the application of neutron scattering to illustrate the kind of information that may be derived. In the case of elastic neutron scattering, the applications are used to derive structural information. The formalism to interpret data is well understood and the questions that are addressed and the systems to which the technique is applied will become increasing sophisticated as the neutron technique is refined. In the case of inelastic neutron scattering the measurements are much more difficult and the interpretation of the data is far from obvious. However, the detailed dynamical information that may be obtained and its potential relevance for understanding biological function is so important, that, despite these present difficulties, an increasing number of such experiments will be undertaken in the future.

7. Acknowledgements

The author wishes to acknowledge many helpful discussions with J. Katsaras and to thank him for permission to quote his unpublished data on bilayer membranes. Figure 1 has been reproduced by permission of the Institute of Physics, Figure 4 by permission of the Canadian Association of Physicists, Figures 8, 9, 10 and 11 by permission of the American Chemical Society, Figures 13 and 15 by permission of the Biophysical Society and Figures 14 and 16 by permission of McMillan Magazines Ltd.

8. References

Babu, Y.S., Sack, J.S., Greenhough, T.C., Bugg, C.E., Means, A.R., and Cook, W.J., (1985) Three-Dimensional Structure of Calmodulin, *Nature*, **315**, 37-40.

Babu, Y.S., Bugg, C.E., and Cook, W.J., (1988) Structure of Calmodulin Refined at 2.2A Resolution, *J. Mol. Biol.*, **204**, 191-204.

Bloom, M., (1992) The Physics of Soft, Natural Materials, *Physics in Canada*, **48**, 7-16.

Boutin, H., and Yip, S., (1968) *Molecular Spectroscopy with Neutrons*, MIT Press, Cambridge.

Cusack, S., and Doster, W., (1990) Temperature Dependence of the Low Frequency Dynamics of Myoglobin, *Biophys. J.*, **58**, 243-251.

Doster, W., Cusack, S., and Petry, W., (1989) Dynamical Transition of Myoglobin revealed by Inelastic Neutron Scattering, *Nature*, **337**, 754-756.

Gupta, V.D., Trevino, S., and Boutin, H., (1968) Vibrational Spectra of Polyglycine, *J. Chem. Phys.* **48**, 3008-3015.

Heidorn, D.B., Seeger, P.A., Rokop, S.E., Blumenthal, D.K., Means, A.R., Crespi, H., and Trewhella, J. (1989) Changes in the Structure of Calmodulin Induced by a Peptide Based on the Calmodulin-Binding Domain of Myosin Light Chain Kinase, *Biochemistry*, **28**, 6757-6764.

Ibel, K., and Stuhrmann, H.B., (1975) Comparison of Neutron and X-ray Scattering of Dilute Myoglobin Solutions, *J. Mol. Biol.*, **93**, 255-265.

Jacrot, B., (1976) The Study of Biological Structures by Neutron Scattering from Solution, *Rept. Prog. Phys.*, **39**, 911-953.

Katsaras, J. and Jeffrey, K.R., (1996) to be published.

Lovesey, S.W., (1984) The Theory of Neutron Scattering from Condensed Matter, in *The International Series of Monographs on Physics*, Clarendon Press, Oxford.

Martel, P., and Lin, H., (1989) Large Scale Oscillatory Motions in Myoglobin, *J. Biol. Phys.*, **17**, 137-144.

Martel, P., (1992) Biophysical Aspects of Neutron Scattering from Vibrational Modes of Proteins, *Prog. Biophys. Molec. Biol.*, **57**, 129-179.

Middendorf, H.D., (1984) Biophysical Applications of Quasi-Elastic and Inelastic Neutron Scattering, *Rev. Biophys. Bioeng.*, **13**, 425-451.

Parsons, D.F. and Akers, C.K., (1969) Neutron Diffraction of Cell Membranes (Myelin), *Science*, **165**, 1016-1018.

Schmatz, W., Kaiser, B., Scherm, R., Schneider, R. and Mayer, A., (1969) Neutron Small-Angle Scattering from Aqueous Solutions of Oxy- and Deoxyhaemoglobin, *J. Mol. Biol.*, **41**, 231-235.

Schoenborn, B.P., (1969) Neutron Diffraction Analysis of Myoglobin, *Nature*, **224**, 143-146.

Schoenborn, B.P., (1976) (ed.), Neutron Scattering for the Analysis of Biological Structures, Brookhaven National Laboratory, (BNL 50453), Upton.

Sears, V.F., (1992) Neutron Scattering Lengths and Cross-Sections, *Neutron News*, **3**, 26-37.

Sköld, K. and Price, D.L. (eds.), (1986) Neutron Scattering, Parts A,B,C., in *Methods of Experimental Physics*, **23**, Academic Press, Orlando.

Squires, G.L., (1978) Introduction to the Theory of Thermal Neutron Scattering, Cambridge University Press, Cambridge.

Whittemore, W.L., (1968) An Investigation of the Polypeptide, Poly-L-Glutamic Acid, using Neutron Inelastic Scattering, *Neutron Inelastic Scattering, vol II*, IAEA, Vienna, pp. 175-184.

Worcester, D.L. and Franks, N.P., (1976) Structural Analysis of Hydrated Egg Lecithin and Cholesterol Bilayers, II. Neutron Diffraction, *J. Mol. Biol.*, **100**, 359-378.

List of Participants

Paul Alewood
3D Centre, University of Queensland, Queensland 4072, Australia
Phone 61 7 365 1270, FAX 61 7 365 1990
E-mail p.alewood@mailbox.uq.oz.au

Dianne Alewood
3D Centre, University of Queensland, Queensland 4072, Australia
Phone 61 7 365 1097, FAX 61 7 365 1990
E-mail d.alewood@mailbox.uq.oz.au

Jon Amster
Dept. of Chemistry, University of Georgia, Athens, GA 30602-2556, USA
Phone 706 542 2001, FAX 706 542 9454,
E-mail amster@sunchem.chem.uga.edu

Robin T. Aplin
The Dyson Perrins Laboratory,Oxford University,
South Parks Road. Oxford, OX1 3QY, U.K.
Phone 44 (0)1865 275 661, FAX 44 (0)1865 275 674

Seppo Auriola
Dept. of Pharmaceutics, University of Kuopio,
Box 1627, SF-70211, Kuopio, Finland
Phone 358 71 162 469, FAX 358 71 162 456
E-mail auriola@messi.uku.fi

Ayeda Ayed
Dept. of Chemistry, University of Manitoba, Winnipeg MB, R3T 2N2, Canada
Phone 204 474 8479, FAX 204 275 0905
E-mail ayed@cc.umanitoba.ca

Brett Bloom
Biology Dept., Brookhaven National Laboratory, Blg.463, Upton, NY 11973, USA
Phone 516 344 7402, FAX
E-mail bloom@bnlux1.bnl.gov

Steven Burley
The Rockefeller University, 1230 York Ave., New York, NY 10021, USA
Phone 212 327 8336, FAX 212 327 8337
E-mail burley@rockvax.rockefeller.edu

Brian Chait
The Rockefeller University, 1230 York Ave, New York, NY 10021, USA
Phone 212 327 8849, FAX 212 327 7547
E-mail chait@rockvax.rockefeller.edu

John Chakel
Hewlett-Packard Laboratories, 3500 Deer Creek Road, PO Box 10350, MS 26U-6
Palo Alto, CA 94303-0867, U.S.A.
Phone 415 857 3130 FAX 415 852 8502
E-mail john_chakel@hpl.hp.com

Igor Chernushevich
Dept. of Physics, University of Manitoba, Winnipeg, MB, R3T2N2, Canada
Phone 204 474 6184 FAX 204 269 8489
E-mail chernus@cc.umanitoba.ca

Caroline Collette
Laboratoire de Spectrometrie de Masse, Institut de Chimie B6
Université de Liège, B4000, Liège, Belgium
Phone 32 41 663 415, FAX 32 41 663 413
E-mail depauw@gw.unipc.ulg.ac.be

Edwin De Pauw
Laboratoire de Spectrometrie de Masse, Institute de Chimie B6
Université de Liège, B4000, Liège, Belgium
Phone 32 41 663 415, FAX 32 41 663 413
E-mail depauw@gw.unipc.ulg.ac.be

Kay Diederichs
Universität Konstanz, Fakultät für Biologie,
Postfach 5560 (M656), D-78434 Konstanz, Germany
Phone 49 7531 88 4049 FAX 49 7531 88 2902
E-mail kay.diederichs@uni-konstanz.de

Lynda Donald
Dept of Chemistry, University of Manitoba, Winnipeg, MB, R3T2N2, Canada
Phone 204 474 8479, FAX 204 275 0905
E-mail ldonald@cc.umanitoba. ca

Harry W. Duckworth
Dept. of Chemistry, University of Manitoba, Winnipeg, MB, R3T2N2, Canada
Phone 204 474 6252 FAX 204 275 0905
E-mail hdckwth@cc.umanitoba.ca

Ragnar Dworschak
Dept. of Physics, University of Manitoba, Winnipeg, MB, R3T2N2, Canada
Phone 204 474 6184, FAX 204 269 8489
E-mail dworsch@cc.umanitoba.ca

John R. Engen
Dept. of Chemistry, University of Nebraska-Lincoln,
Lincoln, NE, 68588-0304, U.S.A.
Phone 402 472 3507 FAX
E-mail jengen@unl.grad1.unl.edu

Werner Ens
Dept. of Physics, University of Manitoba, Winnipeg, MB, R3T 2N2, Canada
Phone 204 474 6178 FAX 204 269 8489
E-mail W_Ens@UManitoba.CA

Michael Fitzgerald
The Scripps Research Institute, MD-CUN-6
10666 N, Torrey Pines Road, La Jolla CA, 92037, USA
Phone 619 784 7051, FAX 619 784 7319
E-mail mfitz@chemical.ligation.com

Michael O. Glocker
Fakultät für Chemie, Universität Konstanz
Postfach 5560 M732, D-78434 Konstanz, Germany
Phone 49 7531 882690 FAX 49 7531 883097
E-mail Michael.Glocker@uni-konstanz.de

Richard Griffey
Isis Pharmaceuticals, 2280 Faraday Ave., Carlsbad, CA 92008, USA
Phone 619 929 2430 FAX 619 929 0036
E-mail rgriffey@isisph.com

Niels H. H. Heegaard
Statens Seruminstitut, bldg. 81, room 528, Artillerivej 5,
DK-2300 Copenhagen S, Denmark
FAX 45 3268 3876
E-mail nheegaard@ai.diag.ssi.dk

Mark E. Hemling
Smithkline Beecham Pharmaceuticals
P.O. Box 1539 UW2940, King of Prussia, PA 19406-0939, USA
Phone 610 270 6549 FAX 610 270 6608
E-mail Mark_E_Hemling@SBPHRD.COM

Franz Hillenkamp
Institut für Medizinische Physik und Biophysik, Universität Münster,
Robert-Koch-Str. 31, D-48149 Münster, Germany
Phone 49 251 83 5103 FAX 49 251 83 5121
E-mail hillenk@uni-muenster.de

Steven Hofstadler
Pacific Northwest National Laboratory, P.O. Box 999 Richland, WA 99352 USA
Phone 509 375 4383 FAX 509 375 6916
E-mail sa_hofstadler@pnl.gov

David Hosfield
Dept. of Chemistry, University of Manitoba, Winnipeg, MB, R3T 2N2, Canada
Phone 204 474 8479 FAX 204 275 0905
E-mail umhosfi1@cc.umanitoba.ca

Carl F. Ijames
Smithkline Beecham Pharmaceuticals
P.O. Box 1539 UW2940, King of Prussia, PA 19406-0939, USA
Phone 610 270 5708 FAX 610 270 6608
E-mail Carl_F_Ijames@SBPHRD.COM

Michael Karas
Institut für Physikalische und Theoretische Chemie
Johann Wolfgang Goethe-Universität
60590 Frankfurt am Main
Phone 49 69 6301 7624 FAX 49 69 631 3984

Robert Karlsson
Pharmacia Biosensor AB, S- 751 82 Uppsala, Sweden
E-mail robert.karlsson@upp.pharmacia.se

Lewis E. Kay
Dept. of Medical Genetics, Medical Sciences Bldg., Univ. of Toronto,
Toronto, ON M5S 1A8, Canada
Phone 416 978 0741 FAX
E-mail Kay@bloch.med.utoronto.ca

Neil Kelleher
Cornell University, Baker Labs #78, Ithaca, NY 14850 - 1301, USA
Phone 607 255 3726 FAX 607 255 7880
E-mail nk18@cornell.edu

Esther C. Kempen
Dept. of Chemistry and Biochemistry, University of Texas, Austin, TX 78712, USA
Phone 512 471 0041 FAX
E-mail esti@utxvms.cc.utexas.edu

Yasuo Konishi
Biotechnology Research Institute, 6100 Royalmount Ave.,
Montreal, Quebec, Canada H4P 2R2
Phone 514 496 6339 FAX 514 496 6232
E-mail konishi@biotech.lan.nrc.ca

Eddy Kragten
CDDT K681.5.03, PO Box 4002, Basel, Switzerland
Phone 41 61 696 5004, FAX 41 61 696 4221
E-mail oostrum@fmi.ch

Andrew Krutchinsky
Dept. of Physics, University of Manitoba, Winnipeg, MB, R3T 2N2, Canada
Phone 204 474 6184 FAX 204 269 8489
E-mail krutchi@cc.umanitoba.ca

Shubhada Kulkarni
Dept. of Chemistry, University of Georgia, Athens, GA 30602, USA
Phone 706 542 2004 FAX 706 542 9454
E-mail kulkarni@sunchem.chem.uga.edu

Gilles Lajoie
Dept. of Chemistry, University of Waterloo, Waterloo, ON, Canada
Phone 819 888 4620, FAX 819 746 0435
E-mail glajoie@chemistry.watstar.uwaterloo.ca

Martin Larsen
Dept. of Molecular Biology, University of Odense
Campusvej 55, DK-5230 Odense M, Denmark
Phone +45 66158696 ext. 2398 FAX +45 65932661
E-mail mrl@mol-bio.ou.dk

Paolo Lecchi
Laboratory of Analytical Chemistry, National Institutes of Health,
NIDDK Bldg 8 room B2A27 - 8 Center Drive, Bethesda, MD 20892, USA
Phone 301 496 7544 FAX 301 402 1967
E-mail paolo@sx102a.niddk.nih.gov

Emmanuelle Leize
Faculté de Chimie, LSMBO, 1 rue Blaise Pascal, 67000 Strasbourg, France
Phone 33 88 416887 FAX 33 88 604687
E-mail vandors@chimie.u-strasbg.fr

Joseph Loo
Parke-Davis Pharmaceutical, 2800 Plymouth Road, Ann Arbor, MI 48105, USA
Phone 313 996 7515 FAX 313 998 2716
E-mail Loo@AA.WL.COM

Andrey Maksimenko
Belozersky Institute of Physico-Chemical Biology
Laboratory of Nucleic Acid Chemistry, Lomonosov Moscow State University
Moscow, 119899 Russia
Phone 007 095 9395407, FAX 007 095 9393181
E-mail maxand@nuclacid.genebee.msu.su

Randall W. Nelson
Dept. of Chemistry and Biochemistry
Arizona State University, Tempe, AZ 85287-1604, USA
Phone 602 965 4244 FAX 602 965 2747
E-mail rnelson@imap2.asu.edu

Hélène Perreault
Dept. of Chemistry, University of Manitoba, Winnipeg, MB, R3T 2N2, Canada
Phone 204 474 7418 FAX 204 275 0905
E-mail perreau@cc.UManitoba.CA

Noelle Potier
Faculté de Chimie, LSMBO, 1 rue Blaise Pascal, 67000 Strasbourg, France
Phone 33 88 416887 FAX 33 88 604687
E-mail vandors@chimie.u-strasbg.fr

Flemming M. Poulsen
Carlsberg Laboratory, Dept. of Chemistry
Gamle Carlsberg Vej 10, DK-2500 Valby, Denmark
Phone +45 33 27 53 48 FAX +45 33 27 47 08
E-mail carlfmp@unidhp.uni-c.dk

Brian Powell
Atomic Energy of Canada Ltd., Chalk River, ON, Canada
E-mail bmpowell@crl4.crl.aecl.ca

Michael Przybylski
Fakultät für Chemie, Universität Konstanz,
Postfach 5560 M732, D-78434 Konstanz, Germany
Phone 49 7531 88 2249 FAX 49 7531 88 3097
E-mail: Michael.Przybylski@uni-konstanz.de

Peter Roepstorff
Dept. of molecular biology, Odense University, Campusvej 55
5230 Odense M, Denmark
Phone 45 66 158600 FAX 45 65 932661
E-mail roe@PR-Group.ou.dk

K.C. Russell
University of Miami, Dept. of Chemistry,
1301 Memorial Dr., Coral Gables, FL 33124 - 0431, USA
Phone 305 284 6856 FAX 305 294 1880
E-mail krussell@impala.ir.miami.edu

K.W. Michael Siu
Institute for National Measurement Standards, National Research Council
Bldg. M-12, Montreal Road, Ottawa, ON, K1A 0R6, Canada
Phone 613 993 5008 FAX (613)993-2451
E-mail michael.siu@nrc.ca

David L. Smith
Dept. of Chemistry, University of Nebraska-Lincoln, Lincoln, NE 68588-0304, USA
Phone 402 472 2794 FAX 402 472 9862
E-mail dls@unlinfo.unl.edu

Kenneth G. Standing
Dept. of Physics, University of Manitoba, Winnipeg, MB, R3T 2N2, Canada
Phone 204 474 9358 FAX 204 269 8489
E-mail standin@cc.umanitoba.ca

Stuart Thomson
Victorian College of Pharmacy, Monash University
381 Royal Parade, Parkville, Victoria, Australia, 3052
Phone 613 9903 9554 FAX 613 9903 9582
E-mail Stuart.Thomson@vcp.monash.edu.au

Kenneth Tomer
NIEHS MD 6-01 PO Box 12233, Research Triangle Park, NC 27709 USA
Phone 919 541 1966 FAX 919 541 7880
E-mail tomer@niehs.nih.gov

Alain Van Dorsselaer
Faculté de Chimie, LSMBO, 1 rue Blaise Pascal, 67000 Strasbourg, France
Phone 33 88 416887 FAX 33 88 604687
E-mail vandors@chimie.u-strasbg.fr

Jan van Oostrum
CDDT, K-681.5.44, Ciba-Geigy, P.O.Box 4002 Basel, Switzerland
Phone 41 61 6961668 FAX 41 61 6964221
E-mail oostrum@fmi.ch

Andreas Wattenberg
Institut Physikalische und Theoretische Chemie,
Pfaffenweg 10, D-61440 Oberursel, Frankfurt, Germany
Phone 49 6171 980854 FAX 49 69 79829560
E-mail wattenberg@chemie.uni-frankfurt.de

Wolfgang Welte
Universität Konstanz, Fakultät für Biologie,
Universitätsstr. 10, D-78434 Konstanz, Germany
Phone 7531 88 2206 FAX 7531 88 3183
E-mail wolfram.welte@uni-konstanz.de

Garrett Westmacott
Dept. of Physics, University of Manitoba, Winnipeg, MB, R3T 2N2, Canada
Phone 204 474 6184 FAX 204 269 8489
E-mail westmac@cc.umanitoba.ca

Julian P. Whitelegge
Department of Chemistry and Biochemistry, 405 Hilgard Ave., UCLA
Los Angeles, CA 90024, USA
Phone 310 206 7886 FAX 310 206 4038
E-mail jpw@xenon.chem.ucla.edu

Evan R. Williams
Dept. of Chemistry, University of California, Berkeley, CA 94720, USA
Phone 510 643 7161 FAX 510 642 8369
E-mail erw@garnet.berkeley.edu

Roman Zubarev
Jonfysik/ISV, Box 535, Uppsala university, 75 121 Uppsala, Sweden
Phone 46 18 183728 FAX +46 -18 - 55-57-36
E-mail zubarev@tsl.uu.se

Index